T0180516

FUNCTIONAL INTEGRALS IN QUANTUM FIELD THEORY
AND STATISTICAL PHYSICS

MATHEMATICAL PHYSICS AND
APPLIED MATHEMATICS

Editors:

M. FLATO, *Université de Dijon, Dijon, France*

R. RACZKA, *Institute of Nuclear Research, Warsaw, Poland*

with the collaboration of :

M. GUENIN, *Institut de Physique Théorique, Geneva, Switzerland*

D. STERNHEIMER, *Collège de France, Paris, France*

VOLUME 8

VIKTOR NIKOLAYEVICH POPOV

FUNCTIONAL INTEGRALS IN QUANTUM FIELD THEORY AND STATISTICAL PHYSICS

Translated from the Russian by J. Niederle and L. Hlavatý

D. REIDEL PUBLISHING COMPANY

A MEMBER OF THE KLUWER  ACADEMIC PUBLISHERS GROUP

DORDRECHT / BOSTON / LANCASTER

Library of Congress Cataloging in Publication Data

Popov, V. N. (Viktor Nikolaevich)
   Functional integrals in quantum field theory
and statistical physics.

   (Mathematical physics and applied mathematics ; v. 8)
   Translation of: Kontinual´nye integraly v teorii
polía i statisticheskoí fizike.
     1. Integration, Functional. 2. Quantum field
theory. 3.´ Statistical physics. I. Title.
QC20.7.F85P6613   1983   530.1´43   83-9477

Published by D. Reidel Publishing Company
P.O. Box 17, 3300 AA Dordrecht, Holland

Sold and distributed in the U.S.A. and Canada
by Kluwer Academic Publishers
190 Old Derby Street, Hingham, MA 02043, U.S.A.

In all other countries, sold and distributed
by Kluwer Academic Publishers Group
P.O. Box 322, 3300 AH Dordrecht, Holland

CONTENTS

FOREWORD

Functional integration is one of the most powerful methods of contemporary theoretical physics, enabling us to simplify, accelerate, and make clearer the process of the theoretician's analytical work. Interest in this method and the endeavour to master it creatively grows incessantly. This book presents a study of the application of functional integration methods to a wide range of contemporary theoretical physics problems.

The concept of a functional integral is introduced as a method of quantizing finite-dimensional mechanical systems, as an alternative to ordinary quantum mechanics. The problems of systems quantization with constraints and the manifolds quantization are presented here for the first time in a monograph.

The application of the functional integration methods to systems with an infinite number of degrees of freedom allows one to uniquely introduce and formulate the diagram perturbation theory in quantum field theory and statistical physics. This approach is significantly simpler than the widely accepted method using an operator approach.

A large part of this book is devoted to the development of nonstandard methods of perturbation theory using specific examples, the first of which is the theory of gauge fields. The method of functional integration with necessary modifications is used for the quantization of the electromagnetic field, the Yang–Mills field, and the gravitational field. Attempts to construct a unified gauge invariant theory of electromagnetic and weak interactions are explored here, too. The next example of functional integration is the derivation of the infrared asymptotic behaviour of the Green's function of quantum electrodynamics. We shall examine the application of functional integration to problems of scattering of high-energy particles and the formulae for a doubly-logarithmic asymptotic and eikonal approximation will be obtained.

The applications of functional integrals to problems of statistical physics begin with examples of superfluidity, superconductivity, and plasma theory. A modified perturbation theory for superfluid Bose and Fermi systems is used in the microscopic approach to the construction of the hydrodynamic Hamiltonian of the system and the equations of superfluid hydrodynamics. For the first time in a monograph, the question of superfluidity of two-dimensional and one-dimensional Bose systems is elucidated. The method describing quantum vortices in Bose and Fermi systems is developed and applied, specifically, to the theory of superconductivity of the second type. The method for using the hydrodynamical Hamiltonian for systems with Coulomb interaction is illustrated in application to the theory of plasma oscillations. The example of a problem which allows such a solution in functional integration formalism is the Ising model. In the last chapters dealing with statistical physics, the Wilson method is studied which utilizes functional integration for the theory of phase transitions. The closing chapter of the book extends the notion of excitations of the quantum vortex type which is customary in statistical physics onto the quantum field theory. This trend, bent on diminishing the number of fundamental fields, is a fairly recent development.

The book need not be read successively. After one becomes acquainted with the definition of the functional integral and the methods of construction of diagram perturbation theory (see (Chapters 1, 2), it is possible to concentrate on those applications of functional integrals to physical systems, which are most interesting. The choice of specific examples is, to a great extent, determined by the scientific interests of the author.

I would like to thank V. Alonso for the help provided during the preparation of the manuscript for press.

CHAPTER 1

FUNCTIONAL INTEGRALS AND
QUANTUM MECHANICS

1. INTRODUCTION

Functional integrals were introduced into mathematics during the twenties by Wiener as a method for solving the problems of diffusion and Brown's motion [1]. In physics, functional integrals were rediscovered in the forties by Feynman and used by him for the reformulation of quantum mechanics. In the late forties, Feynman constructed a new formulation of quantum electrodynamics, based on the method of functional integration, and developed the now-famous diagram technique of perturbation theory [2–4]. This new theory has substantially simplified calculations and has helped to construct the theory of renormalization. The latter has turned out to be an important leap in solving the problem of divergencies which have arisen in quantum electrodynamics since its formulation in the 1929 paper of Heisenberg and Pauli [5]. At the present time, the theory of electromagnetic interactions agrees with experiments up to the seventh decimal digit, and this is one of the merits of the new perturbation theory.

Since the fifties, functional integrals, arising when solving functional equations in quantum field theory (Schwinger equation [6]), have been intensively studied. The functional formulation of quantum field theory has been investigated in the works of Bogoliubov [7], Gel'fand and Minlos [8], Matthews and Salam [9], Khalatnikov [10] and Fradkin [11].

In the sixties, a new field of applications of functional integrals appeared – the quantization of gauge fields. The electromagnetic field, the Einstein gravitational field, the Yang–Mills field and the chiral field can serve as examples of gauge fields. The action functionals of those fields are invariant under gauge transformations which depend on one or several arbitrary functions. From a mathematical point of view, gauge fields are fields of geometrical origin which are connected with fibrations over four-dimensional space-time. The specificity of geometrical fields has to be taken into account when quantizing them; otherwise incorrect results may be obtained.

1

This was noticed for the first time by Feynman in the Yang–Mills and gravitational fields. He has shown that quantization according to a method analogous to the Fermi method in quantum electrodynamics, violates the unitarity condition. Feynman also proposed a method for the removal of the difficulties shown.

Later, as a result of the work of several authors, the problem of gauge-field quantization was studied and the functional integration method [13–18] has turned out to be the most convenient method for solving that problem.

A special place among gauge fields is occupied by the gravitational field. The question of its quantization is connected with the hope that it is a natural physical regularizator cutting the interaction on small distances. The first results supporting this point of view were obtained by De Witt [19] and Khriplovich [20].

At the present time, the method of functional integration is most frequently applied to problems which are some how connected with gauge fields. It is necessary to underline its utilization in attempts to construct a unified theory of electromagnetic and weak interactions [21–23].

There exist many other applications of functional integration to quantum field theory. For example, by using this method it becomes simple to derive various asymptotic formulae for infrared and ultraviolet asymptotics of the Green functions and scattering amplitudes [60, 67–71]. The utilization of functional integrals also turns out to be interesting in dual models.

In the late sixties and early seventies, the theory of automodel (scaling) behaviour of quantum field theory amplitudes at high energies, which has much in common with the theory of phase transitions of the second kind, was developed. Here, the method of functional integration helps to qualitatively describe the picture of high-energy particle scattering and of critical phenomena, and to approximately evaluate the power indices (critical indices).

Lately, a new field of applications of functional integrals has appeared which is connected with the search for excitations in quantum field theory and is analogous to the quantum vortices in statistical physics. The idea that some of the elementary particles can be looked upon as collective excitations of the interacting fundamental fields system allows one to reduce the number of fundamental fields. The method of functional integration is perhaps the only acceptable approach to the solution of the problems appearing here. Some of the results obtained this way are expounded in Chapter 11 of this book.

The application of functional integrals in statistical physics allows one to derive many interesting results which are only obtained with difficulty by other methods. Feynman applied this approach to polaron theory and to the liquid helium theory and he succeeded in accurately evaluating the self-energy of a polaron and in investigating the qualitative features of the λ-transition in liquic helium.

The theory of phase transitions of the second kind, superfluidity, superconductivity, lasers, plasma, Kondo effect, Ising model – this is an incomplete list of problems, for which the application of the functional integration method appears to be very useful. In some of the problems, it allows us to prove results obtained by other methods, clarify the possibilities of their applicability and outline the evaluation of corrections. If there is a possibility of an exact solution, the method of functional integration gives a simple way of obtaining it. In problems far from being exactly solvable (general theory of phase transitions), the application of functional integrals helps to build up the qualitative picture of the phenomenon and to develop the approximative methods of calculations.

Functional integrals are especially useful for the description of collective excitations, such as plasma oscillations in the theory of the system of particles with Coulomb interaction, quantum vortices and long-wave photons in the theory of superfluidity and superconductivity. That is the case when standard perturbation theory should be modified. Functional integrals represent a sufficiently flexible mathematical apparatus, adjusted for such a revision and suggesting the method for its concrete realization.

Functional integration is an 'integral evaluation' adjusted to the needs of contemporary physics. At present, however, the exact mathematical theory and correct definition of functional integrals used in quantum field theory and statistical physics is lacking.

The exact definition and correct mathematical theory can be constructed for functional integrals which give solutions of partial differential equations, including the equations of quantum mechanics and diffusion theory. Mathematical questions of the theory of functional integrals are expounded in the surveys of Gel'fand and Yaglom [26], Kovalchik [27], in the books of Kac [28] and Berezin [29]. Let us also mention the works of Berezin [30], Daletzki [31], Evgraphov [32], Alimov and Buslaiev [33], devoted to the exact definition of some functional integrals.

In the works performed on the physical level of exactness, the functional integral is used as a heuristic means for the construction of perturbation theory and for the transition from one perturbation theory to another. From this point of view the functional integrals are studied in this book.

2. FUNCTIONAL INTEGRALS IN QUANTUM MECHANICS

We present here the definition of the functional integral in quantum mechanics. Feynman in his 1948 article [2] introduced and studied the functional integral in configuration space. For many applications, however, the most suitable form seems to be the expression given by Feynman in 1951 [4], where integration is taken along trajectories in the phase space.

Let us investigate the one-dimensional mechanical system determined by its Hamilton function $H(q, p)$, where q is the coordinate and p is the canonically conjugated momentum. The principle of canonical quantization of such a system consists of replacing the coordinate q and momentum p by operators $\hat{q}$ and $\hat{p}$ according to the rule

$$q \rightarrow \hat{q} \equiv q, \qquad p \rightarrow \hat{p} = -i\hbar \frac{\partial}{\partial q}, \tag{2.1}$$

where $\hbar$ is the Planck constant. In the following we shall use the system of units with $\hbar = 1$. The operators act on a Hilbert space of complex functions $\Psi(q)$. According to (2.1), the effect of the coordinate operator on the function $\Psi(q)$ is a multiplication of that function by the variable q and the operator of momentum is proportional to the differentiation operator $\partial/\partial q$.

The time evolution of a state is determined by the Heisenberg equation

$$i\frac{\partial \Psi}{\partial t} = \hat{H}\Psi, \tag{2.2}$$

where $\hat{H}$ is the energy operator obtained from the classical Hamiltonian function $H(q, p)$ by replacing q and p, according to (2.1), with operators $\hat{q}$ and $\hat{p}$ ordered in a certain way. We can write down the formal solution of equation (2.2) as

$$\Psi(t) = \hat{U}(t, t_0)\Psi(t_0), \tag{2.3}$$

where the evolution operator

$$\hat{U}(t, t_0) = \exp\left(i(t_0 - t)\hat{H}\right) \tag{2.4}$$

is the exponential of the energy operator $\hat{H}$.

The method of functional integration allows one to express the matrix element of the evolution operator as a mean value of the expression

$$\exp(iS[t_0, t]),$$ (2.5)

over trajectories in the phase space, where

$$S[t_0, t] = \int_{t_0}^{t} (p(\tau)\dot{q}(\tau) - H(q(\tau), p(\tau))) \, d\tau$$ (2.6)

is a classical action corresponding to the trajectory $(q(\tau), p(\tau)), (t_0 \leq \tau \leq t)$, in the phase space, $\dot{q}(\tau) \equiv dq(\tau)/d\tau$.

The mean value over trajectories is called the *Feynman functional integral*. Usually this is defined as a limit of finite-dimensional integrals. We shall present here one of the possible definitions.

We divide the interval $[t_0, t]$ with $\tau_1, \ldots, \tau_{N-1}$ points into N equal parts. Let us consider the functions $p(\tau)$, defined on the interval $[t_0, t]$, which are constant on the intervals

$$[t_0, \tau), (\tau_1, \tau_2), \ldots, (\tau_{N-1}, t],$$ (2.7)

and the continuous functions $q(\tau)$ linear on the intervals (2.7). We fix the values of the function $q(\tau)$ at the end points of the interval $[t_0, t]$, putting

$$q(t_0) = q_0; \qquad q(t) = q.$$ (2.8)

The trajectory $(q(\tau), p(\tau))$ is determined by values of the piecewise linear function $q(\tau)$ in points $\tau_1, \ldots, \tau_{N-1}$ (we denote them by $q_1, \ldots, q_{N-1}$) and by values of the piecewise constant function $p(\tau)$ on intervals (τ_k, τ_{k+1}). We denote those values by $p_1, \ldots, p_N$.

Let us consider the finite-dimensional integral

$$(2\pi)^{-N} \int dp_1 \, dq_1 \ldots dq_{N-1} \, dp_N \exp(iS[t_0, t]) \equiv J_N(q_0, q; t_0, t),$$ (2.9)

where $S[t_0, t]$ is the action (2.6) for the described trajectory $(q(\tau), p(\tau))$, defined by the parameters $q_1, \ldots, q_N, p_1, \ldots, p_N$. The basic assertion says that the limit of the integral (2.9) for $N \to \infty$ is equal to the matrix element of the evolution operator

$$\lim_{N \to \infty} J_N(q_0, q; t_0, t) = \langle q | \exp(i(t_0 - t)\hat{H}) | q_0 \rangle.$$ (2.10)

It is not hard to check this statement in those cases when the Hamiltonian H is a function of the coordinate or momentum only.

If $H = H(q)$ (H depends on the coordinate only), then the classical action for the above-mentioned trajectory $(q(\tau), p(\tau))$, takes the form

$$\int_{t_0}^{t} (p\dot{q} - H(q))\, d\tau = p_1(q_1 - q_0) + p_2(q_2 - q_1) + \cdots$$

$$\cdots + p_N(q - q_{N-1}) - \int_{t_0}^{t} H(q(\tau))\, d\tau. \tag{2.11}$$

Integrating in (2.9) over the momenta, we obtain the product of δ-functions:

$$\delta(q_1 - q_0)\delta(q_2 - q_1)\cdots\delta(q - q_{N-1}). \tag{2.12}$$

This product allows us to put the expression $\exp(-i\int_{t_0}^{t} H(q(\tau))\, d\tau)$ to be equal to $\exp(i(t_0 - t)H(q_0))$ and place it in front of the integration symbol. Further integration over $q_1, \ldots, q_{N-1}$ coordinates eliminates all δ-functions except one and leads to the result

$$\delta(q_0 - q)\exp(i(t_0 - t)H(q_0)), \tag{2.13}$$

identical to the matrix element of the evolution operator.

If $H = H(p)$ (H depends on momentum only) then the action takes the form

$$\int_{t_0}^{t} (p(\tau)\dot{q}(\tau) - H(p(\tau)))\, d\tau = p_1(q_1 - q_0) + p_2(q_2 - q_1) + \cdots$$

$$\cdots + p_N(q - q_{N-1}) - \int_{t_0}^{t} H(p(\tau))\, d\tau. \tag{2.14}$$

Integrating in (2.9) first over coordinates $q_1, \ldots, q_{N-1}$ and then over all momenta $p_1, \ldots, p_N$ we obtain the expression

$$\frac{1}{2\pi}\int dp \exp\{ip(q - q_0) + i(t_0 - t)H(p)\}, \tag{2.15}$$

equal to the matrix element of the evolution operator for the Hamiltonian $H = H(\hat{p})$.

The proof of formula (2.10) is more complicated if nontrivial dependence of the Hamiltonian on the coordinate and momentum occurs. In such a case, the prelimit expression (2.9) is not identical to its limit – the evolution operator matrix element. Proof of a formula analogous to (2.10) for the evolution operator of the system described by a parabolic-type equation is given, e.g., in Reference [32]. For the Schrodinger equation, the proof is known only if operator $\hat{H}$ is a sum of a function of coordinates and a function of momenta:

$$H = H_1(q) + H_2(p).\qquad(2.16)$$

Namely, the Hamiltonians of the (2.16) type are used in nonrelativistic quantum mechanics.

We denote the functional integral, defined as the $N \to \infty$ limit of expression (2.9), by the symbol

$$\int_{q(t_0)}^{q(t)} \exp(iS[t_0,t]) \prod_\tau \frac{dp(\tau)\,dq(\tau)}{2\pi}.\qquad(2.17)$$

This form is convenient but it does not reflect the fact that in the prelimit expression (2.9) the number of integrations over momenta is higher by one order than that over the coordinates.

Let us remark that the functional integral, defined by formula (2.10) as a limit of the finite-dimensional one, depends on the method of approximation to the $(q(\tau), p(\tau))$ trajectory. This is connected with the fact that we have no natural prescription for the ordering when replacing the arguments of the function $H(p,q)$ by noncommuting operators $\hat{q}$ and $\hat{p}$. However, the operators with a physical meaning correspond, as a rule, to the functions for which the replacement of arguments by noncommuting operators leads to an unambiguous results. This is true for the energy operator in nonrelativistic quantum mechanics which is equal to the sum of a quadratic function of momenta and a function of coordinates. In such cases the functional integral leads to unambiguous results, too.

We generalize the functional integral formalism to a system with an arbitrary finite number of degrees of freedom.

The action of a mechanical system with n degrees of freedom has the form

$$S[t_0,t] = \int \left(\sum_{i=1}^{n} p_i \dot{q}^i - H(q,p) \right) d\tau.\qquad(2.18)$$

Here q^i is the ith canonical coordinate; p_i is the canonically conjugated momentum;

$$H(q,p) \equiv H(q^1,\ldots,q^n;p_1,\ldots,p_n) \text{ is the Hamiltonian.}$$

By definition, the functional integral for the evolution operator matrix element is a limit of the finite-dimensional integral obtained from (2.9) by the replacement

$$(2\pi)^{-N} \to (2\pi)^{-Nn}; \qquad \mathrm{d}q_k \to \prod_{i=i}^{n} \mathrm{d}q_k^i; \qquad \mathrm{d}p_k = \prod_{i=i}^{n} \mathrm{d}p_{i,k}, \quad (2.19)$$

where q_k^i are the values of the ith coordinate at the point $\tau_k (k = 1, \ldots, N-1)$ and p_{ik} are the values of the ith momentum on (τ_{k-1}, τ_k) interval. It is necessary to keep all the coordinates $q^1, \ldots, q^n$ simultaneously fixed at both ends of the time interval $[t_0, t]$.

We will denote the functional integral defined in such a way by the symbol

$$\int_{q(t')=q'}^{q(t'')=q''} \exp(\mathrm{i}S) \prod_{t} \prod_{i=1}^{n} \frac{\mathrm{d}q^i(t)\,\mathrm{d}p_i(t)}{2\pi}. \tag{2.20}$$

3. QUANTIZATION OF SYSTEMS WITH CONSTRAINTS

In the previous section the quantization of finite-dimensional mechanical systems with the action of Hamilton type (2.18) was studied using the technique of functional integration. Field theory can be looked upon as an infinite-dimensional analogy of a mechanical system with action (2.18). In such an approach the theory of gauge fields is an analogy of mechanical systems with constraints. The quantization of finite-dimensional systems with constraints requires the modification of the functional integral.

A classical action of the finite-dimensional mechanical system with constraints

$$S = \int \left(\sum_{i=1}^{n} p_i \dot{q}^i - H(q,p) - \sum_{a=1}^{m} \lambda_a \varphi^a(q,p) \right) \tag{3.1}$$

also contains, besides the coordinates q and momenta p, the variables λ_a, which come in linearly and play the role of Lagrange multiplicators. The coefficients $\varphi^a(q,p)$ have the meaning of constraints. The variables q,p generate the phase space of dimension $2n$. The number of constraints shall be denoted as m. We suppose that $m < n$ and that the constraints φ^a and Hamiltonian H are in involution, i.e., that they fulfil the conditions

$$\{H, \varphi^a\} = \sum_b c_b^a \varphi^b; \qquad \{\varphi^a, \varphi^b\} = \sum_d c_d^{ab} \varphi^d. \tag{3.2}$$

In these formulae c_b^a, c_d^{ab} are functions of q and p and $\{f, q\}$ is the Poisson bracket:

$$\{f,g\} = \sum_{i=1}^{n} \left(\frac{\partial f}{\partial p_i} \frac{\partial g}{\partial q^i} - \frac{\partial f}{\partial q^i} \frac{\partial g}{\partial p_i} \right). \tag{3.3}$$

The system of equations of motion for action (3.1) contains, besides canonical equations

$$\dot{q}^i = \frac{\partial H}{\partial p_i} + \sum_{a=1}^{m} \lambda_a \frac{\partial \varphi^a}{\partial p_i}; \qquad \dot{p}_i = -\frac{\partial H}{\partial q^i} - \sum_{a=1}^{m} \lambda_a \frac{\partial \varphi^a}{\partial q^i}. \tag{3.4}$$

also the constraint equations

$$\varphi^a(q,p) = 0, \quad a = 1,\ldots, m. \tag{3.5}$$

It is apparent from Equations (3.5) that some of the variables q, p are spurious, i.e., the solution of constraint equations often turns out to be rather difficult. It is therefore desirable to have a formalism where explicit solutions of constraint equations are not required.

Constraint Equations (3.5) define the surface M of the $2n - m$ dimensions in phase space Γ. The involution conditions (3.2) guarantee, for arbitrary functions $\lambda_a(t)$, the fulfilment of constraint conditions (3.5), provided those equations are satisfied for initial conditions. In other words, a trajectory which starts on the manifold M does not leave it.

We shall regard as observables on the manifold M the variables which are not influenced by arbitrariness in the choice of $\lambda_a(t)$. This requirement is fulfilled by the functions $f(q,p)$, which obey the conditions

$$\{f, \varphi^a\} = \sum_b d_b^a \varphi^b. \tag{3.6}$$

Indeed, in the equations of motion for those functions

$$\dot{f} = \{H, f\} + \sum_a \lambda_a \{\varphi^a, f\} \tag{3.7}$$

the λ_a-depending terms vanish on M.

The function $f(q,p)$ defined on M and satisfying conditions (3.6) does not in fact depend on all variables. Conditions (3.6) can be looked upon as a system of m differential equations of the first order on M for which Equations (3.2) are conditions of integrability. The function f is therefore unambiguously defined by its values on a submanifold of the system's initial conditions which has the dimension $(2n - m) - m = 2(n - m)$. It is convenient to take as such a manifold a surface Γ^*, defined by constraint equations (3.5) and m additional conditions:

$$\chi_a(q,p) = 0, \ a = 1,\ldots, m. \tag{3.8}$$

The functions χ_a must satisfy the condition

$$\det \| \{\chi_a, \varphi^b\} \| \neq 0, \tag{3.9}$$

because only in that case can Γ^* play the role of an initial surface for Equation (3.6). It is convenient to suppose that χ_a mutually commute*:

$$\{\chi_a, \chi_b\} = 0. \tag{3.10}$$

In such a case it is possible to introduce canonical variables onto the manifold Γ^*. Indeed, if condition (3.9) is satisfied, then, using canonical transformation in Γ, we can introduce a new set of variables where χ_a take a simple form:

$$\chi_a(q, p) = p_a, \quad a = 1, \ldots, m, \tag{3.11}$$

where $p_a(a = 1, \ldots, m)$ is a subset of canonical momenta of the new system of variables. Condition (3.9) can be written, in terms of the new variables, as

$$\det \left\| \frac{\partial \varphi^a}{\partial q^b} \right\| \neq 0, \tag{3.12}$$

and the constraint Equation (3.6) can therefore be solved with respect to q^a. Finally, the surface Γ^* is given by the equations

$$p_a = 0, \qquad q^a = q^a(q^*, p^*), \tag{3.13}$$

on Γ, so that q^* and p^* are independent, canonically constructed, variables on Γ^*.

Let us study now what the functional integral for the finite-dimensional mechanical system with constraints looks like. We shall introduce additional conditions $\chi_a(q, p)$ so that relations (3.9) and (3.10) are satisfied. The basic assertion is that the evolution operator matrix element is given by the functional integral

$$\int \exp \left\{ i \int_{t_0}^{t} \left(\sum_{i=1}^{n} p_i \dot{q}^i - H(q, p) \right) d\tau \right\} \prod_{\tau} d\mu(q(\tau), p(\tau)), \tag{3.14}$$

where the integration measure is given by the formula

$$d\mu(\tau) = (2\pi)^{m-n} \det \| \{\chi_a, \varphi^b\} \| \prod_a \delta(\chi_a)\delta(\varphi^a) \prod_{i=1}^{n} dq^i(\tau)dp_i(\tau). \tag{3.15}$$

---

*As from now, we shall mean by a commutator of functions f and g on phase space the Poisson bracket $\{f, g\}$ (3.3). The functions commute if the Poisson bracket is equal to zero.

To prove the assertion we transform integral (3.14) with measure (3.15) to integral (2.20) where integration is taken along the trajectories in the physical phase space Γ^*. Now we use the above-mentioned coordinates q^a, q^*, p_a, p^*. Integral (3.14) then transforms into the integral with the measure

$$d\tilde{\mu} = (2\pi)^{m-n} \det \left\| \frac{\partial \varphi^a}{\partial q^b} \right\| \prod_a \delta(p_i) \delta(\varphi^a) \prod_{i=1}^n dq^i \, dp_i, \tag{3.16}$$

which can be rewritten as

$$\prod_a \delta(p_a) \delta(q^a - q^a(q^*, p^*)) \, dq^a \, dp_a \prod_{i=1}^{n-m} \frac{dq^{*j} \, dp_j^*}{2\pi}. \tag{3.17}$$

One need not integrate over q and p thanks to the δ-functions. As a result, the integral takes the following form

$$\int \exp \left\{ i \int_{t_0}^t \left(\sum_j p_j^* \dot{q}^{*j} - H^*(q^*, p^*) \right) d\tau \right\} \prod_\tau \prod_{j=1}^{n-m} \frac{dq^{*j} \, dp_j^*}{2\pi}, \tag{3.18}$$

which coincides with (2.20). It is therefore possible to regard formulae (3.14) and (3.15) as proved.

Let us bear in mind that integral (3.14) can be rewritten as

$$\int \exp \left\{ i \int_{t_0}^t \left(\sum_i p_i \dot{q}^i - H - \sum_a \lambda_a \varphi^a \right) d\tau \right\} \prod_\tau \det \| \{\chi_a, \varphi^b\} \| (2\pi)^{m-n} \times$$

$$\times \prod_a \delta(\chi^a) \prod_{i=1}^n dq^i \, dp_i \prod_b \frac{\Delta\tau \, d\lambda_b}{2\pi}. \tag{3.19}$$

The symbol $\prod_b \Delta\tau(d\lambda_b/2\pi)$ shows that in the prelimit expression there are integrals over $\lambda_b(\tau_i)$ (τ_i are dividing points of the interval $[t_0, t]$) of the type

$$\int \exp \left\{ -i \sum_{i,a} \lambda_a(\tau_i) \varphi^a(q(\tau_i), p(\tau_i)) \Delta(\tau) \right\} \prod_{i,b} d\tau \frac{d\lambda_b}{2\pi}. \tag{3.20}$$

Expression (3.20) is equal to the product of δ-functions

$$\prod_{i,a} \delta[\varphi^a(q(\tau_i), p(\tau_i))]. \tag{3.21}$$

It means that in integral (3.19) we can carry out the integration over λ_b and return again to integral (3.14).

We shall show that functional integral (3.14) does not depend on the choice of additional conditions. Let $\delta\chi_a$ be an infinitesimally small change of those conditions. Up to the linear combination of constraints, it is possible to represent $\delta\chi_a$ as a result of infinitesimal canonical transformation in Γ, the generator of which is a linear combination of constraints. Indeed, $\delta\chi_a$ can be expressed as

$$\delta\chi_a = \{\Phi, \chi_a\} + \sum_b c_{ab}\varphi^b, \tag{3.22}$$

where

$$\Phi = \sum_a h_a\varphi^a, \tag{3.23}$$

and the solution of the system of equations

$$\sum_b \{\chi_a, \varphi^b\}h_b = -\delta\chi_a. \tag{3.24}$$

can be used as h_a.

This system has an unambiguous solution due to conditions (3.9).

Using the canonical transformation described above, we can replace the constraints by their linear combinations

$$\delta\varphi^a = \sum_b A_b^a\varphi^b, \tag{3.25}$$

where

$$A_b^a = \{h_b, \varphi^a\} - \sum_c h_c c_b^{ac}. \tag{3.26}$$

The variables of integral (3.14) and measure (3.15) transform as follows

$$\chi_a \to \chi_a + \delta\chi_a, \qquad \varphi^a + \sum_b A_b^a\varphi^b, \qquad H \to \dot{H};$$

$$\prod_a \delta(\varphi^a) \to \prod_a \delta(\varphi^a + \delta\varphi^a) = \left(1 + \sum_a A_a^a\right)\prod_a \delta(\varphi^a);$$

$$\det\|\{\chi_a, \varphi^b\}\| \to \det\|\{\chi_a + \delta\chi_a, \varphi^b + \delta\varphi^b\}\|$$

$$= \det\|\{\chi_a + \delta\chi_a, \varphi^b\}\|\det\left\|\frac{\partial(\varphi^a + \delta\varphi^a)}{\partial\varphi^b}\right\|$$

$$= \det\|\{\chi_a + \delta\chi_a, \varphi^b\}\|\left(1 + \sum_a A_a^a\right).$$

As a result of the canonical transformation, the integration measure differs from measure (3.15) only by the substitution of $\chi_a \to \chi_a + \delta\chi_a$. This also proves the independence of integral (3.14) on the choice of additional conditions.

In the following section the functional integrals for finite-dimensional mechanical systems obtained in the preceding sections are generalized to the field theory which describes systems with an infinite number of degrees of freedom.

4. FUNCTIONAL INTEGRALS AND QUANTIZATION ON MANIFOLDS

In this section we shall examine two examples of functional integral calculation enabling us to carry out the quantization of mechanical systems on manifolds. The spectrum of observables of a dynamical system can be calculated using the method of functional integration and, if the manifold is a group, the method is useful for finding irreducible group representation. The examples shown below differ in the structure of the phase space Γ.

1. $\Gamma = u(1) \times R$ (cylinder). We shall examine the functional integral.

$$\langle \varphi' | \exp(-i\hat{H}t) | \varphi \rangle = \int \exp\left\{ i \int (\pi \, d\varphi - H \, dt) \right\} \prod \frac{d\varphi \, d\pi}{2\pi} \qquad (4.1)$$

over the phase space of a mechanical system, the coordinate part of which is a unit circumference (interval $(0, 2\pi)$) with joint end points.

Functional integral (4.1) with the Hamiltonian depending on momenta belongs to a class of exactly calculable ones (Section 1). The compactibility of coordinate space leads, however, to an important change when compared with the example studied in Section 1, where coordinate space was a straight line.

For evaluation of integral (4.1) we replace the exponential argument with the prelimit expression

$$i \sum \pi_i(\varphi_{i+1} - \varphi_i) - i \sum H(\pi_i) \Delta t, \qquad (4.2)$$

where π_i are values of π on the intervals $(t_{i+1} - t_i)$, $i = 1, 2, \ldots, N$. We integrate first over variables φ_i – the values of variable $\varphi(t)$ at the end points of the interval $t_{i+1} - t_i$, different from $t_0 = t, t_N = t'$. It is easy to see that to perform it we have to integrate over variables $\varphi_i (i = 1, \ldots, N - 1)$ along the whole real axis and then to sum over all functions $\varphi(t)$ satisfying the conditions

$$\varphi(t) = \varphi, \qquad \varphi(t') = \varphi' + 2\pi m, \qquad (4.3)$$

where m is an integer. Integration over the momentum variables has to be carried out along the whole real axis. It is therefore necessary to evaluate the expression

$$(2\pi)^{-N} \sum_{m=-\infty}^{\infty} \int \prod_{i=1}^{N} d\pi_i \prod_{i=1}^{N-1} d\varphi_i \exp\left\{ i\sum_i \pi_i(\varphi_{i+1} - \varphi_i) + \right.$$

$$\left. + i\pi_N 2\pi m - i\sum_i H(\pi_i)\Delta t \right\}. \tag{4.4}$$

Integration over φ_i yields

$$(2\pi)^{-N} \int \exp\left\{ i\sum_{i=1}^{N} \pi_i(\varphi_{i+1} - \varphi_i) \right\} \prod_{i=1}^{N-1} d\varphi_i$$

$$= (2\pi)^{-1} \prod_{i=1}^{N-1} \delta(\pi_{i+1} - \pi_i) \exp i(\pi_N \varphi' - \pi_1 \varphi). \tag{4.5}$$

The product of $N-1$ δ-functions allows us to integrate over all momentum variables except one. Expression (3.4) takes the form

$$(2\pi)^{-1} \int d\pi' \sum_{m=-\infty}^{\infty} \exp\left\{ i\pi'(\varphi' - \varphi) + 2\pi m - i(t' - t)H(\pi') \right\}. \tag{4.6}$$

Summing over m

$$\sum_m \exp(i\pi' 2\pi m) = \sum_{m=-\infty}^{\infty} \delta(\pi' - m) \tag{4.7}$$

and integrating over π', we obtain

$$(2\pi)^{-1} \sum_{m=-\infty}^{\infty} \exp[im(\varphi' - \varphi)] \exp[i(t' - t)H(m)]. \tag{4.8}$$

This result can be interpreted as follows: The expression $\exp(im\varphi)$ is a one-dimensional irreducible representation of the group $U(1)$. Then, (4.8) gives the sum over all such representations.

As can be seen from (4.8) when quantizing a system on the one-dimensional group of rotations, the momentum operator spectrum is discrete (with integer values). Analogous formulae can be obtained for other groups, too, in particular for the group of three-dimensional rotations $O(3)$.

In the following example, the quantization of a dynamical system on a torus – a compact phase space of finite volume – is studied. The volume of phase space has to be a multiple of 2π.

2. $\Gamma = U(1) \times U(1)$ (torus; $0 \leq x \leq a$, $0 \leq y \leq b$). We shall represent as a continual integral the matrix element

$$\langle x'|\exp(-iHt)|x\rangle. \tag{4.9}$$

We define the functional integral as a limit of finite-dimensional integrals over values $y_1, \ldots, y_N$ of a piecewise constant function $y(t)$ and over values $x_1, \ldots, x_{N-1}$ of a piecewise linear function $x(t)$ at dividing points $\tau_1, \ldots, \tau_{N-1}$. Due to the continuity of the function $x(t)$, one has to integrate over variables $x_1, \ldots, x_{N-1}$ along the whole real axis and then, identically with the quantization on a cylinder, to sum over trajectories incoming into the points $x - ma$ (m is an integer). As for integrals over y-variables, it is straightforward to realize that integration over one of them (e.g., y_1), has to be carried out at an interval $[0, b]$ and over the others ($y_2, \ldots, y_N$), along the whole real axis.

The prelimit expression

$$(2\pi)^{-N} \sum_m \int dy_1 \, dx_1 \ldots dx_{N-1} \, dy_N \exp\Big\{ iy_1(x_1 - x) +$$
$$+ iy_2(x_2 - x_1) + \cdots + iy_N(x' + 2\pi m - x_{N-1}) - i\int H \, dt\Big\} \tag{4.10}$$

can be calculated exactly if H depends on y only. Without any loss of generality, H can be taken as a periodic function with a period b. The integral over the x-variables has the form

$$(2\pi)^{-N} \int dx_1 \ldots dx_{N-1} \times$$
$$\times \exp\{iy_1(x_1 - x) + iy_2(x_2 - x_1) + \cdots$$
$$\cdots + iy_N(x' + 2\pi m - x_N)\} \tag{4.11}$$
$$= (2\pi)^{-1}\delta(y_1 - y_2)\delta(y_2 - y_3)\ldots\delta(y_{N-1} - y_N) \times$$
$$\times \exp\{iy_N(x' + 2\pi m) - iy_1 x\}.$$

Integration over $y_1 \ldots, y_N$ removes all δ-functions and we receive the expression

$$(2\pi)^{-1}\sum_m \int_0^b dy \exp\{iy(x' - x + ma) - iH(y)(t' - t)\}. \tag{4.12}$$

Using formula

$$\sum_m \exp(imay) = \sum_n \delta\left(\frac{ay}{2\pi} - n\right) = \frac{2\pi}{a}\sum_n \delta\left(y - \frac{2\pi n}{a}\right), \qquad (4.13)$$

we obtain the expression

$$a^{-1}\int_0^b dy \exp\{iy(x'-x) - iH(y)(t'-t)\}\sum_n \delta\left(y - \frac{2\pi n}{a}\right) \qquad (4.14)$$

instead of (4.12).

Function (4.13) has an important property of b-periodicity if $ab/2\pi$ is an integer k. We obtain the condition of 'volume quantization':

$$ab = 2\pi k, \qquad (4.15)$$

where k is an integer. If the volume quantization condition is satisfied, expression (4.14) takes the form

$$a^{-1}\sum_{s=0}^{k-1} \exp\left\{i\frac{2\pi s}{a}(x'-x)\right\}\exp\left\{-i(t'-t)H\left(\frac{s}{k}b\right)\right\}. \qquad (4.16)$$

It follows from there that the possible values are

$$y_s = \frac{s}{k}b, \quad s = 0, 1, \ldots, k-1. \qquad (4.17)$$

The analogous possible x-values are

$$x_s = \frac{s}{k}a, \quad s = 0, 1, \ldots, k-1. \qquad (4.18)$$

The volume quantization condition (4.15) is equivalent to the condition of quantization of the Planck constant derived in Berezin's paper [148]. To convince ourselves of it, we introduce, instead of x, y, the 'cylindrical variables' x_1, x_2 defined as

$$x = \frac{a}{2\pi}x_1, \qquad y = \frac{b}{2\pi}y_1 \qquad (4.19)$$

and denote the factor $ab/4\pi^2$ emerging before the action S as h^{-1}.

CHAPTER 2

FUNCTIONAL INTEGRALS IN QUANTUM FIELD THEORY AND STATISTICAL PHYSICS

5. FUNCTIONAL INTEGRALS AND PERTURBATION THEORY IN QUANTUM FIELD THEORY

Field theory can be regarded as the theory of a mechanical system with an infinite number of degrees of freedom. The functional integral in field theory can be constructed using various methods. First, it is possible to start with field action, written in the Hamiltonian form, and construct the functional integral over the phase space of a system with an infinite number of degrees of freedom. Second, it is possible to start with the action not written explicitly in Hamiltonian form and study the functional integral over all fields. This approach enables us to construct explicitly relativistic theory. In the Hamiltonian approach, relativistic invariance is often not explicit and requires special proof.

The explanation and basic features of the method of integration over all fields may be given if a transformation of resulting functional integrals in to integrals of a Hamiltonian type is possible.

We shall examine the definition and the rules of dealing with the functional integral using as an example the theory of scalar field with action:

$$S = \int d^4x \left(\frac{1}{2} g_0^{\mu\nu} \frac{\partial \varphi}{\partial x^\mu} \frac{\partial \varphi}{\partial x^\nu} - \frac{m^2}{2} \varphi^2 - \frac{g}{3!} \varphi^3 \right). \tag{5.1}$$

Here $\varphi(x)$ are field functions depending on a point $x = (x^0, x^1, x^2, x^3)$ of pseudoeuclidean space V_4 and $g_0^{\mu\nu}$ is the diagonal Minkowski tensor $[1, -1, -1, -1]$. The action is the sum of the functional S_0 – the action of the free field quadratic in φ, and the integral over $-(g/3!)\varphi^3$ that describes the self-interaction with the coupling constant g. The factor $1/3!$ before $g\varphi^3$ has been chosen for convenience.

To define the functional integral over all fields the finite-dimensional approximation is frequently used.

In the space V_4 we take a big cubic volume V, divided into N^4 equal small cubes $v_i(i = 1, \ldots, N^4)$. We approximate function $\varphi(x)$ in the volume

17

V by a function constant in the volumes v_i and the first derivatives $\partial\varphi/\mathrm{d}x_\mu$ by the finite differences:

$$\frac{1}{\Delta l}[\varphi(x_\nu + \delta_{\mu\nu}\Delta l) - \varphi(x_\nu)], \tag{5.2}$$

where Δl is the length of the edge of the cube v_i. Approximating piecewise constant function $\varphi(x)$ is defined through its values in volumes v_i.

Let us consider the finite-dimensional integral

$$\int \exp(\mathrm{i}S) \prod_{\substack{i=1 \\ x\in v_i}}^{N} n(x)\,\mathrm{d}\varphi(x) \tag{5.3}$$

over the values of function $\varphi(x)$ in volume v_i. Here, S is the action integral for approximating function $\varphi(x)$ (with (5.2) as an approximation for its first derivatives) and $n(x)$ is independent of $\varphi(x)$, (it is chosen so that for $V \to \infty$, $v_i \to 0$ integral (5.3) would acquire the form $\exp(cV)$ with c being a V-independent constant). Usually $n(x)$ is of the type

$$n(x) = a(\Delta l)^\alpha \tag{5.4}$$

with constants (not depending on x) a and α.

Finite-dimensional integrals of type (5.3) are present in the prelimit expressions used for the definition of functional integrals encountered in field theory. We shall write down the Green function as a functional integral.

The *Green function* is an expectation value of the product of two or more field functions weighted by $\exp(\mathrm{i}S)$, e.g., the two-point function is defined by the formula

$$G(x, y) \equiv -\mathrm{i}\langle\varphi(x)\varphi(y)\rangle =$$

$$= -\mathrm{i}\lim_{\substack{V\to\infty \\ v_i\to 0}} \frac{\int \exp(\mathrm{i}S)\varphi(x)\varphi(y)\Pi_{i=1,x\in v_i}^{N^4} n(x)\,\mathrm{d}\varphi(x)}{\int \exp(\mathrm{i}S)\Pi_{i=1,x\in v_i}^{N^4} n(x)\,\mathrm{d}\varphi(x)}. \tag{5.5}$$

We denote the limit at the right-hand side of that formula by the symbol

$$\frac{\int \exp(\mathrm{i}S)\varphi(x)\varphi(y)\Pi_x n(x)\,\mathrm{d}\varphi(x)}{\int \exp(\mathrm{i}S)\Pi_x n(x)\,\mathrm{d}\varphi(x)}. \tag{5.6}$$

Green functions are considered to be known if we know the *generating functional*

$$Z[\eta] = \frac{\int \exp\mathrm{i}(S + \int\eta(x)\varphi(x)\,\mathrm{d}^4x)\Pi_x n(x)\,\mathrm{d}\varphi(x)}{\int \exp(\mathrm{i}S)\Pi_x n(x)\,\mathrm{d}\varphi(x)}. \tag{5.7}$$

Specifically, the two-point Green function is given by the formula

$$G(x, y) = i\frac{\delta}{\delta\eta(x)}\frac{\delta}{\delta\eta(y)}Z[\eta]|_{\eta=0}. \tag{5.8}$$

It is not difficult to evaluate the functional $Z[\eta]$ for free field theory. To perform it, we apply the shift

$$\varphi(x) \rightarrow \varphi(x) + \varphi_0(x), \tag{5.9}$$

when integrating over φ in the numerator of formula (5.7), where $\varphi_0(x)$ is chosen so that the terms linear in φ cancel in the argument of the exponential. This leads to the equation

$$-(\Box + m^2)\varphi_0(x) = -\eta(x) \tag{5.10}$$

for $\varphi_0(x)$. A solution of Equation (5.10) can be expressed in terms of the Green function $D(x, y)$ of the operator $(-\Box - m^2)$ by the formula

$$\varphi_0(x) = -\int D(x, y)\eta(y)\,d^4y. \tag{5.11}$$

The Green function is the solution of the equation

$$(-\Box_x - m^2)D(x, y) = \delta(x - y) \tag{5.12}$$

with the δ-function on the right-hand side.

When shift (5.9) is applied, the integral in the numerator of the right-hand side of (5.7) is reduced to multiplication of the denominator integral by the factor

$$\exp\left\{-\frac{i}{2}\int\eta(x)D(x, y)\eta(y)\,d^4x\,d^4y\right\}, \tag{5.13}$$

which gives the form of the generating functional in the case of free field theory. In this theory the two-point function (denoted $G_0(x, y)$), calculated according to formula (5.8), is the Green function of operator $(-\Box - m^2)$:

$$G_0(x, y) = D(x, y). \tag{5.14}$$

The definition of this function by Equation (5.12) is not unique. It is defined only up to an additive part – the solution of the homogeneous equation

$$(-\Box - m^2)f = 0. \tag{5.15}$$

There exists a most natural choice for the function $D(x, y)$. This choice can be justified by many arguments. We bring here one of them.

The expression $\exp(iS)$ is an oscillating functional of $\varphi(x)$. We shall study instead of it the functional $\exp(iS_\varepsilon)$, where S_ε is a complex action depending on a nonnegative parameter ε.

$$S_\varepsilon = \tfrac{1}{2} \int \varphi(-\Box - m^2 + i\varepsilon)\varphi \, d^4x, \tag{5.16}$$

chosen so that the absolute value of the functional $\exp(iS_\varepsilon)$ is less than one and vanishes for $\int \varphi^2 d^4x \to \infty$.

The results obtained are unambiguous if the corrected action S_ε is used for the definition of the Green function and if the limit $\varepsilon \to +0$ is performed. Specifically $D(x, y)$ becomes a limit of the Green function of the operator $(-\Box - m^2 + i\varepsilon)$. This function is unambiguously defined. It depends on the difference $(x - y)$ and is given by

$$D(x - y) = \frac{1}{(2\pi)^4} \int \frac{d^4k \, e^{ik(x-y)}}{k^2 - m^2 + i\varepsilon}. \tag{5.17}$$

The limit of this function for $\varepsilon \to +0$ is called the *propagator* or *Feynman's Green function* and is denoted by $D_F(x - y)$.

In the theory of free fields we have

$$\langle \varphi(x)\varphi(y) \rangle = iD_F(x - y). \tag{5.18}$$

for the expectation value of the product of fields $\varphi(x)$ and $\varphi(y)$.

The expectation value of the product of an arbitrary but finite number of functions $\varphi(x_1), \ldots, \varphi(x_n)$ is in the case of free field theory obtained by taking the nth functional derivative of expression (5.13) and putting $\eta = 0$.

The expectation value of an odd number of the functions φ is obviously equal to zero. For the expectation value of an even number of functions it is easy to derive a statement known as the *Wick theorem*.

The expectation value of the product of an even number of functions $\varphi(x_1), \ldots, \varphi(x_{2n})$ is equal to the sum of the products of the expectation values of all possible pairs. For instance,

$$\begin{aligned}
\langle \varphi(x_1)\varphi(x_2)\varphi(x_3)\varphi(x_4) \rangle &\\
= \langle \varphi(x_1)\varphi(x_2) \rangle \langle \varphi(x_3)\varphi(x_4) \rangle &+ \\
+ \langle \varphi(x_1)\varphi(x_3) \rangle \langle \varphi(x_2)\varphi(x_4) \rangle &+ \\
+ \langle \varphi(x_1)\varphi(x_4) \rangle \langle \varphi(x_2) \times \varphi(x_3) \rangle. &
\end{aligned} \tag{5.19}$$

The proof for the expectation value of the $2n$ functions can be accomplished by differentiating $2n$ times functional (5.13) and putting $\eta = 0$.

Wick's theorem can be used for the construction of the formal perturbation theory and the corresponding diagram technique. We shall construct the perturbation theory for the scalar field with Lagrangian (5.1). Let us represent the functional $\exp(iS)$ as

$$\exp(iS) = \exp(iS_0)\exp(iS_1), \tag{5.20}$$

where S_0 is the free field action and the term

$$S_1 = -\frac{g}{3!}\int \varphi^3(x)\,d^4x \tag{5.21}$$

describes the self interaction. The perturbation theory is based on the expansion of $\exp(iS_1)$ into the series in g

$$\exp(iS_1) = \sum_{n=0}^{\infty} \frac{(-ig)^n}{n!(3!)^n}\int \varphi^3(x_1),\ldots,\varphi^3(x_n)\,d^4x_1,\ldots,d^4x_n \tag{5.22}$$

and an integration of the resulting series term by term, e.g., for a two-point function we receive an expression in the form of a ratio of two series

$$G(x, y) = -i\frac{\sum_{n=0}^{\infty}\dfrac{(-ig)^n}{n!(3!)^n}\int \exp(iS_0)\varphi(x)\varphi(y)\int \varphi^3(x_1),\ldots}{\sum_{n=0}^{\infty}\dfrac{(-ig)^n}{n!(3!)^n}\int \exp(iS_0)\varphi^3(x_1),\ldots} \longrightarrow$$

$$\longrightarrow \frac{\ldots,\varphi^3(x_n)\,d^4x_1,\ldots,d^4x_n\,\Pi_x\,n(x)\,d\varphi(x)}{\ldots,\varphi^3(x_n)\,d^4x_1,\ldots,d^4x_n\,\Pi_x\,n(x)\,d\varphi(x)}. \tag{5.23}$$

Dividing the numerator and the denominator of the right-hand side by the integral

$$\int \exp(iS_0)\prod_x n(x)\,d\varphi(x), \tag{5.24}$$

we transfer the problem to the calculation of the expectation values

$$\langle \varphi^3(x_1),\ldots,\varphi^3(x_n)\rangle_0 \equiv \frac{\int \exp(iS_0)\varphi^3(x_1),\ldots,\varphi^3(x_n)\Pi_x\,n(x)\,d\varphi(x)}{\int \exp(iS_0)\Pi_x\,n(x)\,d\varphi(x)} \tag{5.25}$$

in the denominator of the right-hand side of (5.23) and to the calculation of the expectation values

$$\langle \varphi(x)\varphi(y)\varphi^3(x_1),\ldots,\varphi^3(x_n)\rangle_0 \tag{5.26}$$

in the numerator. Here we use Wick's theorem, which expresses the expectation values $\langle \ldots \rangle_0$ as a sum of all possible pairs expectation values allowing one to evaluate any term of the series in (5.23). Feynman has shown that a picture diagram can be assigned to each term of the series considered. The perturbation theory in which every term corresponds to a diagram is called the *diagram technique*. In the scalar field theory with self interaction studied here it is possible to construct diagrams in the following way.

We assign a diagram made of n points (each having three legs) depicting points $x_1,\ldots,x_n$ of pseudoeuclidean space to expectation value (5.25). Such a diagram has for $n = 4$ the form

$$\tag{5.27}$$

To the expectation value (5.26) we assign the diagram obtained from the corresponding diagram for expectation value (5.25) by adding two points (each having one leg) that connect points x and y in V_4, e.g., for $n = 4$ we get

$$\tag{5.28}$$

Diagrams introduced in such a way shall be called *prediagrams* to distinguish them from those introduced later on. Prediagrams are symmetric with respect to the permutation of $x_1,\ldots,x_n$ points and also with respect to the permutation of legs in each point. It is therefore possible to speak about the symmetry group G_n of an n-point prediagram of the order

$$R_n = n!(3!)^n. \tag{5.29}$$

The symmetry of prediagrams reflects the symmetry of corresponding expectation values (5.25) and (5.26) with do not change under a permutation of arguments and under permutation in any of the triplets of the field functions $\varphi(x_i)\varphi(x_i)\varphi(x_i) = \varphi^3(x_i)$ in the expression for the expectation value.

Let us notice that the expression R_n^{-1} (together with $(-ig)^n$) plays the role of a multiplicative factor of the expectation value $\langle \dots \rangle_0$ in series (5.23).

According to Wick's theorem, the expectation values (5.25) and (5.26) are sums of the products of all possible expectation values of field function pairs. Connecting each pair of the points x_i, x_j with a line, one can assign a diagram to every type of formation of the pair expectation values, if among those values the expectation value $\langle \varphi(x_i)\varphi(x_j) \rangle_0$ is present. The number of lines is equal to the number of pairs, i.e., half the number of field functions.

All diagrams arising from prediagram (5.27) are shown in (5.30) and those arising from (5.28) in (5.31)

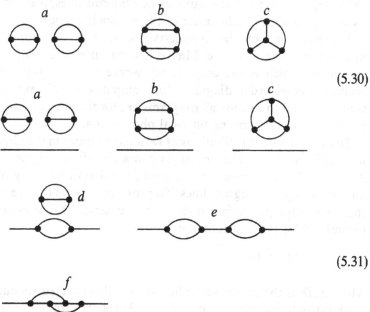

$$(5.30)$$

$$(5.31)$$

The expression corresponding to a diagram is obtained when the product of the pair expectation values is integrated over $x_1, \dots, x_n$, multiplied by $(-ig)^n R_n^{-1}$ and by the number of ways through which a given diagram can be constructed from a prediagram. It is not difficult to observe that the number of ways is equal to the ratio $R_n/r_{n,d}$ where R_n is the order of the prediagram symmetry group and $r_{n,d}$ is the order of the symmetry group of the diagram which is obtained from the prediagram by connecting its vertices with lines. As a result we get the factor before the integral over $x_1, \dots, x_n$ equal to $(-ig)^n r_{n,d}^{-1}$.

The rules of correspondence can be formulated as follows. To each line connecting the points x_i and x_j, the Green function $D_F(x_i - x_j)$ is assigned which differs from the expectation value $\langle \varphi(x_i)\varphi(x_j) \rangle_0$ by factor i, and each vertex has its counterpart in the coupling constant g:

$$\underline{\hspace{3cm}}_{x_i \hspace{2cm} x_j} \quad D_F(x_i - x_j) \qquad\qquad g \qquad\qquad (5.32)$$

The expression for a diagram is obtained when the product of expressions that correspond to elements of diagrams – to vertices and lines – is integrated over the coordinates of vertices and multiplied by $(i)^{l-n-1} r_{n,d}^{-1}$, where l is the number of diagram lines, n is the number of its vertices and $r_{n,d}$ is the order of the symmetry group.

The existence of the symmetry factor $r_{n,d}^{-1}$ has not always been mentioned in the literature. Maybe it is because in the quantum electrodynamics (which is the only theory where it is necessary to take into account higher-order diagrams for comparison with experiments) this factor is equal to one for all diagrams except the vacuum ones which can be neglected in describing physical phenomena.

To evaluate the contributions of various diagrams in the Green function, it is sufficient to restrict oneself to connected diagrams, i.e., those where it is possible to go from any vertex of the diagram to any other vertex, moving along the diagram lines. To prove it we have to take into account that the diagrams corresponding to the series in the denominator of formula (5.23) give the total contribution

$$\exp \sum_i D_i^c, \qquad\qquad (5.33)$$

where $\Sigma_i D_i^c$ is the sum of contributions of all connected vacuum diagrams (without external lines). Formula (5.33) follows from the fact that the diagram consisting of n_1 connected components of a given type, n_2 connected components of another type and so on, has the following symmetry factor

$$r^{-1} = \prod_i ((n_i)! r_i^{n_i})^{-1}, \qquad\qquad (5.34)$$

where r_i is the order of the symmetry group of a connected component of the ith kind. The factors $((n_i)!)^{-1}$ reflect the symmetry of a diagram with respect to a permutation of the same components and lead to the exponential (5.33). It is worthwhile remarking that the sum of contributions of diagrams in the numerator (5.23) can be reduced to the product of the sum of contributions of connected components with the factor (5.33).

In quantum theory textbooks (see, e.g., [35–37]), the diagram technique is usually constructed using an operator approach. Derivation of this technique by means of the functional integral approach, as shown here, is more natural. Feynman derived his diagrams using the functional integration technique.

For particular calculations, the diagram technique in momentum space is more useful. It is obtained when we pass to the Fourier transforms $\tilde{\varphi}(k)$ of the field functions $\varphi(x)$:

$$\varphi(x) = \frac{1}{(2\pi)^4} \int \exp(ikx)\,\tilde{\varphi}(k)\,d^4k \tag{5.35}$$

and expectation values of the type

$$\langle \tilde{\varphi}(k_1),\ldots,\tilde{\varphi}(k_n) \rangle. \tag{5.36}$$

play the role of Green functions.

The expressions, corresponding to the elements of a diagram – vertices and lines – take the form

$$
\begin{aligned}
&\overset{k_1 \qquad k_2}{\rule{3cm}{0.4pt}} \qquad \delta(k_1 + k_2)(k_1^2 - m^2 + i\varepsilon)^{-1} \\[2mm]
&\qquad\qquad\qquad\qquad\qquad\qquad g\delta(k_1 + k_2 + k_3).
\end{aligned}
\tag{5.37}
$$

In the momentum diagram technique, the contribution of a particular diagram is obtained if the product of expressions corresponding to its elements according to (5.37) is integrated over all internal momenta and multiplied by $r_{n,d}^{-1}(i/(2\pi))^{l-n-1}$, where n is the number of vertices, l is the number of lines and $r_{n,d}$ is the order of the diagram symmetry group.

Let us remark that the Green function in momentum space contains as a multiplicative factor the δ-function $\delta(\Sigma_i k_i)$ which ensures the conservation of the 4-momentum

$$G(k_i) = M(k_i)\delta(\Sigma_i k_i). \tag{5.38}$$

Knowing the Green function, the elements of the S-matrix can be calculated according to the formula

$$S(k_1,\ldots,k_n) = \lim_{k_i^2 \to m_i^2} M(k_i) \cdot$$

$$\cdot \left\{ \prod_{i=1}^{n} (k_i^2 - m_i^2)\theta(\pm k_i^0)|2k_i^0|^{-1/2}(2\pi)^{-3/2} \right\}. \tag{5.39}$$

The proof of formula (5.39) is not presented here, it can be found in many textbooks on quantum field theory.

The method of functional integration over all fields can be explained and substantiated if it is possible to transform the functional integrals obtained here into integrals of a Hamiltonian form which represent a field theory generalization of the integrals obtained above, in the quantization of the finite-dimensional mechanical systems.

Continuing the examination of the scalar field example, we shall write down in Hamiltonian form the functional integral

$$\int \exp{(iS)} \prod_x n(x)\, d\varphi(x). \tag{5.40}$$

To proceed, we consider the integral

$$\int \exp{(iS[\varphi, \pi])} \prod_x n(x)\, d\varphi(x)\, d\pi(x), \tag{5.41}$$

where the expression

$$S[\varphi, \pi] = \int \left(\pi\, \partial_0 \varphi - \frac{1}{2}\pi^2 - \frac{1}{2}(\nabla\varphi)^2 - \frac{m^2}{2}\varphi^2 - \frac{g}{3!}\varphi^3 \right) d^4x \tag{5.42}$$

coincides with action (5.1) provided $\partial_0\varphi(x)$ is substituted for $\pi(x)$.

Action (5.42) is of Hamiltonian form and the corresponding function is

$$H = \int d^3x \left(\frac{1}{2}\pi^2 + \frac{1}{2}(\nabla\varphi^2) + \frac{m^2}{2}\varphi^2 + \frac{g}{3!}\varphi^3 \right), \tag{5.43}$$

where the functions $\varphi(x)$, $\pi(x)$ have the meaning of coordinate and conjugated momentum densities, respectively. We show that integral (5.41) over variables φ and π results in integral (5.40) over all fields. To achieve this, it is sufficient to notice that the integral over π in formula (5.41) can be expressed explicitly if the shift

$$\pi(x) \to \pi(x) + \partial_0\varphi(x), \tag{5.44}$$

is performed which causes the integral to transform into the product of integral (5.40) over φ and the integral over π

$$\int \exp\left(-\frac{i}{2}\int \pi^2(x)\, d^4x \right) \prod_x d\pi(x), \tag{5.45}$$

leading to the product of normalization factors. In formulae for Green's functions – expectation values of a product of several fields φ – integrals of the type (5.45) enter both the numerator and the denominator and are therefore cancelled.

In such a way, we have succeeded in expressing the functional integral of scalar field theory through the Hamiltonian form, artificially introducing an integral over a new variable – the canonical momentum. Such an approach will be used in the sequel for the proof of the Hamiltonicity of given systems of quantum field theory and statistical physics.

The scheme of functional integration over all fields produces a method of quantization of Bose fields. In operator formalism such a quantization leads to the replacement of field functions with operators obeying Bose commutation relations.

Quantization of Fermi fields can be performed using the functional integral over anticommuting variables (for details see Berezin's book [29]). The following basic facts are necessary to accomplish it.

The integral over Fermi fields (over an infinite Grassman algebra with involution) is defined as a limit of the integral on an algebra with a unit element and a finite even number of generators $x_i, x_i^* (i = 1, 2, \ldots, n)$ obeying commutation relations

$$x_i x_j + x_j x_i = 0; \qquad x_i^* x_j^* + x_j^* x_i^* = 0; \qquad x_i x_j^* + x_j^* x_i = 0. \quad (5.46)$$

Any of the elements of the algebra $f(x, x^*)$ is a polynom of the form

$$f(x, x^*) = \sum_{a_i, b_i = 0, 1} c_{a_1, \ldots, a_n b_1, \ldots, b_n} x_2^{a_1}, \ldots, x_n^{a_n} (x_1^*)^{b_1}, \ldots, (x_n^*)^{b_n} \quad (5.47)$$

with the coefficients $c_{a_1, \ldots, a_n b_1, \ldots, b_n}$ being complex numbers. Due to commutation relations (5.42), we have $x_i^2 = (x_i^*)^2 = 0$ for $i = j$ so that the powers of generators higher than one vanish. It is sufficient to limit oneself to the ordering of factors accepted in (5.47) because any other type of ordering can be changed into it using commutation relations (5.46) for $i \neq j$.

Let us define the operation of involution, acting on element (5.47), as

$$f \to f^* = \sum_{a_i, b_i = 0, 1} \bar{c}_{a_1, \ldots, a_n b_1, \ldots, b_n} (x_n)^{b_n}, \ldots, (x_1)^{b_1} (x_n^*)^{a_n}, \ldots, (x_1^*)^{a_1}. \quad (5.48)$$

On the algebra, we can introduce the integral

$$\int f(x, x^*) \, dx^* \, dx \equiv \int f(x_1, \ldots, x_n, x_1^*, \ldots, x_n^*) \, dx_1^* \, dx_1, \ldots, dx_n^* \, dx_n, \quad (5.49)$$

This integral is defined through the relations

$$\int dx_i = 0; \qquad \int dx_i^* = 0; \qquad \int x_i \, dx_i = 1; \qquad \int x_i^* \, dx_i^* = 1 \quad (5.50)$$

and we demand that the symbols dx_i, dx_i^* anticommute with each other and with the generators if the natural condition of linearity is imposed

$$\int (c_1 f_1 + c_2 f_2) \, dx^* \, dx = c_1 \int f_1 \, dx^* \, dx + c_2 \int f_2 \, dx^* \, dx. \quad (5.51)$$

When we integrate the sum (5.47), only the contribution of the term for which $a_i = b_i = 1$ for all $i = 1, \ldots, n$ is different from zero.

The following two formulae will be useful later on:

$$\int \exp(-x^* A x) \, dx^* \, dx = \det A; \qquad\qquad (5.52)$$

$$\frac{\int \exp(-x^* A x + \eta^* x + x^* \eta) \, dx^* \, dx}{\int \exp(-x^* A x) \, dx^* \, dx} = \exp(\eta^* A^{-1} \eta), \qquad (5.53)$$

where

$$x^* A x = \sum_{i,k} a_{ik} x_i^* x_k \qquad\qquad (5.54)$$

is a quadratic form of the generators x_i, x_i^* corresponding to the matrix A.

The expressions

$$\eta^* x = \sum_i \eta_i^* x_i; \qquad x^* \eta = \sum_i x_i^* \eta_i \qquad (5.55)$$

are linear forms of the generators x_i, x_i^* whose coefficients η_i, η_i^* anticommute with each other and with the generators. The elements η_i, η_i^*, together with the generators x_i, x_i^*, can be regarded as generators of a larger algebra. The expression $\eta^* A^{-1} \eta$ in formula (5.51) is a quadratic form of the matrix A^{-1} inverse to the matrix A.

The exponential in the integrand of formulae (5.52) and (5.53) can be expressed through the expansion into series, in which, due to commutation relations (5.46), only several first terms are different from zero.

Formula (5.52) is not difficult to prove taking into account that only the nth term of the expansion of the exponential gives a contribution to the integral. Formula (5.53) can be proved using the shift $x \to x + \tilde{\eta}$, $x^* \to x^* + \tilde{\eta}^*$ which cancels the linear form of x, x^* in the exponent of the integrand.

Detailed proof and some generalizations of formulae (5.52), and (5.53) can be found in the above-mentioned Berezin's monograph [29].

6. FUNCTIONAL INTEGRALS AND THE TEMPERATURE DIAGRAM TECHNIQUE IN STATISTICAL PHYSICS

In this section the temperature diagram technique for perturbation theory in quantum statistical physics is derived by means of the method of functional integration.

There are several modifications of Green's functions for quantum mechanical systems at temperatures different from zero – temperature, temperature-time and so on. The diagram technique for perturbation theory can be directly obtained only for temperature Green's functions. They are convenient for calculation of the thermodynamical characteristics of the system and they also contain information on the quasiparticle spectrum and weakly nonequilibrium kinetical phenomena.

We shall explore the functional integral formalism for the system of Bose particles placed into a cubic volume $V = L^3$ with periodic boundary conditions. The functional integral for such a case is an integral in the space of complex functions ('fields') $\psi(\mathbf{x}, \tau)$, $\bar{\psi}(\mathbf{x}, \tau)$, $\mathbf{x} \in V$, periodical in the parameter τ ('time') with period $\beta = (kT)^{-1}$, where k is the Boltzmann constant and T is the absolute temperature.*

We define the Green's functions as expectation values (in the above-defined space) of the products of several functions ψ, $\bar{\psi}$ with different arguments weighted with $\exp S$, where S is the functional of ψ, $\bar{\psi}$, having the meaning of action

$$S = \int_0^\beta \mathrm{d}\tau \int \mathrm{d}^3 x \, \bar{\psi}(\mathbf{x}, \tau) \, \partial_\tau \psi(\mathbf{x}, \tau) - \int_0^\beta H'(\tau) \, \mathrm{d}\tau; \qquad (6.1)$$

$H'(\tau)$ is the Hamiltonian of the form

$$H'(\tau) = \int \mathrm{d}^3 x \left[\left(\frac{1}{2\pi} \bar{\nabla}\bar{\psi}(\mathbf{x}, \tau), \nabla\psi(\mathbf{x}, \tau) \right) - \lambda \bar{\psi}(\mathbf{x}, \tau)\psi(\mathbf{x}, \tau) \right] +$$
$$+ \frac{1}{2} \int \mathrm{d}^3 x \, \mathrm{d}^3 y \, u(\mathbf{x} - \mathbf{y}) \bar{\psi}(\mathbf{x}, \tau)\bar{\psi}(\mathbf{y}, \tau)\psi(\mathbf{y}, \tau)\psi(\mathbf{x}, \tau). \qquad (6.2)$$

---

*In the following we shall use the system of units with $\hbar = k = 1$, where $\hbar$ is the Planck constant and k is the Boltzmann constant.

Here λ is the chemical potential of the system and $u(\mathbf{x} - \mathbf{y})$ is the pair interaction potential of two Bose particles.

Due to the periodicity, the functions ψ, $\bar{\psi}$ can be decomposed into the Fourier series

$$\psi(\mathbf{x},\tau) = (\beta V)^{-1/2} \sum_{\mathbf{k},\omega} \exp(-\mathrm{i}(\mathbf{k}\mathbf{x} - \omega\tau))a(\mathbf{k},\omega);$$

$$\bar{\psi}(\mathbf{x},\tau) = (\beta V)^{-1/2} \sum_{\mathbf{k},\omega} \exp(\mathrm{i}(\mathbf{k}\mathbf{x} - \omega\tau))a^+(\mathbf{k},\omega), \tag{6.3}$$

where

$$\omega \equiv \omega_n = 2\pi n/\beta; \qquad k_i = 2\pi n_i/L; \tag{6.4}$$

n, n_i being integers.

The expansions of type (6.3) are valid for Green's functions. For example, we can represent the one-particle Green's function

$$G(\mathbf{x},\tau;\mathbf{x}_1,\tau_1) = \langle \psi(\mathbf{x},\tau)\bar{\psi}(\mathbf{x}_1,\tau_1) \rangle$$

$$\equiv -\frac{\int e^S \psi(\mathbf{x},\tau)\psi(\mathbf{x}_1,\tau_1)\,\mathrm{d}\bar{\psi}\,\mathrm{d}\psi}{\int e^S\,\mathrm{d}\bar{\psi}\,\mathrm{d}\psi} \tag{6.5}$$

as a ratio of two functional integrals, denoting the measure by the symbol $\mathrm{d}\bar{\psi}\,\mathrm{d}\psi$.

Function (6.5) depends, evidently, on differences $\mathbf{x} - \mathbf{x}_1'$, $\tau - \tau_1$. Substituting for ψ, $\bar{\psi}$ their Fourier expansions (6.3) we come to the conclusion that function (6.5) can be expressed as an expectation value

$$G(\mathbf{k},\omega) = -\langle a(\mathbf{k},\omega)a^+(\mathbf{k},\omega) \rangle. \tag{6.6}$$

We shall examine the perturbation theory diagram technique arising from the calculation of integrals of type (6.5). The quadratic form in ψ, $\bar{\psi}$ in action (6.1) will be taken as a nonperturbed action S_0 and a fourth-order form denoted as S_1 will be regarded as a perturbation.

In the action, we shall pass from the functions ψ, $\bar{\psi}$ to their Fourier coefficients a, a^+. We get the expression

$$S = \sum_{\mathbf{k},\omega} \left(\mathrm{i}\omega - \frac{k^2}{2m} + \lambda \right)a^+(\mathbf{k},\omega)a(\mathbf{k},\omega) - \frac{1}{2\beta V} \times$$

$$\times \sum_{\substack{\mathbf{k}_1 + \mathbf{k}_2 = \mathbf{k}_3 + \mathbf{k}_4 \\ \omega_1 + \omega_2 = \omega_3 + \omega_4}} v(\mathbf{k}_1 - \mathbf{k}_3)a^+(\mathbf{k}_1,\omega_1)a^+(\mathbf{k}_2,\omega_2)a(\mathbf{k}_3,\omega_3)a(\mathbf{k}_4,\omega_4) \tag{6.7}$$

Here $v(\mathbf{k})$ is the Fourier transformation of the potential $u(\mathbf{x})$ defined by the formula

$$u(\mathbf{x}) = \frac{1}{V} \sum_{\mathbf{k}} \exp(i\mathbf{k} \cdot \mathbf{x}) v(\mathbf{k}). \tag{6.8}$$

The use of Fourier coefficients $a(\mathbf{k}, \omega)$, $a^{+}(\mathbf{k}, \omega)$ gives an exceptionally simple form to the functional integral as well as to diagram perturbation theory. The measure in functional integral (6.5), denoted symbolically by $d\bar{\psi} \, d\psi$, can be written as

$$d\bar{\psi} \, d\psi = \prod_{\mathbf{k}, \omega} da^{+}(\mathbf{k}, \omega) \, da(\mathbf{k}, \omega). \tag{6.9}$$

The functional integral can be regarded as a limit of the finite-dimensional integral over the variables $a^{+}(\mathbf{k}, \omega)$, $a(\mathbf{k}, \omega)$ with $|\mathbf{k}| < k_0$, $|\omega| < \Omega_0$ for $k_0, \Omega_0 \to \infty$.

We shall construct the perturbation theory diagram technique applying the scheme used for the scalar field in section 5. The unperturbed Green function can be easily found knowing the generating functional

$$Z_0[\eta, \eta^{+}] = \frac{\begin{array}{c} \int \exp\left[S_0 + \Sigma_{\mathbf{k}, \omega}(\eta^{+}(\mathbf{k}, \omega)a(\mathbf{k}, \omega) + \eta(\mathbf{k}, \omega)a^{+}(\mathbf{k}, \omega))\right] \times \to \\ \to \times \Pi_{\mathbf{k}, \omega} \, da^{+}(\mathbf{k}, \omega) \, da(\mathbf{k}, \omega) \end{array}}{\int \exp(S_0) \Pi_{\mathbf{k}, \omega} \, da^{+}(\mathbf{k}, \omega) \, da(\mathbf{k}, \omega)}. \tag{6.10}$$

We evaluate this functional using the shift which removes linear forms in η, η^{+} from the exponent of the integrand. As a result we get

$$Z_0[\eta, \eta^{+}] = \exp\left\{ -\sum_{\mathbf{k}, \omega} \eta^{+}(\mathbf{k}, \omega) \eta(\mathbf{k}, \omega) \left(i\omega - \frac{k^2}{2m} + \lambda \right)^{-1} \right\}. \tag{6.11}$$

Differentiating first according to $\eta(\mathbf{k}, \omega)$, then according to $\eta^{+}(\mathbf{k}, \omega)$ and putting $\eta = \eta^{+} = 0$, we receive an expression for the unperturbed Green's function:

$$G_0(\mathbf{k}, \omega) = -\frac{\int a(\mathbf{k}, \omega) a^{+}(\mathbf{k}, \omega) e^{S_0} \Pi_{\mathbf{k}, \omega} \, da^{+} \, da}{\int e^{S_0} \Pi_{\mathbf{k}, \omega} \, da^{+} \, da}$$

$$= \left(i\omega - \frac{k^2}{2m} + \lambda \right)^{-1}. \tag{6.12}$$

From formula (6.10), Wick's theorem for expectation values of the type

$$\left\langle \prod_{i=1}^{n} a(\mathbf{k}_i, \omega_i) \prod_{i=1}^{n} a^{+}(\mathbf{k}_j, \omega_j) \right\rangle_0, \tag{6.13}$$

follows, where

$$\langle f(a^+, a)\rangle_0 \equiv \frac{\int f(a^+, a)\,e^{S_0}\Pi_{\mathbf{k},\omega}\,da^+\,da}{\int e^{S_0}\Pi_{\mathbf{k},\omega}\,da^+\,da}. \tag{6.14}$$

According to Wick's theorem, expectation value (6.13) is equal to a sum of the products of all possible pair expectation values of the following type

$$\langle a(\mathbf{k}_i, \omega_i)a^+(\mathbf{k}_j, \omega_j)\rangle = -\delta_{\mathbf{k}_i \mathbf{k}_j}\delta_{\omega_i \omega_j}G_0(\mathbf{k}_i, \omega_i). \tag{6.15}$$

We shall construct the perturbation theory expanding the functional $\exp S = \exp S_0 \exp S_1$ standing in the integrand of the functional integral into series in powers of S_1. We express Green's function as a ratio of two series

$$G(\mathbf{k}, \omega) = -\frac{\sum_{n=0}^{\infty}(1/n!)\langle a(\mathbf{k}, \omega)a^+(\mathbf{k}, \omega)S_1^n\rangle_0}{\sum_{n=0}^{\infty}(1/n!)\langle S_1^n\rangle_0}. \tag{6.16}$$

Each of the S_1 inside $\langle ...\rangle$ is a fourth-order form in the integration variables a, a^+. Thus, we deal with the expectation values

$$\left\langle \prod_{i=1}^{n} a^+(\mathbf{k}_{1j}\omega_{1j})a^+(\mathbf{k}_{2j}\omega_{2j})a(\mathbf{k}_{3j}\omega_{3j})a(\mathbf{k}_{4j}\omega_{4j})\right\rangle_0 \tag{6.17}$$

in the denominator (6.16). The following expectation values

$$\left\langle a(\mathbf{k}, \omega)a^+(\mathbf{k}, \omega)\prod_{i=1}^{n} a^+(\mathbf{k}_{1j}\omega_{1j})a^+(\mathbf{k}_{2j}, \omega_{2j})a(\mathbf{k}_{3j}, \omega_{3j})a(\mathbf{k}_{4j}, \omega_{4j})\right\rangle_0 \tag{6.18}$$

appear in the numerator. We assign a prediagram made of n vertices of fourth order to the expectation value (6.17). For $n = 2$ the prediagram has the form

$$\tag{6.19}$$

The arrows incoming into the vertex correspond to the variable a, the outgoing ones to a^+. We assign the following prediagram

$$\tag{6.20}$$

to expectation value (6.18). In accordance with Wick's theorem, the expectation values (6.17) and (6.18) are the sums of all possible pair expectation values. To each fixed way of creating pair expectation values we assign a diagram connecting each pair of the vertices i, j with a line if, among pair expectation values, the following one can be found

$$\langle a(\mathbf{k}_i, \omega_i) a^+ (\mathbf{k}_j, \omega_j) \rangle_0.$$

We enumerate for $n = 2$ all diagrams arising from prediagrams (6.19)

$$(6.21)$$

and also all diagrams arising from prediagram (6.20)

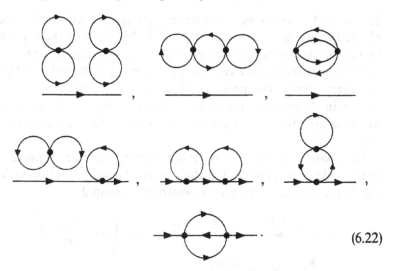

$$(6.22)$$

Expressions corresponding to a given diagram can be obtained if the product of pair expectation values is multiplied by the expression

$$\frac{(-1)^n}{n!} \prod_{i=1}^{n} \frac{v(\mathbf{k}_{1i} - \mathbf{k}_{3i})}{2\beta V},$$

$$(6.23)$$

and also by the number of ways through which a given diagram can be obtained from the prediagram. Then the summation over all momenta and frequencies of internal lines is carried out.

It is convenient to reformulate slightly the resulting rules of correspondence. The perturbation S_1 can be written down as

$$-\frac{1}{4\beta V} \sum_{\substack{k_1+k_2=k_3+k_4 \\ \omega_1+\omega_2=\omega_3+\omega_4}} (v(k_1-k_3)+v(k_1-k_4))a^+(k_1,\omega_1) \times$$

$$\times a^+(k_2,\omega_2)a(k_3,\omega_3)a(k_4,\omega_4), \qquad (6.24)$$

where the coefficient of $a^+(k_1,\omega_1)a^+(k_2,\omega_2)a(k_3,\omega_3)a(k_4,\omega_4)$ is symmetric with respect to the replacements

$$(k_1,\omega_1) \rightleftarrows (k_2,\omega_2), (k_3,\omega_3) \rightleftarrows (k_4,\omega_4).$$

The demonstrated symmetry allows one to speak about the prediagram symmetry group of the order

$$R_n = 4^n n!. \qquad (6.25)$$

The factor 4^n corresponds to symmetry groups at each fourth-order vertex; $n!$ – to the group of vertex permutations. Multiplying the expression R_n^{-1} by the number of ways through which a given diagram can be obtained from the prediagram, we receive $r_{n,d}^{-1}$ where $r_{n,d}$ is the order of the diagram symmetry group. The presence of the symmetry factor $r_{n,d}^{-1}$ – a common feature in diagram perturbation technique – was discussed in Section 5 where we investigated perturbation theory for relativistic scalar field theory with self interaction.

The above arguments allow us to formulate the rules of correspondence in the following manner. We shall assign Green's function $G_0(k,\omega)$ to a line of the diagram and the symmetrized potential

$$G_0(k,\omega) = \left(i\omega - \frac{k^2}{2m} + \lambda \right)^{-1} \qquad (6.26)$$

$$v(k_1-k_3)+v(k_1+k_4).$$

to a vertex. The expression corresponding to the diagram for the denominator in (6.16) can be obtained by summing over independent momenta and frequencies of the product of expressions which correspond to lines and vertices of the diagram and then multiplying the result by

$$\frac{1}{r_{n,d}} \left(\frac{-1}{\beta V} \right)^{l-n}, \qquad (6.27)$$

where l is the number of lines, n is the number of vertices and $r_{n,d}$ is the order of the diagram symmetry group. For the diagram corresponding to the numerator the factor in front of the diagram is equal to

$$\frac{1}{r_{n,d}} \left(\frac{-1}{\beta V} \right)^{l-n-1} , \tag{6.28}$$

where $l - n - 1 = c$ is the number of independent diagram loops.

A contribution into Green's function is given only by connected diagrams of the numerator in (6.16). The denominator in (6.16) has the form

$$\sum_{n=0}^{\infty} \frac{1}{n!} \langle S_1^n \rangle_0 = \exp \left(\sum_i D_i^c \right), \tag{6.29}$$

where $\sum_i D_i^c$ is the sum of all connected vacuum diagrams. The numerator in (6.16) is a product of factor (6.29) with the sum of all connected diagrams. The transition to connected diagrams is a common fact as well as the above-mentioned appearance of the factor $r_{n,d}^{-1}$.

Perturbation theory constructed for Bose systems coincides with the temperature diagram technique explained, e.g., in Reference [38]. Its derivation in functional integral formalism is simpler than in operator formalism. Moreover, the method of functional integration is more suitable for reformulation of perturbation theory which is applied in those cases when the standard perturbation theory constructed here is not applicable. From the functional integration point of view, such a reformulation is an alternative method of asymptotic evaluation of the functional integral.

Now, we shall stop briefly at the functional integral and diagram technique for Fermi systems. The quantization of a Fermi system can be obtained as a result of integration over the space of anticommuting functions (the elements of an infinite Grassmann algebra). To obtain the correct result it is necessary to impose the conditions of antiperiodicity in τ:

$$\psi(\mathbf{x}, \tau + \beta) = -\psi(\mathbf{x}, \tau), \qquad \bar{\psi}(\mathbf{x}, \tau + \beta) = -\bar{\psi}(\mathbf{x}, \tau). \tag{6.30}$$

on $\psi, \bar{\psi}$. As a result, the Green functions $\psi, \bar{\psi}$ are expanded into the Fourier series

$$\psi(\mathbf{x}, \tau) = (\beta V)^{-1/2} \sum_{\mathbf{k}, \omega} \exp \left[-i(\mathbf{k}\mathbf{x} - \omega \tau) \right] a(\mathbf{k}, \omega);$$

$$\bar{\psi}(\mathbf{x}, \tau) = (\beta V)^{-1/2} \sum_{\mathbf{k}, \omega} \exp \left[i(\mathbf{k}\mathbf{x} - \omega \tau) \right] a^+(\mathbf{k}, \omega), \tag{6.31}$$

where

$$\omega = 2\pi(n + 1/2)/\beta; \qquad k_i = 2\pi n_i/L; \qquad (6.32)$$

and n, n_i are integers. In contrast to the case of Bose systems, the Fermi frequencies ω are proportional to the half-integers $n + 1/2$.

The action S of the Fermi system has formally the same form (6.1) as for the Bose system. Passing to Fourier coefficients, it is convenient to write S as

$$S_1 = \frac{1}{4\beta V} \sum_{\substack{\mathbf{k}_1 + \mathbf{k}_2 = \mathbf{k}_3 + \mathbf{k}_4 \\ \omega_1 + \omega_2 = \omega_3 + \omega_4}} (v(\mathbf{k}_1 - \mathbf{k}_3) - v(\mathbf{k}_1 - \mathbf{k}_4))a^+(\mathbf{k}_1, \omega_1) \times$$

$$\times a^+(\mathbf{k}_2, \omega_2)a(\mathbf{k}_3, \omega_3)a(\mathbf{k}_4, \omega_4), \qquad (6.33)$$

having simultaneously antisymmetrized the potential. The derivation of the diagram technique is completely analogous to that performed above for Bose systems. The elements of the diagram technique

$$G_0(\mathbf{k}, \omega) = \left(i\omega - \frac{k^2}{2m} + \lambda \right)^{-1};$$

$$v(\mathbf{k}_1 - \mathbf{k}_3) - v(\mathbf{k}_1 - \mathbf{k}_4) \qquad (6.34)$$

differ from the corresponding elements for the Bose system by the fact that the frequencies are multiples of half-integers $\omega = 2\pi(n + 1/2)/\beta$ and by antisymmetrization of the potential.

The expression corresponding to the diagram can be obtained summing the product of expressions corresponding to diagram elements over independent momenta and frequencies and multiplying the result by the factor

$$(-1)^F \frac{1}{r_{n,d}} \left(-\frac{1}{\beta V} \right)^{l-n} \qquad (6.35)$$

for vacuum diagrams and by

$$(-1)^F \frac{1}{r_{n,d}} \left(-\frac{1}{\beta V} \right)^{l-n-1} \qquad (6.36)$$

for diagrams corresponding to a one-particle Green's function.

Expansions (6.35) and (6.36) differ from analogous expressions for (6.27) and (6.28) valid for the Bose system by the factor $(-1)^F$, where F is the number of closed Fermi loops of the diagram. The presence of the factor $(-1)^F$ is a consequence of anticommutativity of $\psi, \bar{\psi}$ fields.

The perturbation theory for Green's functions, constructed here, allows modifications connected with a partial summation of diagrams. It has been shown above that vacuum diagrams do not give a contribution to the Green's function, which is represented as a sum of the contributions of connected diagrams which can be passed from the entry to exit moving along diagram lines.

It is known that the full Green's function can be expressed through the irreducible self-energy part $\Sigma(\mathbf{k}, \omega)$, which is a sum of the contributions of all diagrams with two legs such that the diagram entry cannot be separated from the exit by cutting one line of the diagram. The partial summation of diagrams allowing one to express the function $G(\mathbf{k}, \omega)$ through $G_0(\mathbf{k}, \omega)$ and $\Sigma(\mathbf{k}, \omega)$ can be inferred from the graphical equation

The analytical form of that equation is

$$G(\mathbf{k}, \omega) = G_0(\mathbf{k}, \omega) + G_0(\mathbf{k}, \omega)\Sigma(\mathbf{k}, \omega)G_0(\mathbf{k}, \omega) +$$
$$+ G_0(\mathbf{k}, \omega)\Sigma(\mathbf{k}, \omega)G_0(\mathbf{k}, \omega)\Sigma(\mathbf{k}, \omega)G_0(\mathbf{k}, \omega) + \cdots \quad (6.38)$$
$$= G_0(\mathbf{k}, \omega) + G_0(\mathbf{k}, \omega)\Sigma(\mathbf{k}, \omega)G(\mathbf{k}, \omega).$$

The solution of Equation (6.38) has the form

$$G(\mathbf{k}, \omega) = (G_0(\mathbf{k}, \omega)^{-1} - \Sigma(\mathbf{k}, \omega))^{-1}. \quad (6.39)$$

Thus, to evaluate Green's function $G(\mathbf{k}, \omega)$, it is sufficient to find a corresponding self-energy part.

We shall now examine the modification of perturbation theory called the *skeleton diagram technique*. We come to it by making the partial summation of diagrams which leads to the replacement of light lines of the diagram, corresponding to an unperturbed Green's function, by bold ones to which the full Green's functions are assigned. In contrast with the ordinary diagram technique, however, there is no need to consider diagrams with self-energy parts inserted into internal lines. The expressions for the full functions are not known in advance so that for their definition a system of equations is employed. The first of them is Equation (6.39), the second one is the equation for the self-energy part $\Sigma(\mathbf{k}, \omega)$:

$$\Sigma(\mathbf{k}, \omega) = \rightarrow\!\!\left(\Sigma\right)\!\!\rightarrow \;=\; \rightarrow\!\!\bigcirc\!\!\rightarrow + \cdots, \qquad (6.40)$$

representing it as a sum of an infinite number of diagrams of the skeleton diagram technique, the elements of which are full Green's functions.

The skeleton diagram technique appears to be especially suitable in theories with anomalous Green's functions which are identically equal to zero when calculated by a standard perturbation theory. Such a situation is encountered in the theory of superconductivity (see chapter 7). The system of equations, generalizing systems (6.39) and (6.40) for the case when anomalous Green's functions occur has, besides trivial solutions, also nontrivial ones emerging below the phase transition point. The possibility of partial summation is a common feature of the diagram perturbation technique, valid for statistical physics as well as for relativistic quantum field theory.

To conclude this section we shall show expressions for several physical quantities, e.g., mean number of particles, momentum, kinetic energy, and pressure.

Let us first study the Bose system. The mean number of particles N of the system can be received averaging the functional

$$\int \bar\psi(\mathbf{x}, \tau)\psi(\mathbf{x}, \tau)\, d^3x. \qquad (6.41)$$

This gives the formula

$$N = \frac{\int \exp(S)\left[\int \bar\psi(\mathbf{x}, \tau)\psi(\mathbf{x}, \tau)\, d^3x\right] d\bar\psi\, d\psi}{\int \exp(S)\, d\bar\psi\, d\psi} = \int d^3x \langle \bar\psi(\mathbf{x}, \tau)\psi(\mathbf{x}, \tau)\rangle. \qquad (6.42)$$

In a translationally-invariant system with action (6.1), Green's function (6.5) depends on $\mathbf{x} - \mathbf{x}_1$, $\tau - \tau_1$. Equation (6.42) can therefore be written as

$$N = V \lim_{\substack{\mathbf{x}_1 \to \mathbf{x} \\ \tau_1 \to \tau}} \langle \psi(\mathbf{x}, \tau)\bar\psi(\mathbf{x}_1, \tau_1)\rangle. \qquad (6.43)$$

This expression depends on the way of approaching the limit. For instance, if we put $x_1 = x$, we receive different limits for $\tau_1 \to \tau - 0$ and $\tau_1 \to \tau + 0$. We shall demonstrate this using as an example an ideal Bose system with Green's function (6.26); in that case

$$\langle \psi(x,\tau)\bar{\psi}(x_1,\tau_1)\rangle = \frac{1}{\beta V}\sum_{k,\omega}\frac{\exp\left[i(k(x-x_1)-\omega(\tau-\tau_1))\right]}{i\omega - \dfrac{k^2}{2m} + \lambda}. \quad (6.44)$$

We put here $x_1 = x$, $\tau_1 = \tau + \varepsilon$ $(\varepsilon > 0)$. The evaluation of the sum over frequencies gives

$$T\sum_{\omega}\exp\left(i\omega\varepsilon\right)\left(i\omega - \frac{k^2}{2m} + \lambda\right)^{-1} = \frac{\exp\left[\varepsilon\varepsilon(k)\right]}{\exp\left[\beta\varepsilon(k)\right] - 1}, \quad (6.45)$$

where $\varepsilon(k) = (k^2/2m) - \lambda$. In the limit $\varepsilon \to 0$, the Bose partition function is obtained

$$\left(\exp\left[\beta\varepsilon(k)\right] - 1\right)^{-1}. \quad (6.46)$$

Putting $\varepsilon < 0$ in (6.44), we receive in the limit $\varepsilon \to -0$ the expression

$$\left(\exp\left[\beta\varepsilon(k)\right] - 1\right)^{-1} + 1. \quad (6.47)$$

instead of (6.46). The correct answer is received for $\tau_1 - \tau \to +0$. In that case we find the well-known Bose partition formula

$$\rho = N/V = V^{-1}\sum_{k}\left[\exp\beta\left(\frac{k^2}{2m} - \lambda\right) - 1\right]^{-1} \quad (6.48)$$

for the density $\rho = N/V$. The approaching $\tau_1 \to \tau + 0$ gives the correct answer for nonideal system, too. As a result, the following formula is obtained

$$N/V = \rho = -\lim_{\varepsilon \to +0}\frac{T}{V}\sum_{k,\omega}\exp\left(i\omega\varepsilon\right)G(k,\omega). \quad (6.49)$$

It expresses the density through the Green's function $G(k,\omega)$ in (k,ω) representation.

The expectation value for the momentum and kinetic energy can be received averaging the functionals

$$\frac{i}{2}\int(\nabla\bar{\psi}\psi - \bar{\psi}\nabla\psi)\,d^3x; \quad (6.50)$$

$$\frac{1}{2m}\int \nabla\bar{\psi}\nabla\psi\, d^3x. \tag{6.51}$$

It leads to the formulae

$$K/V = -(T/V)\sum_{k,\omega}\exp(i\omega\varepsilon)\mathbf{k}G(\mathbf{k},\omega), \tag{6.52}$$

$$H_{kin}/V = -(T/V)\sum_{k,\omega}\exp(i\omega\varepsilon)\frac{k^2}{2m}G(\mathbf{k},\omega) \tag{6.53}$$

in which $\varepsilon \to +0$.

We shall derive further the expression for pressure, starting from a formula for the ratio of statistical sums of ideal and nonideal systems expressed as the sum of vacuum diagrams:

$$\frac{Z}{Z_0} = \frac{\exp(-\beta\Omega)}{\exp(-\beta\Omega_0)} = \frac{\int\exp(S)\,d\bar{\psi}\,d\psi}{\int\exp(S_0)\,d\bar{\psi}\,d\psi} = \exp\sum_i D_i^c. \tag{6.54}$$

Here Ω_0, Ω are Ω potentials of ideal and nonideal systems, respectively. We have

$$\Omega_0 = -p_0 V, \qquad \Omega = -pV, \tag{6.55}$$

where p_0, p is the pressure of ideal and nonideal systems. From (6.54) and (6.55), the formula

$$p = p_0 + \frac{1}{\beta V}\sum_i D_i^c, \tag{6.56}$$

follows. It expresses the pressure p through the pressure p_0 and the sum of contributions of vacuum diagrams.

Formula (6.56) is valid for Fermi systems, too. Formulae for the mean number of particles, for momentum and kinetic energy per unit volume differ for the Fermi system from the corresponding formulae for the Bose system (6.49), (6.52), (6.53) by the sign and by replacement of the Bose frequencies $\omega = 2\pi n T$ with Fermi ones $nT(2n+1)$.

CHAPTER 3

GAUGE FIELDS

7. QUANTIZATION OF GAUGE FIELDS

Gauge fields have geometrical origin. They can be regarded as connections on a fibre bundle, its base being the space-time V_4 and its fibre a finite-dimensional space carrying a representation of a group.

The geometrical nature of gauge fields also demonstrates itself when constructing the corresponding quantum theory. Attempts to quantize geometrical fields by standard methods lead to the difficulties and contradictions which have already been encountered in the problem of electromagnetic field quantization – the simplest one from a geometrical point of view. In that case, the problems can be avoided if the Fermi method with the use of indefinite metric is applied to its quantization. It has been shown, however, that uncritical application of the Fermi method, justified in quantum electrodynamics, to more complicated systems can cause the violation of unitarity of the theory. This was revealed for the first time in 1963 by Feynman [12] for the case of the Yang – Mills field and the gravitational field. Feynman outlined how the indicated difficulty can be disposed of. He showed that the unitarity of a diagram of a closed-loop type could be restored by subtraction of another closed-loop type diagram, which describes the propagation of a ghost particle.

Feynmans method did not provide the possibility of generalization to more complicated diagrams. A solution of the problem for an arbitrary diagram was given in 1967 by De Witt [13] and also by Fadeev and Popov [14, 15], who used substantially different approaches. Both approaches have in common the use of the functional integration method which provides the scheme of covariant perturbation theory for gauge fields.

The method developed in [14, 15] is based on the following idea. The fields, obtained from other fields through gauge transformation (e.g., A_μ and $A_\mu + \partial_\mu \Lambda$ in electrodynamics), describe the same physical (geometrical) situation and are therefore physically indistinguishable. This leads to an idea that the classes of those fields which can be obtained from other fields through gauge transformations should be the basic objects of the theory. In such a way all fields of the type $A_\mu + \partial_\mu \Lambda$ are unified into one class.

The action in the theory of gauge fields is the same for all fields obtained through gauge transformations. In other words, the action is a functional defined on classes.

In functional integral formalism it is possible to obtain a theory whose basic objects are classes if we can write down the functional integral as an integral over all classes. It can be accomplished, e.g., if the integration is taken over the surface in the manifold of all fields whose elements intersect each of the classes once. Then each class will have exactly one representative on that surface. The integration measure arising on such surfaces changes with the variation of the surface, but all physical results must be independent of the choice of the surface.

We now formulate the rules for the quantization of gauge fields in the formalism of functional integration over all fields.

We shall denote the gauge fields by A, its components being A_μ^a where $\mu = 0, 1, 2, 3$ is a space-time index and a is an isotropic index.

The gauge group G is a direct product of the groups G_0, operating at every point x of space-time:

$$G = \prod_x G_0(x). \tag{7.1}$$

Let Ω be an element of the gauge group which is a function on V_4 with functional values in G_0. We denote by an A^Ω the result of action of the element Ω on the field A. The set of fields A^Ω with A fixed, Ω running through gauge group G, is called the gauge group orbit.

It has been mentioned in Section 5 that quantization of a field with action S leads to the averaging of the functional $\exp iS$ over all fields. In the theory of gauge fields the action $S[A]$ is gauge invariant, i.e., the same for all fields obtained from each other through gauge transformations

$$S[A^\Omega] = S[A]. \tag{7.2}$$

To construct the functional integral, it is necessary to choose the measure on the manifold of all fields A. The measure

$$d\mu[A] = \prod_x \prod_{\mu,a} dA_\mu^a(x) \tag{7.3}$$

turns out to be the simplest one. The symbol $\prod_x$ is defined in Section 5. We shall call this measure local. In particular, examples of the theory of gauge-field measure (7.3) exhibit the property of gauge invariance

$$d\mu[A^\Omega] = d\mu[A].$$ (7.4)

It follows from the invariance of action $S[A]$ and the measure $d\mu[A]$ with respect to gauge transformations $A \to A^\Omega$ that the corresponding functional integral

$$\int \exp(iS[A]) d\mu[A]$$ (7.5)

becomes proportional to the 'orbit volume', i.e., to the functional integral

$$\int \prod_x d\Omega(x)$$ (7.6)

over gauge group G. Here, $\Pi_x\, d\Omega(x)$ is an invariant measure on group G which is equal to the product of measures on G_0, operating at every point of space-time V_4.

The approach to the integration over classes, developed here, consists of the explicit factorization of that factor from the functional integral. Such a factorization can be realized by several methods. The idea of one of them consists of the transition from integral (7.5) over all fields to the integral over the surface in the manifold of all fields whose elements intersect with each of the gauge-group orbits just once. Let the equation of the surface be

$$f(A) = 0.$$ (7.7)

The equation $f(A^\Omega) = 0$ should have a unique solution with respect to $\Omega(x)$ for any $A(x)$.

We introduce the functional $\Delta_f[A]$, defined by the condition

$$\Delta_f[A] \int \prod_x \delta(f(A^\Omega(x))) d\Omega(x) = 1.$$ (7.8)

Here, the integration is carried out over gauge group G of the infinite-dimensional δ function $\Pi_x \delta(f(A^\Omega(x)))$. Such a δ function is a functional defined by the specification of the rules for its integration with other functionals. In the following, we shall demonstrate several specific examples of the evaluation of integrals of type (7.8). Let us notice that the functional $\Delta_f[A]$ is gauge invariant, i.e.,

$$\Delta_f[A^\Omega] = \Delta_f[A].$$ (7.9)

To factorize factor (7.6) from functional integral (7.5) we insert the left hand side of formula (7.8) (which is equal to one) into the integral and make the substitution $A^\Omega \to A$. The measure $d\mu[A]$ and functionals $S[A], \Delta_j[A]$ are invariant under such a substitution. Integral (7.5) leads to the multiplication of the group's volume $\Pi_x \, d\Omega(x)$ by the integral

$$\int \exp(iS[A])\Delta_f[A][\prod_x \delta(f(A))] \, d\mu[A]. \tag{7.10}$$

Just this integral provides a starting point for the quantum theory of gauge fields.

We shall demonstrate that integral (7.10), formally depending on the choice of the surface $f[A] = 0$, is in fact invariant with respect to the choice of the surface. To prove it, we insert into the integrand (7.10) 'another unit element':

$$1 = \Delta_g[A] \int_0 \prod_x \delta[g(A^\Omega(x))] \, d\Omega(x), \tag{7.11}$$

where $g(A) = 0$ is an equation of another surface which, like the surface $f(A) = 0$, intersects each of the orbits of group G only once. Interchanging the integration over A and Ω, performing then the shift $A^\Omega \to A$ and finally interchanging again the integration over A and Ω we can express integral (7.10) as

$$\int \exp(iS[A])\Delta_g[A]\left[\prod_x \delta(g(A))\right] d\mu[A]. \tag{7.12}$$

The method described allows one to pass in the functional integral from one surface to another or, we can say, from one gauge to another. Especially, such a method is suitable for the transition from the Hamiltonian form of the functional integral to the integral in the relativistic gauge. In the following, we examine such a transition using specific examples. It is possible to show a method for the factorization of the volume of a gauge group from the functional integral which is more general than the method just described. Let us study the gauge invariant functional $F[A]$. We define a functional $\Phi[A]$ by the equation

$$\Phi[A] \int F[A^\Omega] \prod_x d\Omega(x) = 1. \tag{7.13}$$

It is gauge invariant. It is necessary, however, to require that the functional integral at the left-hand side of expression (7.13) really exists. Inserting the left-hand side of (7.13) into integral (7.5) and then performing the shift $A^{\Omega} \to A$ we obtain the product of the group volume (7.6) with the integral

$$\int \exp(iS[A])\Phi[A]F[A]d\mu[A]. \tag{7.14}$$

Integral (7.10) is a special case of (7.14). The independence of integral (7.14) on the choice of functional can be proved in the same way as the independence of integral (7.10) on the choice of the surface $f(A) = 0$.

In the theory of gauge fields, Green's function is defined as an expectation value of the product of field functions at different points of space-time V_4. The generating functional of Green's functions has the form

$$\bar{Z}[\eta] = \frac{\int \exp\{iS[A] + i\int \eta A\, d^4x\}F[A]\Phi[A]\, d\mu[A]}{\int \exp(iS[A])F[A]\Phi[A]\, d\mu[A]}, \tag{7.15}$$

where $S[A]$ is the action of the field A, $d\mu[A]$ is the local gauge invariant measure; the functionals F and Φ are defined above. The linear functional

$$\int \left(\sum_{\mu,a} \eta_a^{\mu}(x)A_{\mu}^a(x) \right) d^4x, \tag{7.16}$$

is denoted by $\int \eta A\, d^4x$ where $\eta_a^{\mu}(x)$ are arbitrary test functions. Green's functions – functional derivatives of functional (7.15) – depend on the choice of gauge, i.e., on the choice of the functional $F[A]$. Physical results, obtained by averaging gauge-invariant functionals, however, do not depend on the choice of gauge.

Let us notice the characteristic difference between perturbation theory in the theory of gauge fields and that of section 5, developed for scalar field theory.

Let action S contain a small parameter ε. We shall consider action S_0, corresponding to the zero value of parameter ε, to be a quadratic form of field functions. This is typical for many examples in field theory.

When constructing the perturbation theory in the functional integral formalism of Section 5, the functional $\exp iS$ is expanded into the series

$$\exp(iS) = \exp(iS_0)\exp(iS_1) = \exp(iS_0) \sum_{n=0}^{\infty} \frac{\varepsilon^n}{n!} a_n \tag{7.17}$$

in powers of ε. The resulting functional series should be integrated term by term and individual terms of the perturbation theory series should be evaluated using Wikc's theorem, which follows from the fact that S_0 is a nondegenerated quadratic form.

In the theory of gauge fields, both the product of functionals $F\Phi$ and the functional $\exp iS$ appear in the integrand.

This product is also expanded in to a series in the perturbation theory. If F is the δ functional, its expansion into a series in ε is complicated. It is therefore desirable to choose the equation of the surface $f(A) = 0$ in such a manner that it does not contain parameter ε. This is true, e.g., for equations that define the Lorentz and Coulomb gauge in electrodynamics and also in the Yang–Mills theory. In gravitation theory, the role of equations with analogous properties is assumed by harmonical conditions.

Let us bear in mind that the functions $\Delta_j[A]$ depend on parameter ε even if equation $f(A) = 0$ does not contain this parameter. To demonstrate this, it is necessary to expand the functional $\Delta_j[A]\exp(iS[A])$ into the series

$$\exp(iS_0[A]) \sum_{n=0}^{\infty} \frac{\varepsilon^n}{n!} b_n[A]. \tag{7.18}$$

In general, the product

$$F_0 \Phi_0 \exp(iS_0) \equiv M_0, \tag{7.19}$$

where F_0 and Φ_0 are the principal terms of the functionals F and Φ, should be such that Wick's theorem is fulfilled when integrating a product of fields with M_0. This property is satisfied if M_0 is an exponential of a nondegenerated quadratic form or the product of a degenerated quadratic form exponential with a δ functional that corresponds to the flat surface, orthogonal to the zero directions of that form.

We shall study the construction of the perturbation theory that arises from expansion (7.18) using particular examples. In these examples the rule ensuring the validity of Wick's theorem will be satisfied.

8. QUANTUM ELECTRODYNAMICS

Using specific examples, we shall demonstrate how the general scheme of gauge-field quantization works. Let us investigate an electromagnetic field, the one with the simplest geometrical structure.[*]

---

[*]From now on we shall write the vector indices below, making no difference between k_0 and contravariant components. We use the summation according to which repeated Greek indices are summed using pseudoeuclidean metric and repeated Latin indices are summed over values 1, 2, 3. For instance, $k^2 \equiv k_\mu k_\mu = k_0^2 - \mathbf{k}^2$, $\partial_\mu A_\mu = \partial_0 A_0 - \partial_1 A_1 - \partial_2 A_2 - \partial_3 A_3$, $k_i k_i = k_1^2 + k_2^2 + k_3^2$. We shall denote by $\delta_{\mu\nu}$ the Minkowski tensor (with nonzero components $\delta_{00} = -\delta_{11} = -\delta_{22} = -\delta_{33} = 1$).

The action of a free electromagnetic field

$$S = -\frac{1}{4}\int(\partial_\mu A_\nu - \partial_\nu A_\mu)^2 \, d^4 x \tag{8.1}$$

is invariant under the Abelian group of gauge transformations

$$A_\mu(x) \to A_\mu(x) + \partial_\mu \lambda(x). \tag{8.2}$$

It follows from the previous section that the quantization of gauge fields is realized using the functional integral of the functional

$$\Phi F \exp(iS), \tag{8.3}$$

where S is the action of the system, F is an arbitrary gauge-invariant functional, and the functional Φ^{-1} is the expectation value of F averaged over the gauge group. The local integration measure, having in this case the form

$$d\mu[A] = \prod_x \prod_{\mu=0}^{3} dA_\mu(x), \tag{8.4}$$

is evidently gauge-invariant. Functionals of the type

$$F_1[A] = \prod_x \delta(\partial_\mu A_\mu(x));$$

$$F_2[A] = \prod_x \delta(\operatorname{div} \mathbf{A}(x));$$

$$\tag{8.5}$$

$$F_3[A] = \exp\left\{ -\frac{i}{2d_l} \int (\partial_\mu A_\mu(x))^2 \, d^4 x \right\}.$$

turn out to be the most suitable for the following construction of the perturbation theory. The functionals F_1 and F_3 lead to an explicitly relativistic quantization and the use of F_2 is convenient when passing to the Hamilton theory. The corresponding gauge-invariant functionals are given by the formulae:

$$\Phi_1^{-1}[A] = \int \prod_x \delta(\partial_\mu(A_\mu(x) + \partial_\mu \lambda(x))) \, d\lambda(x);$$

$$\Phi_2^{-1}[A] = \int \prod_x \delta(\operatorname{div}(\mathbf{A}(x) + \nabla\lambda(x))) \, d\lambda(x); \tag{8.6}$$

$$\Phi_3^{-1}[A] = \int \exp\left\{ -\frac{i}{2d_l} \int (\partial_\mu(A_\mu(x) + \partial_\mu \lambda(x)))^2 \, d^4 x \right\} \prod_x d\lambda(x).$$

All these functionals do not, in fact, depend on the field $A_\mu(x)$, as can be seen if we perform the shift $\lambda \to \lambda - \square^{-1} \partial_\mu A_\mu$ in the first and third functionals and $\lambda \to \lambda - \square^{-1} \operatorname{div} A$ in the second one. Thus, with precision up to an (infinite) constant factor, we can suppose that

$$\Phi_1 = \Phi_2 = \Phi_3 = 1. \tag{8.7}$$

Now the form of the functional integral is defined in all three cases.

The use of the functional F_2 means the integration over the fields satisfying the equation

$$\operatorname{div} \mathbf{A} = 0. \tag{8.8}$$

This is a well-known Coulomb gauge condition. We shall show how the integral with the functional F_2 can be transformed into the integral of an explicitly Hamilton type. Such a transformation is possible if the integral over auxiliary fields is introduced. In our case we are led to the functional integral of the form

$$\int \exp\left(iS[A_\mu, F_{\mu\nu}]\right) \prod_x \delta(\operatorname{div} \mathbf{A}(x)) \prod_\mu dA_\mu(x) \prod_{\mu < \nu} dF_{\mu\nu}(x) \tag{8.9}$$

with the action

$$S[A_\mu, F_{\mu\nu}] = \int \left(\frac{1}{4} F_{\mu\nu} F_{\mu\nu} - \frac{1}{2} F_{\mu\nu}(\partial_\mu A_\nu - \partial_\nu A_\mu) \right) d^4x, \tag{8.10}$$

depending not only on the vector $A_\mu(x)$, but also on the antisymmetric tensor $F_{\mu\nu}(x)$. In the classical theory, $F_{\mu\nu}$ are electromagnetic field strengths:

$$F_{\mu\nu} = \partial_\mu A_\nu - \partial_\nu A_\mu. \tag{8.11}$$

To pass to quantum theory it is necessary to regard the functions $A_\mu, F_{\mu\nu}$ as independent and integrate over them as over independent variables. The integral over $F_{\mu\nu}$ in (8.9) can be evaluated exactly. To accomplish this, it is sufficient to perform a shift

$$F_{\mu\nu} \to F_{\mu\nu} + \partial_\mu A_\nu - \partial_\nu A_\mu, \tag{8.12}$$

which transforms integral (8.9) into the product of the integral over A_μ and the integral over $F_{\mu\nu}$ of the type

$$\int \exp\left(\frac{i}{4} \int F_{\mu\nu} F_{\mu\nu} d^4x \right) \prod_x \prod_{\mu < \nu} dF_{\mu\nu}(x). \tag{8.13}$$

This integral is just a normalization constant.

We rewrite action (8.10) using three-dimensional rotations:

$$\int (\mathbf{E}\partial_0\mathbf{A} - \frac{1}{2}E^2 + \frac{1}{2}H^2 - (\mathbf{H}, \text{rot }\mathbf{A}) + A_0 \, \text{div }\mathbf{E})\, d^4x, \qquad (8.14)$$

where

$$E_i = F_{0i}; \qquad H_1 = F_{23}; \qquad H_2 = F_{31}; \qquad H_3 = F_{12}.$$
$$(8.15)$$

We shall integrate over $\mathbf{H}$ in (8.9). This leads to the substitution $\mathbf{H} \to \text{rot }\mathbf{A}$ in action (8.14). Then we integrate over A_0, which yields the functional

$$\prod_x \delta\,(\text{div }\mathbf{E}(x)). \qquad (8.16)$$

We obtain

$$\int \exp\,(iS[\mathbf{A}, \mathbf{E}]) \prod_x \delta\,(\text{div }\mathbf{A}(x))\delta(\text{div }\mathbf{E}(x)) \prod_{i=1}^{3} dA_i(x)\, dE_i(x)$$
$$(8.17)$$

With a Hamiltonian-type action

$$\int\left(\mathbf{E}\partial_0\mathbf{A} - \frac{1}{2}E^2 - \frac{1}{2}(\text{rot }\mathbf{A})^2\right)d^4x. \qquad (8.18)$$

Integral (8.17) turns out to be an analog of integrals studied in Section 3 when we quantized finite-dimensional systems with constraints. Here the role of the constraint is played by div $\mathbf{E}$, the role of the additional condition by the Coulomb gauge equation (8.8). It is possible to use the transverse (in the three-dimensional sense) components of vectors $\mathbf{A}$ and $\mathbf{E}$ – vector potential and electric field strength, respectively – as independent variables.

The necessity for modification of a functional integral in quantum electrodynamics was pointed out by Białynicki – Birula [39] who, apparently for the first time, studied the functional integral with δ-functional $\Pi_x\delta(\partial_\mu A_\mu)$ (without, however, mentioning the possibility that an additional factor of the Δ_j type may appear during the transition to the nonabelian gauge group).

The Lagrangian of spinorial quantum electrodynamics

$$L(x) = \bar\psi(x)(i\gamma_\mu(\partial_\mu - ieA_\mu(x)) - m)\psi(x) -$$
$$- \tfrac{1}{4}(\partial_\mu A_\nu(x) - \partial_\nu A_\mu(x))^2, \qquad (8.19)$$

where γ_μ are Dirac matrices, also contains besides the electromagnetic potential $A_\mu(x)$, the four-component spinors $\psi(x)$, $\bar\psi(x)$ describing the Fermi electron-positron field. The coefficient of the trilinear term $\bar{\psi}\gamma_\mu\psi A_\mu$ describes the interaction of an electromagnetic field with an electron-positron field. Lagrangian (8.19) is invariant under the Abelian group of gauge transformations:

$$A_\mu(x) \to A_\mu(x) + \partial_\mu\lambda(x); \tag{8.20}$$
$$\psi(x) \to \exp(ie\lambda(x))\psi(x); \qquad \psi(x) \to \bar{\psi}(x)\exp(-ie\lambda(x)).$$

In the functional integration scheme we shall consider the components of spinors $\psi_\alpha(x), \bar{\psi}_\alpha(x)$ to be anticommuting elements of Grassmann algebra, and we shall integrate the functional $\exp iS$ over the measure

$$\prod_x \left(\prod_\mu dA_\mu(x) \prod_\alpha d\bar{\psi}_\alpha(x)\, d\psi_\alpha(x) \right) \equiv \prod_x dA\, d\bar{\psi}\, d\psi. \tag{8.21}$$

We shall outline here the construction of the perturbation theory in the parameter e. As the functionals F_1, F_2, F_3 do not depend on e, we are led to the term-by-term integration of the series

$$\exp(ie\Delta S) = \sum_{n=0}^{\infty} \frac{(ie)^n}{n!}(\Delta S)^n. \tag{8.22}$$

The form of Green's functions depends on the choice of functional $F[A]$. We shall find generating functionals for unperturbed functions corresponding to the zero value of the parameter e. Using the choice $F = F_1$ we evaluate the integral

$$Z_0[\eta] = \frac{\int \exp\{i(S_0 + \int(\bar{\eta}\psi + \bar{\psi}\eta + \eta_\mu A_\mu)d^4x)\}\Pi_x\delta(\partial_\mu A_\mu)dA\, d\bar{\psi}\, d\psi}{\int \exp(iS_0)\Pi_x\delta(\partial_\mu A_\mu)dA\, d\bar{\psi}\, d\psi}, \tag{8.23}$$

where $\bar{\eta}, \eta, \eta_\mu$ are 'sources' of the fields $\psi, \bar{\psi}, A_\mu$. The integration in (8.23) is taken over fields $A_\mu(x)$, which fulfil the Lorentz condition

$$\partial_\mu A_\mu(x) = 0. \tag{8.24}$$

The integrals with the functional $F_1[A]$ are therefore called *integrals in the Lorentz gauge*. The functional (8.23) is evaluated by a standard method using the shift

$$\psi \to \psi + \psi_0; \bar{\psi} \to \bar{\psi} + \bar{\psi}_0; \qquad A_\mu \to A_\mu + A_\mu^{(0)}, \tag{8.25}$$

which cancels the terms linear in $\psi, \bar{\psi}, A_\mu$ in the exponent of the integrand in the numerator (8.23). So that shift (8.25) should not change the δ functional $F_1[A]$, it is sufficient to choose the field $A_\mu^{(0)}(x)$ to be transverse, i.e., fulfilling condition (8.24).

The expression for the functional $Z_0[\eta]$ turns out to be equal to

$$\exp\left(-i\int\bar{\eta}(x)G(x-y)\eta(y)\,d^4x\,d^4y\,-\right.$$

$$\left.-\frac{i}{2}\int\eta_\mu(x)D_{\mu\nu}^{\text{tr}}(x-y)\eta_\nu(y)\,d^4x\,d^4y\right), \tag{8.26}$$

where

$$G(x-y)=(2\pi)^{-4}\int d^4p\exp\left[i(p,x-y)\right](\hat{p}+m)/(p^2-m^2+i0);$$

$$D_{\mu\nu}^{\text{tr}}(x-y)=(2\pi)^{-4}\int d^4k\exp\left[i(k,x-y)\right](-k^2\delta_{\mu\nu}+k_\mu k_\nu)/(k^2+i0)^2. \tag{8.27}$$

Here $(p,x-y)=p_\mu(x-y)_\mu,\hat{p}=\gamma_\mu p_\mu.$

The necessity of the replacement $k^2\to k^2+i0$ in integrand expressions for Green's functions was explained in Section 5. The correctness of Wick's theorem follows from formula (8.27). It leads to the diagram technique where to each electron (photon) line the function $G(D_{\mu\nu}^{\text{tr}})$ is assigned and a coupling constant e corresponds to each vertex.

Analogous schemes of the perturbation theory, which differ only in the form of the photon Green's function, emerge when using the functionals F_2 and F_3. The integrals with the functional F_2 will be called *the integrals in the Coulomb gauge*. The unperturbed Green's function in the Coulomb gauge is given by the formula

$$D_{\mu\nu}^q(x-y)=(2\pi)^{-4}\int d^4k\exp\left[i(k,x-y)\right]\tilde{D}_{\mu\nu}^q(k); \tag{8.28}$$

$$\tilde{D}_{00}^q(k)=\frac{1}{\mathbf{k}^2};\qquad\tilde{D}_{0i}^q=0;\qquad\tilde{D}_{ij}^q(k)=(k^2+i0)^{-1}\left(\delta_{ij}-\frac{k_ik_j}{\mathbf{k}^2}\right).$$

The unperturbed Green's function $D_{\mu\nu}$ for the functional F_3 has the form

$$(k^2+i0)^{-2}(-k^2\,\delta_{\mu\nu}+(1-d_t)k_\mu k_\nu). \tag{8.29}$$

Formula (8.29) takes a simple form for $d_t=1$. The corresponding Green's function $\delta_{\mu\nu}(-k^2-i0)^{-1}$ is usually called the *function in the Feynman gauge*.

In such a way we have obtained three schemes of perturbation theory, which differ in the form of the photon Green's functions. These schemes are well known and lead to the same results for physical quantities.

The method of functional integration is also convenient for the derivation of exact relations which appear to be the consequences of gauge invariance. As an illustration we derive the Ward identity and also the relation between Green's functions in different gauges. We shall consider the electron Green's function in the Lorentz gauge:

$$G_L(x-y) = -\mathrm{i}\langle \psi(x)\bar{\psi}(y)\rangle_L$$
$$\equiv -\mathrm{i}\frac{\int \exp(\mathrm{i}S)\psi(x)\bar{\psi}(y)\Pi_x\delta(\partial_\mu A_\mu)\,\mathrm{d}A\,\mathrm{d}\bar{\psi}\,\mathrm{d}\psi}{\int \exp(\mathrm{i}S)\Pi_x\delta(\partial_\mu A_\mu)\,\mathrm{d}A\,\mathrm{d}\bar{\psi}\,\mathrm{d}\psi}. \tag{8.30}$$

The transformation of spinor fields $\psi(x)\to\exp(\mathrm{i}ec(x))\psi(x)$, $\bar{\psi}(x)\to\bar{\psi}(x)\exp(-\mathrm{i}ec(x))$ in the integral of numerator (8.30) leads to the emergence of the factor

$$\exp\mathrm{i}e\left(c(x)-c(y)+\int c(z)\partial_\mu j_\mu(z)\,\mathrm{d}^4z\right), \tag{8.31}$$

in the integrand, where $j_\mu = \bar{\psi}\gamma_\mu\psi$. Differentiating with respect to $c(z)$ and putting then $c\equiv 0$, we get the formula

$$G_L(x-y)[\delta(x-z)-\delta(y-z)]$$
$$= \mathrm{i}\langle \psi(x)\bar{\psi}(y)d_\mu j_\mu(z)\rangle_L, \tag{8.32}$$

from which, after the transition to the momentum representation, the Ward identity [40] follows

$$G^{-1}(\hat{p}) - G^{-1}(\hat{q}) = (p-q)_\mu\Gamma_\mu(p,q), \tag{8.33}$$

which connects the electron Green function $G(\hat{p})$ and the irreducible vertex part $\Gamma_\mu(p,q)$. It can be seen that this identity is valid at any gauge of the photon function because the change of variables in the functional integral in its derivation only affect the spinor fields.

We shall now investigate the transition from the Coulomb gauge to the Lorentz gauge using the example of a one-electron function:

$$G_R(x-y) = -\mathrm{i}\langle \psi(x)\bar{\psi}(y)\rangle_R$$
$$\equiv -\mathrm{i}\frac{\int \exp(\mathrm{i}S)\psi(x)\bar{\psi}(y)\Pi_x\delta(\mathrm{div}\,\mathbf{A})\,\mathrm{d}A\,\mathrm{d}\bar{\psi}\,\mathrm{d}\psi}{\int \exp(\mathrm{i}S)\Pi_x\delta(\mathrm{div}\,\mathbf{A})\,\mathrm{d}A\,\mathrm{d}\bar{\psi}\,\mathrm{d}\psi}. \tag{8.34}$$

Into the numerator and the denominator of the right-hand side of this formula we insert the integral $\int \Pi_x\delta(\partial_\mu A_\mu - \square\lambda)\,\mathrm{d}\lambda(x)$, which does not depend on $A_\mu(x)$, and then we perform transformation (8.20) not changing the action. The δ function $\delta(\partial_\mu A_\mu - \square\lambda)$ changes into $\delta(\partial_\mu A_\mu)$, $\delta(\mathrm{div}\,\mathbf{A})$ into

$\delta(\text{div }\mathbf{A} + \Delta\lambda)$. In the numerator the factor $\exp ie(\lambda(x) - \lambda(y))$ emerges, where $\lambda(x)$ can be replaced by the solution $c(x)$ of the equation

$$\Delta c(x) + \text{div }\mathbf{A}(x) = 0, \tag{8.35}$$

i.e., with the function

$$\frac{1}{4\pi} \int |\mathbf{x} - \mathbf{z}|^{-1} \delta(x_0 - z_0) \text{div }\mathbf{A}(z) d^4 z$$

$$\equiv \int l_i(x - z) A_i(z) d^4 z, \tag{8.36}$$

where

$$l_i(x - z) = \delta(x_0 - z_0) \frac{\partial}{\partial x_i} (4\pi|\mathbf{x} - \mathbf{z}|)^{-1}. \tag{8.37}$$

After that, the integrals over $\lambda(x)$ in the numerator and denominator mutually cancel. The resulting formula

$$G_R(x - y)$$

$$= -i\langle \psi(x)\bar\psi(y) \exp ie \int (l_i(x - z) - l_i(y - z)) A_i(z) d^4 z \rangle_L \tag{8.38}$$

expresses the Green's function in the Coulomb gauge after the expansion into the powers of e as a series in Green's functions

$$\langle \psi(x)\bar\psi(y) \prod_{k=1}^{n} A_{ik}(z_k) \rangle_L$$

in the Lorentz gauge.

9. YANG–MILLS FIELDS

The theory of Yang–Mills fields [41] is the simplest example of the theory with a nonabelian gauge group.

The vector Yang–Mills field, connected with a simple compact Lie group G can be conveniently described by the matrices $B_\mu(x)$ acquiring values in the Lie algebra of that group

$$B_\mu(x) = \sum_{a=1}^{n} b_\mu^a(x)\tau_a. \tag{9.1}$$

Here τ_a are linearly-independent matrices in the adjoint representation of the Lie algebra, normalized by the conditions

$$\mathrm{tr}\,\tau_a\tau_b = -2\delta_{ab}, \tag{9.2}$$

n is the number of group parameters; $b_\mu^a(x)$ is a c-number function with the vector index μ and the 'isotopic' index a. As is well known, in the adjoint representation it is possible to also use the latter index for the enumerating of matrix elements so that

$$(B_\mu)_{ab} = (\tau_c)_{ab}b_\mu^c = t_{abc}b_\mu^c, \tag{9.3}$$

where t_{abc} are group-structure constants, antisymmetric in all three indices.

The Lagrangian of the Yang–Mills field

$$Z(x) = \tfrac{1}{8}\,\mathrm{tr}\,F_{\mu\nu}F_{\mu\nu}, \tag{9.4}$$

where

$$F_{\mu\nu} = \partial_\nu B_\mu - \partial_\mu B_\nu + \varepsilon[B_\mu, B_\nu] \tag{9.5}$$

is invariant under the gauge transformations

$$B_\mu \to \Omega B_\mu \Omega^{-1} + \frac{1}{\varepsilon}\partial_\mu\Omega\Omega^{-1} \tag{9.6}$$

with the matrix Ω acting in the adjoint representation of the group.

The analogues of the functionals $F[A]$, used in section 8 for quantum electrodynamics, turn out to be suitable for quantization in the functional integral formalism. Now, they have the form:

$$F_1[B] = \prod_x \delta(\partial_\mu B_\mu(x)) \equiv \prod_x \prod_a \delta(\partial_\mu b_\mu^a(x));$$

$$F_2[B] = \prod_x \delta(\mathrm{div}\,B(x)) \equiv \prod_x \prod_a \delta(\mathrm{div}\,\mathbf{b}^a(x)); \tag{9.7}$$

$$F_3[B] = \exp\left(\frac{\mathrm{i}}{4d_l}\int \mathrm{tr}\,(\partial_\mu B_\mu(x))^2\,\mathrm{d}^4x\right)$$

$$= \exp\left(-\frac{\mathrm{i}}{2d_l}\int \sum_a (\partial_\mu b_\mu^a(x))^2\,\mathrm{d}^4x\right).$$

Here, the functionals F_1 and F_2 select, among all fields, those satisfying the conditions

$$f_L[B] = \partial_\mu B_\mu = 0 \quad \text{for } F_1; \qquad f_R[B] = \mathrm{div}\,\mathbf{B} = 0 \quad \text{for } F_2. \tag{9.8}$$

Each of these equations is a matrix equation and, in fact, it represents n additional conditions (according to the number of parameters of the group G).

The factor Φ_1 corresponding to the functional F_1 will be denoted by Δ_L. In the functional integral, this factor stands before the δ function of $\partial_\mu B_\mu$ and it is therefore sufficient to know its value only for the transverse fields $(\partial_\mu B_\mu = 0)$. In such a case, the whole contribution to the integral

$$\Delta_L^{-1}[B] = \int \prod_x \delta(\partial_\mu B_\mu^\Omega(x))\,d\Omega(x) \tag{9.9}$$

comes from a neighbourhood of the unit element where the substitution

$$\Omega(x) = 1 + \varepsilon\mu(x), \tag{9.10}$$

can be performed ($u(x)$ is an element of the Lie algebra) and only the terms linear in u remain

$$\begin{aligned} \partial_\mu B_\mu^\Omega &= \partial_\mu(B_\mu + \varepsilon[u, B_\mu] + \partial_\mu u) \\ &= \square u - \varepsilon[B_\mu, \partial_\mu u] \equiv \hat{A}u, \end{aligned} \tag{9.11}$$

where $\square = \hat{A}_0$ is the d'Alambert operator. Instead of the matrices $u(x)$, we introduce the column

$$u(x) = \sum_{a=1}^n \tau_a u_a(x), \tag{9.12}$$

which the operator $\hat{A}$ acts on according to the rule

$$\begin{aligned} (\hat{A}u)_a &= (\square u - \varepsilon[B_\mu, \partial_\mu u])_a \equiv (\square\delta_{ac} - \varepsilon(B_\mu)_{ac}\,\partial_\mu)u_c \\ &= \square u_a - \varepsilon t_{abc} b_\mu^b\,\partial_\mu u_c. \end{aligned} \tag{9.13}$$

Integral (9.9) can be written as

$$\Delta_L^{-1}[B] = \int \prod_x \prod_a \delta((\hat{A}u)_a)\,du_a. \tag{9.14}$$

Formally, $\Delta_L[B]$ is the determinant of the operator $\hat{A}$. Putting out the trivial (infinite) factor $\det \square$, it is possible to expand the logarithm $\Delta_L[B]$ into series in ε:

$$\begin{aligned} \ln\Delta_L[B] &= \ln(\det\hat{A}/\det\hat{A}_0) = \mathrm{Sp}\ln(1 - \varepsilon\square^{-1}B_\mu\,\partial_\mu) \\ &= -\sum_{n=2}^\infty \frac{\varepsilon^n}{n}\int d^4x_1,\ldots,d^4x_n\,\mathrm{tr}\,(B_{\mu_1}(x_1),\ldots,B_{\mu_n}(x_n)) \times \\ &\quad \times \partial_{\mu_1}D(x_1 - x_2),\ldots,\partial_{\mu_n}D(x_n - x_1). \end{aligned} \tag{9.15}$$

$D(x)$ is the Green's function of the d'Alambert operator (5.17). The symbol Sp in (9.15) and thereafter denotes the trace in an operator sense in contradistinction to tr for the trace of a matrix.

The corresponding factor in the Coulomb gauge is denoted by Δ_R. Analogous evaluation leads to the formula

$$\ln \Delta_R[B] = \text{Sp} \ln (1 - \varepsilon \Delta^{-1} B_i \partial_i)$$

$$= - \sum_{n=2}^{\infty} \frac{\varepsilon^n}{n} \int d^4 x_1, \ldots, d^4 x_n \, \text{tr} \, (B_{i_1}(x_1), \ldots, B_{i_n}(x)) \times \qquad (9.16)$$

$$\times \partial_{i_1} \tilde{D}(x_1 - x_2), \ldots, \partial_{i_n} \tilde{D}(x_n - x_1),$$

where

$$\tilde{D}(x) = - \frac{1}{(2\pi)^4} \int \frac{d^4 k}{\mathbf{k}^2} \exp i(k, x) = - \delta(x_0)(4\pi|\mathbf{x}|)^{-1}. \qquad (9.17)$$

The indices $i_1, \ldots, i_n$ in (9.16) acquire the values $1, 2, 3$.

The factor $\Phi_3[B]$ is given by the formula

$$\Phi_3^{-1}[B] = \int \exp \left(\frac{i}{4d_l} \int \text{tr} \, (\partial_\mu B_\mu^\Omega(x))^2 \, d^4 x \right) \prod_x d\Omega(x). \qquad (9.18)$$

For the right-hand side of (9.18), we have not succeeded in obtaining a closed expression analogous to formulae (9.15) and (9.16) representing the factors Δ_L, Δ_R as determinants, but, as will be shown below, this fact does not interfere in developing a simple scheme of the perturbation theory.

We shall first construct the perturbation theory in the Lorentz gauge. It arises as a result of the expansion of the functional

$$\Delta_L[B] \exp (iS[B]) = \exp (iS + \ln \Delta_L[B]). \qquad (9.19)$$

into series in ε. The expression $\ln \Delta_L$ is conveniently interpreted as an addition to the action S. The term of nth order in the expansion of $\ln \Delta_L$ into series in ε leads to the diagram vertex with n outgoing lines. The explicit expression for this term, which follows from (9.15), suggests the interpretation of the vertex as a circle with n outgoing lines along which the *ghost scalar particle* propagates, interacting with the vector field according to $\sim \varepsilon \, \text{tr} \, (\varphi B_\mu \partial_\mu \varphi)$. This statement can be interpreted exactly if we write down the determinant as an integral over the anticommuting variables

$$\det (\square - \varepsilon B_\mu \partial_\mu)$$

$$= \int \exp \left(i \int L(B_\mu, \bar{\eta}, \eta) \, d^4 x \right) \prod_x \prod_a d\bar{\eta}^a(x) \, d\eta^a(x), \qquad (9.20)$$

where

$$L(B_\mu, \bar\eta, \eta) = -\tfrac{1}{2}\,\mathrm{tr}\,\bar\eta(\square - \varepsilon B_\mu\,\partial_\mu)\eta$$
$$= \bar\eta^a\,\square\,\eta^a - \varepsilon t_{abc}b_\mu^c\bar\eta^a\,\partial_\mu\eta^b. \qquad (9.21)$$

Formula (9.20) is an infinite integral of (5.52). It shows that our system can be looked upon as a system of Bose fields $b_\mu^a(x)$, interacting with each other and with scalar Fermi fields $\eta^a(x), \bar\eta^a(x)$.

The construction of the perturbation theory and diagram technique is in many ways analogous to the method outlined in Section 8 for quantum electrodynamics. The elements of the diagram technique in the Yang–Mills theory are lines of two types, corresponding to the transverse vector and ghost scalar particles, and also the vertices describing the interaction of vector particles with the scalar ones and with each other.

We shall represent the vector particles with solid lines and the ghost scalar particles with dashed lines. The elements of the diagrams are the vertices and lines of the form

$$(9.22)$$

We shall write down the expressions for the momentum representation of the diagram elements depicted in (9.22)

$$G_{\mu\nu}^{ab}(p) = -\delta_{ab}(p^2\delta_{\mu\nu} - p_\mu p_\nu)(p + i0)^{-2};$$
$$G^{ab}(p) = -\delta_{ab}(p^2 + i0)^{-1};$$
$$V_{\mu,\nu\rho}^{abc} = i\varepsilon t_{abc}(p_{i\nu}\delta_{\mu\rho} - p_{i\rho}\delta_{\mu\nu}); \qquad (9.23)$$
$$V_{\mu\nu\rho\sigma}^{abcd} = \varepsilon^2 t_{abe}t_{cde}(\delta_{\mu\rho}\delta_{\nu\sigma} - \delta_{\mu\sigma}\delta_{\nu\rho});$$
$$V_\mu^{abc} = \frac{i\varepsilon}{2}t_{abc}(p_3 - p_2)_\mu.$$

To find the contribution of a given diagram, it is necessary to integrate over independent 4-momenta the product of expressions which correspond

to all its elements, sum over all independent discrete indices and multiply
the result by

$$r^{-1}\left(\frac{i}{(2\pi)^4}\right)^{l-v-1}(-2)^s,\tag{9.24}$$

where v is the number of diagram vertices, l is the number of its internal
lines, s is the number of closed loops of ghost scalar particles and r is the
order of the symmetry group of the diagram. Let us remark that
$l-v-1=c$ is the number of independent contours of the diagram.

We show that in the perturbation theory developed above, the transverse
Green's function $G_{\mu\nu}^{ab}$ can be replaced, without changing the physical
results, by a function with an arbitrary longitudinal part d_l:

$$G_{\mu\nu}^{ab}(d_l,p)=-\delta_{ab}(p^2+i0)^{-2}(p^2\delta_{\mu\nu}+(d_l-1)p_\mu p_\nu).\tag{9.25}$$

The original proof of that fact, given by De Witt [13], is rather
cumbersome. We shall present here the proof suggested by t'Hooft [18].
We shall examine a set of conditions of the form

$$\partial_\mu B_\mu(x)-c(x)=0.\tag{9.26}$$

The mean value of the gauge-invariant functional X over fields with
gauge condition (9.26) can be written in the form

$$\langle X\rangle=\frac{\int X[B]\exp(iS[B])\Delta_c[B]\Pi_x\delta(\partial_\mu B_\mu-c)\,dB}{\int\exp(iS[B])\Delta_c[B]\Pi_x\delta(\partial_\mu B_\mu-c)\,dB},\tag{9.27}$$

where $\Delta_c[B]$ is the factor that corresponds to condition (9.26). The
numerator and denominator of the right-hand side of (9.27) do not depend
on c. Using this fact, we rewrite expression (9.27) as

$$\langle X\rangle=\frac{\int\exp\left(\frac{i}{4d_l}\int\operatorname{tr}c^2\,d^4x\right)dc\int X[B]\exp(iS[B])\Delta_c[B]\Pi_x\delta(\partial_\mu B_\mu-c)\,dB}{\int\exp\left(\frac{i}{4d_l}\int\operatorname{tr}c^2\,d^4x\right)dc\int\exp(iS[B])\Delta_c[B]\Pi_x\delta(\partial_\mu B_\mu-c)\,dB},\tag{9.28}$$

i.e., as a ratio of functional integrals over the variable $c(x)=\Sigma_a\tau^a c^a(x)$.
Integrating over c, we obtain for the mean value $\langle X\rangle$ the expression

$$\langle X\rangle=\frac{\int X[B]\tilde\Delta[B]\exp(i(S[B]+(1/4d_l)\int\operatorname{tr}(\partial_\mu B_\mu)^2\,d^4x))\Pi_x dB}{\int\tilde\Delta[B]\exp(i(S[B]+(1/4d_l)\int\operatorname{tr}(\partial_\mu B_\mu)^2\,d^4x))\Pi_x dB}.\tag{9.29}$$

The factor

$$\tilde{\Delta}[B] = \Delta_c[B]|_{c=\partial_\mu B_\mu} \tag{9.30}$$

is given by the equality

$$\Delta_c^{-1}[B] = \int \prod_x \delta(\partial_\mu B_\mu^\Omega - c)\, d\Omega(x). \tag{9.31}$$

The evaluation of the factor $\Delta_c[B]$ is analogous to the evaluation of Δ_L (9.9). The whole contribution to integral (9.31) is given by a neighbourhood of unit elements where replacement (9.10) can be made and only the terms linear in n left in $(\partial_\mu B_\mu - c)$:

$$\partial_\mu B_\mu^\Omega - c = \partial_\mu(\varepsilon[u, B_\mu] + \partial_\mu u) = \hat{A}u. \tag{9.32}$$

The operator $\hat{A}$ acts on the column $u(x)$ defined above by formula (9.12), according to the rule

$$(\hat{A}u)_a = \square u_a - \varepsilon t_{abc}\, \partial_\mu(b_\mu^b u_c). \tag{9.33}$$

The operator $\hat{A}$ is conjugated to $\hat{\tilde{A}}$ and therefore the determinants of the operators $\hat{A}$ and $\tilde{A}$ are equal:

$$\tilde{\Delta}[B] = \det \hat{A} = \det \hat{\tilde{A}} = \Delta_L[B]. \tag{9.34}$$

The evaluation of $\langle X \rangle$ as a ratio of functional integrals (9.29) in terms of the perturbation theory, leads to the diagram technique with Green's function of the vector Yang–Mills particle (9.25). The contribution of the factor $\Delta_L[B] = \tilde{\Delta}[B]$ can again be integrated as a contribution of additional diagrams describing the interaction of vector particles with ghost scalar particles.

The perturbation theory discussed above is not the only possible one. Another form of perturbation theory and diagram technique emerges in the so-called *first-order formalism*. This formalism can be obtained if Lagrangian (9.4) is written as

$$L(x) = -\tfrac{1}{8} \operatorname{tr} F_{\mu\nu} F_{\mu\nu} +$$
$$+ \tfrac{1}{4} \operatorname{tr} F_{\mu\nu}(\partial_\nu B_\mu - \partial_\mu B_\nu + \varepsilon[B_\mu, B_\nu]) \tag{9.35}$$

and the integration over $B_\mu, F_{\mu\nu}$, as over independent variables, is performed. The expression 'first-order formalism' means that the symbol of the derivative enters Lagrangian (9.35) in an order not higher than the first.

Using the Lorentz gauge we obtain the functional integral of the form

$$\int \exp\left(iS[B, F]\right)\Delta_L[B] \prod_x \delta(\partial_\mu B_\mu)\, dB\, dF, \tag{9.36}$$

where the expression

$$dB(x)\, dF(x) = \prod_a \left(\prod_\mu db_\mu^a(x) \right) \prod_{\mu < \nu} df_{\mu\nu}^a(x), \tag{9.37}$$

as well as $dB(x)$, is gauge invariant.

If we make the expansion of the functional $\Delta_L[B] \exp(iS[B, F])$ into the series in ε in the integrand of (9.36), we obtain a new alternative of the diagram perturbation theory with three lines corresponding to the functions $\langle BB \rangle$, $\langle BF \rangle$, $\langle FF \rangle$ and one vertex describing the trilinear interaction εFBB. The elements of the diagram technique have the form

$$
\begin{array}{ccc}
\underline{\mu a \quad p \quad \nu b} & \underline{\mu\nu a \qquad \rho b} & \underline{\mu\nu a \quad p \quad \rho\sigma b} \\
G_{\mu\nu}^{ab}(p) & G_{\mu\nu,\rho}^{ab}(p) & G_{\mu\nu,\rho\sigma}^{ab}(p)
\end{array}
$$

$$\tag{9.38}$$

$$
p_1\mu\nu a \quad \diagdown \begin{array}{l} p_2\mu b \\[6pt] p_3\nu c \end{array} \qquad V_{\mu\nu}^{abc}
$$

where the field B is depicted by a single line and the field F by a double line. The expressions which correspond to elements (9.38) are given by the formulae

$$
\begin{aligned}
G_{\mu\nu}^{ab} &= \delta_{ab}(-p^2\delta_{\mu\nu} + p_\mu p_\nu)(p^2 + i0)^{-2}; \\
G_{\mu\nu,\rho}^{ab} &= i\delta_{ab}(p_\nu\delta_{\mu\rho} - p_\mu\delta_{\nu\rho})(p^2 + i0)^{-1}; \\
G_{\mu\nu\rho\sigma}^{abcd} &= \delta_{ab}(\delta_{\mu\rho}\delta_{\nu\sigma} - \sigma_{\mu\sigma}\delta_{\nu\rho}) - (p^2 + i0)^{-1} \times \\
&\quad \times (\delta_{\mu\rho}p_\nu p_\sigma + \delta_{\nu\sigma}p_\mu p_\rho - \delta_{\mu\sigma}p_\nu p_\rho - \delta_{\nu\rho}p_\mu p_\sigma); \\
V_{\mu\nu}^{abc} &= \varepsilon t_{abc}.
\end{aligned}
\tag{9.39}
$$

The lines and vertices describing the propagation of ghost scalar particles and their interaction with vector particles remain the same as in the formalism of the second order, because the factor Δ_L, which depends on B but not on F, remains the same.

The formalism of the first order is convenient for the passage to the canonical quantization. We shall examine such a transition starting from the integral over B_μ in the Coulomb gauge with sources, which is the generating functional of the Green's functions:

$$Z[\eta] = \frac{\int \exp\{iS[B,F] + i\int(\eta_{\mu a}b_{\mu a} + \tfrac{1}{2}\eta_{\mu v a}f_{\mu v a})\,d^4x\}\Delta_R[B]}{\int \exp(iS[B,F])\Delta_R[B]} \cdots$$

$$\cdots \frac{\Pi_x\delta(\text{div }\mathbf{B})\,dB\,dF}{\Pi_x\delta(\text{div }\mathbf{B})\,dB\,dF}.$$ (9.40)

Much the same way as in electrodynamics, we shall use the transverse (in the three-dimensional sense) components of the fields B_i, F_{0i} ($i = 1, 2, 3$) as dynamical variables. We shall suppose that the sources stand only at certain dynamical variables, i.e., that the following conditions are satisfied

$$\eta_{0a} = \eta_{ika} = \partial_i\eta_{0ia} = \partial_i\eta_{ia} = 0.$$ (9.41)

In three-dimensional notation, Lagrangian (9.35) acquires the form

$$\begin{aligned}L(x) = \text{tr}\{&-\tfrac{1}{8}F_{ik}F_{ik} + \tfrac{1}{4}F_{0i}F_{0i} + \\ &+ \tfrac{1}{4}F_{ik}(\partial_k B_i - \partial_i B_k + \varepsilon[B_i, B_k]) - \tfrac{1}{2}F_{0i}\partial_0 B_i - \\ &- \tfrac{1}{2}B_0(\partial_i F_{0i} - \varepsilon[B_i, F_{0i}])\}.\end{aligned}$$ (9.42)

Because there are no sources at B_0, F_{ik}, we can integrate over these variables in (9.40), which leads to the appearance of δ functional

$$\prod_x \delta(\partial_i F_{0i} - \varepsilon[B_i, F_{0i}])$$ (9.43)

and to the replacement of F_{ik} by

$$H_{ik} \equiv \partial_k B_i - \partial_i B_k + \varepsilon[B_i, B_k]$$ (9.44)

in the integral over the remaining variables B_i, F_{0i}. We shall insert into integral (9.40) the factor

$$\int \prod_x \delta(\Delta c + \partial_i F_{0i})\,dc(x),$$ (9.45)

which in fact does not depend on F_{0i}, and then perform the shift $F_{0i} \to F_{0i} - \partial_{ic}$. The functional $\Pi\delta(\Delta c + \partial_i F_{0i})$ is transformed into $\Pi_x\delta(\partial_i F_{0i})$ and the functional $\Pi_x\delta(\partial_i F_{0i} - \varepsilon[B_i, F_{0i}])$ into the expression $\Pi_x\delta(\Delta c - \partial_i F_{0i} - \varepsilon[B_i, \partial_{ic}] + \varepsilon[B_i, F_{0i}])$ which is equal to $\Pi_x\delta(\Delta c - \varepsilon[B_i, \partial_{ic}] + \varepsilon[B_i, F_{0i}])$ due to $\partial_i F_{0i} = 0$.

Let $c_0(x)$ be a solution of the equation

$$\Delta c - \varepsilon[B_i, \partial_{ic}] = -\varepsilon[B_i, F_{0i}],$$ (9.46)

which can be expressed in terms of the Green's function depending on

$$c_0(x) = -\varepsilon \int D(x, y; B)[B_i(y), F_{0i}(y)] \, d^3y. \tag{9.47}$$

After the shift $c \to c + c_0$, the functional $\Pi_x \delta(\Delta c - \varepsilon[B_i, \partial_{ic}]$ originates and the function $c(x)$ can be put as equal to zero everywhere except in the argument of the δ functional. The integral

$$\int \prod_x \delta(\Delta c - \varepsilon[B_i, \partial_i c]) \, dc(x) \tag{9.48}$$

cancels with the factor $\Delta_R[B]$. As a result, functional (9.40) takes the form

$$Z[\eta] = \frac{\int \exp\{iS[B_i, F_{0i}] + i\int(\eta_i^a b_i^a + \eta_{0i}^a f_{0i}^a)\} \, d^4x}{\int \exp\{iS[B_i, F_{0i}]\}} \cdots$$

$$\cdots \frac{\Pi_x \delta(\partial_i B_i) \, \delta(\partial_i F_{0i}) \, dB \, dF}{\Pi_x \delta(\partial_i B_i) \, \delta(\partial_i F_{0i}) \, dB \, dF}, \tag{9.49}$$

where

$$S[B_i, F_{0i}] = \int dx_0 \left(\int f_{0i}^a \partial_0 b_i^a \, d^3x - H \right); \tag{9.50}$$

$$H = \int d^3x (\tfrac{1}{4} h_{ik}^a h_{ik}^a + \tfrac{1}{2} f_{0i}^a f_{0i}^a + \tfrac{1}{2} \partial_i c_0^a \partial_i c_0^a). \tag{9.51}$$

In those formulae, $[B_i, F_{0i}]$ is the action corresponding to the Hamiltonian H, where the transverse fields b_i, f_{0i} have the meaning of canonically conjugated coordinates and momenta.

It has been shown in Section 5, using the example of a scalar field, that the formalism of functional integration over canonically conjugated coordinates and momenta is equivalent to the canonical quantization. When the canonical quantization is applied to the system described by Hamiltonian (9.51), it results in the replacement of functions b_i^a, f_{0j}^b, through which h_{ik}^a, c_0^a are expressed, by the operators $\hat{b}_i^a(x), \hat{f}_{0j}^b(y)$, satisfying commutation relations

$$[\hat{b}_i^a(x), \hat{f}_{0j}^b(y)] = i\delta_{ab} \delta_{ij}^{tr}(x - y)$$

$$\equiv \frac{i\delta_{ab}}{(2\pi)^3} \int d^3k \exp i(k, x - y) \left(\delta_{ij} - \frac{k_i k_j}{k^2} \right). \tag{9.52}$$

Hamiltonian (9.51) becomes a self-conjugated and positive definite energy operator. Such a quantization of the Yang–Mills field has been suggested by Schwinger [42]. It has been shown how the functional integral formalism leads to Schwinger's canonical quantization. Let us emphasize that the existence of the factor $\Delta_R[B]$ in the original integral (9.40) is important in bringing the integral to an explicitly Hamiltonian form.

We shall examine the construction of the S matrix for the Yang–Mills field. It is natural to start from the functional integral in the Coulomb gauge. Namely, in such a gauge the integral is brought to the integral over canonical variables. The unitarity of the S matrix is evident here, in any case if its elements are calculated in the perturbation theory.

The element of the S matrix, describing the conversion of m ingoing particles into $(n - m)$ outgoing ones, is expressed in terms of the Fourier transformed function $G^{a_1,\ldots,a_n}_{(p_1,\ldots,p_n)}$ of the Green's function in the Coulomb gauge

$$\langle b_{i_1}^{a_1}(x_1),\ldots,b_{i_n}^{a_n}(x_n)\rangle_R$$

$$\equiv \frac{\int \exp{(iS)}\Delta_R[B]b_{i_1}^{a_1}(x_1),\ldots,b_{i_n}^{a_n}(x_n)\Pi_x\delta(\operatorname{div}\mathbf{B}(x))\,dB(x)}{\int \exp{(iS)}\Delta_R[B]\Pi_x\delta(\operatorname{div}\mathbf{B}(x))\,dB(x)}$$

(9.53)

by the formula

$$S^{a_1,\ldots,a_n}_{i_1,\ldots,i_n}(p_1,\ldots,p_n)$$

$$= \lim_{p_k^2 \to 0}\left(\prod_{k=1}^{n} Z_R^{-1/2}u_k(\hat{e}_k)_{i_kj_k}\right)G^{a_1,\ldots,a_n}_{j_1,\ldots,j_n}(p_1,\ldots,p_n)$$

(9.54)

where

$$(\hat{e}_k)_{ij} = \delta_{ij} - \frac{(p_k)_i(p_k)_j}{\mathbf{p}_k^2}$$

(9.55)

is the (transverse) polarization operator and the factors u_k are equal to $p_k^2\theta(p_k^0)|2p_k^0|^{-1/2}(2\pi)^{-3/2}$ for ingoing particles and to $p_k^2\theta(-p_k^0)\times$ $\times|2p_k^0|^{-1/2}(2\pi)^{-3/2}$ for outgoing ones. Finally, Z_R is the residue of the full one-particle Green's function in the Coulomb gauge at $p^2 \to 0$. Suppose that for $p^2 \to 0$ the one-particle Green's function has the form

$$G_{ij}^{ab} = \frac{Z_R\delta_{ab}}{p^2 + i0}\left(\delta_{ij} - \frac{p_ip_j}{\mathbf{p}^2}\right).$$

(9.56)

The expressions for the S-matrix elements in the Coulomb gauge are not explicitly relativistically invariant. We transform them into the relativistic Lorentz gauge.

We insert into the functional integrals in the numerator and the denominator the factor

$$\Delta_L[B] \int \prod_x \delta(\partial_\mu B_\mu^\Omega) \, d\Omega(x), \tag{9.57}$$

equal to one. Then we perform the shift $B^\Omega \to B; B \to B^{\Omega^{-1}}$. The integral over the gauge group

$$\int (B^{\Omega^{-1}}(x_1))_{i_1}^{a_1}, \ldots, (B^{\Omega^{-1}}(x_n))_{i_n}^{a_n} \prod_x \delta(\text{div } \mathbf{B}^{\Omega^{-1}}) \, d\Omega(x), \tag{9.58}$$

is obtained, in which the product $B^{\Omega^{-1}}(x_1)_{i_1}^{a_1}, \ldots, B^{\Omega^{-1}}(x_n)_{i_n}^{a_n}$ can be put before the integration sign. The remaining integral over Ω is cancelled by $\Delta_R[B]$.

Expression (9.53) for the Green's function takes the form

$$\frac{\int \exp{(iS)} \Delta_L[B] (B^{\Omega^{-1}}(x_1))_{j_1}^{a_1}, \ldots, (B^{\Omega^{-1}}(x_n))_{i_n}^{a_n} \Pi_x \delta(\partial_\mu B_\mu) \, dB}{\int \exp{(iS)} \Delta_L[B] \Pi_x \delta(\partial_\mu B_\mu) \, dB}. \tag{9.59}$$

In (9.59) the matrix Ω^{-1} is chosen such that the condition of three-dimensional transversality of the field $B^{\Omega^{-1}}(\text{div } B^{\Omega^{-1}} = 0)$ is satisfied. Expanding $B^{\Omega^{-1}}$ in powers of ε, we obtain the series

$$B^{\Omega^{-1}}(x) = B^{\text{tr}} - \frac{\varepsilon}{2}[\Delta^{-1} \text{div } \mathbf{B}, B + B^{\text{tr}}]^{\text{tr}} + \cdots \tag{9.60}$$

The first two terms of the expansion are presented here and the index tr marks the three-dimensional transverse part of the corresponding vector. We can evaluate integral (9.59) using the perturbation theory if the functions $B^{\Omega^{-1}}$ are expanded into series (9.60) and if to every term of the expansion depending on the product of m fields B, a vertex with m outgoing lines is assigned.

The transition to the Lorentz gauge of the Green's functions turns out to be rather complicated. To contruct the S-matrix it is, however, sufficient to know the Green's functions only on the mass shell (all $p_k^2 \to 0$). In that case the factors μ_k vanish and the transition to the Lorentz gauge reduces to the insertions

$$\frac{G_R}{\quad} = \underset{\sigma}{\bigcirc}\underset{G_L}{\quad\quad}\underset{\sigma}{\bigcirc}$$

$$(9.61)$$

into the external legs. The contribution to the insertion (we denote it σ is given by all the diagrams, beginning in the vertex generated by expansion (9.60) and terminating in the vertex connected with the rest of the diagram only by one line. Especially in the diagrams of the above structure, the vanishing of the factors μ_k is compensated by the poles of the one-particle functions G_L in the Lorentz gauge (with residues Z_L). The contribution of all other diagrams vanishes when going to the mass shell and the Green's function (9.59) in the Coulomb gauge differs from the corresponding function in the Lorentz gauge by the factor σ^n. It follows from the comparison of one-particle functions (the first one from diagram equations (9.61)) that $\sigma = (Z_R/Z_L)^{1/2}$, i.e., that it is expressed via the relation of residues. As a result, it turns out that on the mass shell it is possible to pass to the Green's functions in the Lorentz gauge, to replace Z_R with Z_L, and consequently to write down an S matrix element in an explicitly Lorentz-invariant form.

To conclude this section, we shall investigate corrections of the second order of perturbation theory to the Green's function. They are interesting because they indicate a situation opposite to the notorious situation of *null-charge* in quantum electrodynamics [37].

The photon Green's function in the transverse gauge has for $k^2 \gg m^2$ the form

$$D_{\mu\nu}^{\text{tr}} = (k^2 + i0)^{-1}(k^2\delta_{\mu\nu} - k_\mu k_\nu)[k^2 + i0 - P]^{-1}$$

$$\approx (k^2 + i0)^{-2}(k^2\delta_{\mu\nu} - k_\mu k_\nu)\left(1 - \frac{e^2}{12\pi^2}\ln\left(\frac{k^2}{m^2}\right)\right)^{-1}, \quad (9.62)$$

if we restrict ourselves to a one-loop second-order diagram for the self-energy part considering its asymptotic form at $k^2 \gg m^2$. The approximation (9.62) contains an unphysical 'ghost' pole for $(e^2/12\pi^2)$

$\ln(-k^2/m^2) \approx 1$, whose incidental existence would lead to serious contradictions with many principles of the theory [35], [37].

The formula for the Green's function of the Yang–Mills field for $k^2 \gg m^2$

$$G_{\mu\nu} = (k^2 + i0)^{-2} (k^2 \delta_{\mu\nu} - k_\mu k_\nu) \left(1 + \frac{13\varepsilon^2}{16\pi^2} \ln\left(-\frac{k^2}{M^2} \right) \right)^{-1}, \quad (9.63)$$

where M^2 is the renormalization constant, has the $+$ sign before the logarithmic term in the denominator and the problems with the unphysical pole and null-charge do not arise. This fact has been studied in References [43, 44]. It has been shown later that the situation does not change if higher diagrams of the perturbation theory are taken into account.

Formula (9.63) is easy to prove in the 'first-order formalism' where it is necessary to evaluate the following diagrams:

$$(9.64)$$

In these diagrams three types of internal lines for the Yang–Mills field and also the dashed lines corresponding to ghost scalar particles appear.

The expressions for self-energy parts (9.63) can be evaluated by means of the rules formulated above and turn out to be equal to the following expressions:

$$\Sigma_{\mu\nu}^{ab}(k) = \frac{\delta_{ab}\varepsilon^2}{12\pi^2} \left[(\delta_{\mu\nu}k^2 - k_\mu k_\nu) \ln\left(-\frac{k^2}{k_0^2} \right) - (ak^2 + b)\delta_{\mu\nu} + ck_\mu k_\nu \right];$$

$$\Sigma_{\mu\nu,\sigma}^{ab}(k) = \frac{\delta_{ab}\varepsilon^2 3i}{16\pi^2} (k_\mu \delta_{\nu\sigma} - k_\nu \delta_{\mu\sigma}) \left(\ln\left(-\frac{k^2}{k_0^2} \right) + d \right);$$

$$(9.65)$$

$$\Sigma_{\mu\nu,\sigma}^{ab}(k) = \frac{\delta_{ab}\varepsilon^2}{16\pi^2} \left\{ (\delta_{\mu\rho}\delta_{\nu\sigma} - \delta_{\mu\sigma}\delta_{\nu\rho}) \left(\ln\left(-\frac{k^2}{k_0^2} \right) + e \right) + \right.$$

$$\left. + \frac{1}{2(k^2 + i0)} (k_\mu k_\rho \delta_{\nu\sigma} + k_\nu k_\sigma \delta_{\mu\rho} - k_\mu k_\sigma \delta_{\nu\rho} - k_\nu k_\rho \delta_{\mu\sigma}) \right\}.$$

Here, k_0 is some fixed four-momentum fulfilling the condition $k_0^2 > 0$; a, b, c, d, e are renormalization constants. As a matter of fact, the first derivatives of $\Sigma_{\mu\nu,\rho\sigma}^{ab}(k)$, second derivatives of $\Sigma_{\mu\nu,\sigma}^{ab}(k)$ and third derivatives of $\Sigma_{\mu\nu}^{ab}(k)$ are uniquely defined. To ensure the transversability of $\Sigma_{\mu\nu}^{ab}(k)$, it is necessary to have $b = 0$, $a = -c$. Knowing the self-energy part (9.65), we evaluate the Green's function in the second order of the perturbation theory according to the formula

$$
G_{\mu\nu}^{ab} + G_{\mu\sigma}^{ac} \Sigma_{\sigma\rho}^{cd} G_{\rho\nu}^{db} + G_{\mu\sigma}^{cc} \Sigma_{\sigma,\rho\lambda}^{cd} G_{\rho\lambda,\nu}^{db} +
$$

$$
+ G_{\mu,\rho\lambda}^{ac} \Sigma_{\rho\lambda,\sigma}^{cd} G_{\sigma\nu}^{db} + G_{\mu,\rho\lambda}^{ac} \Sigma_{\rho\lambda,\sigma\tau}^{cd} G_{\sigma\tau,\nu}^{db}
$$

$$
= (k^2 + i0)^{-2} (k^2 \delta_{\mu\nu} - k_\mu k_\nu) \left[1 - \frac{11\varepsilon^2}{12\pi^2} \left(\ln\left(-\frac{k^2}{k_0^2} \right) + f \right) \right]
$$

$$
\approx (k^2 + i0)^{-2} (k^2 \delta_{\mu\nu} - k_\mu k_\nu) \left[1 + \frac{11\varepsilon^2}{12\pi^2} \left(\ln\left(-\frac{k^2}{k_0^2} \right) + f \right) \right]^{-1}
$$

(9.66)

where the renormalization constant f is a linear combination of a, d, e ($f = a - d - e$). The Green's function $G_{\mu\nu,\sigma}^{ab}(k)$ and self-energy part $\Sigma_{\mu\nu,\sigma}^{ab}(k)$ differ by the sign from the corresponding $G_{\sigma,\mu\nu}^{ab}(k)$, $\Sigma_{\sigma,\mu\nu}^{ab}(k)$. Formula (9.63) is proved.

10. QUANTIZATION OF A GRAVITATIONAL FIELD

As has been already noted in the introduction, when the quantum theory of gravitation is constructed, it is supposed that the gravitational field may become a natural 'physical regulator' which removes the quantum field theory divergencies. The gravitational field can be considered a special case of the gauge field and can be quantized according to the general scheme of gauge field quantization. The gauge transformations are the coordinate transformations that leave the space infinity untouched and the gauge group is a Poincaré group.

Specialities of the gravitational field quantization are related above all to the self-interaction of the field. Therefore, we shall deal mainly with a pure self-interacting gravitational field.

Among many others, there are two most common ways of gravitational field parametrizations: via metric tensor or via tetrads. We shall present both of them.

In metric field formalism, the gravitational field is described by the potentials $g_{\mu\nu}(x)$ and by the Christoffel symbols $\Gamma_{\mu\nu}^\rho(x)$. The latter may be considered either independent variables (Palatini's formalism) or functions of $g_{\mu\nu}$:

$$\Gamma^\rho_{\mu\nu} = \tfrac{1}{2}g^{\rho\sigma}(\partial_\mu g_{\nu\sigma} + \partial_\nu g_{\mu\sigma} - \partial_\sigma g_{\mu\nu}).\tag{10.1}$$

The contravariant matrix $g^{\mu\nu}$ is the inverse to $g_{\mu\nu}$; g is the determinant of the matrix $g_{\mu\nu}$.

We shall investigate asymptotically flat fields only. In that case, the space-time manifold is topologically equivalent to the four-dimensional Euclidean space and may be parametrized by the global coordinates $x_\mu(-\infty < x_\mu < +\infty, \mu = 0, 1, 2, 3)$. We suppose that the choice of coordinates is consistent with the conditions at the space infinity

$$g_{\mu\nu} = \eta_{\mu\nu} + O\left(\frac{1}{r}\right); \qquad \Gamma^\rho_{\mu\nu} = O\left(\frac{1}{r^2}\right),\tag{10.2}$$

where $r = \sqrt{(x^1)^2 + (x^2)^2 + (x^3)^2}$; $\eta_{\mu\nu}$ is the Minkowski tensor. The signature $(+ - - -)$ is used here.

The action functional

$$S = \frac{1}{2\kappa^2}\int[-\Gamma^\rho_{\mu\rho}\partial_\nu(\sqrt{-g}\,g^{\mu\nu}) + \Gamma^\rho_{\mu\nu}\partial_\rho(\sqrt{-g}\,g^{\mu\nu}) +$$
$$+ \sqrt{-g}\,g^{\mu\nu}(\Gamma^\rho_{\mu\sigma}\Gamma^\sigma_{\rho\nu} - \Gamma^\rho_{\mu\nu}\Gamma^\sigma_{\rho\sigma})]\,\mathrm{d}^4x,\tag{10.3}$$

where κ is the Newton constant, is invariant under the coordinate transformations acting on the quantities $g^{\mu\nu}, \Gamma^\rho_{\mu\nu}$ according to the rules

$$\delta g^{\mu\nu} = -\eta^\lambda\partial_\lambda g^{\mu\nu} + g^{\mu\nu}\partial_\lambda\eta^\nu + g^{\nu\lambda}\partial_\lambda\eta^\mu;$$
$$g\Gamma^\rho_{\mu\nu} = -\eta^\lambda\partial_\lambda\Gamma^\rho_{\mu\nu} - \Gamma^\rho_{\mu\lambda}\partial_\nu\eta^\lambda - \Gamma^\rho_{\nu\lambda}\partial_\mu\eta^\lambda + \Gamma^\lambda_{\mu\nu}\partial_\lambda\eta^\rho.\tag{10.4}$$

They are the formulae for infinitesimal transformations; η^μ are infinitesimally small components of the vector field that generate the coordinate transformations

$$\delta x^\mu = \eta^\mu(x).\tag{10.5}$$

Variations of the action with respect to $\Gamma^\rho_{\mu\nu}$ give equations whose solutions are functions (10.1). In this sense $\Gamma^\rho_{\mu\nu}$ can be regarded as independent variables.

When the explicit expression (10.1) of the Christoffel symbols in terms of the metric tensor is inserted into (10.3) the action acquires the form

$$S = \frac{1}{4\kappa^2} \cdot \int (h^{\rho\sigma} \partial_\rho h^{\mu\nu} \, \partial_\nu h_{\mu\sigma} - \tfrac{1}{2} h^{\rho\sigma} \, \partial_\rho h^{\mu\nu} \, \partial_\sigma h_{\mu\nu} +$$
$$+ \tfrac{1}{4} h^{\rho\sigma} \, \partial_\rho \ln h \partial_\sigma \ln h) \, \mathrm{d}^4 x \tag{10.6}$$

where the contravariant density

$$h^{\mu\nu} = \sqrt{-g} g^{\mu\nu}, \qquad h = \det h^{\mu\nu} \tag{10.7}$$

is introduced for convenience.

In the tetrad formalism, the gravitational field is described by the tetrad components $e^{\mu a}(x)$ and torsion coefficients $\omega_{\mu,ab}(x) = -\omega_{\mu,ba}(x)$. The family $e^{\mu a}(x)$ forms a matrix with a positive determinant $e(x)$. The action functional

$$S = \frac{1}{2\kappa^2} \int [\omega_{\mu ab} \, \partial_\mu (e^{-1} e^{\mu a} e^{\nu b}) - \omega_{\mu ab} \, \partial_\nu (e^{-1} e^{\mu a} e^{\nu b}) +$$
$$+ e^{-1} e^{\mu a} e^{\nu b} (\omega_{\mu ac} \omega^c_{\nu b} - \omega_{\nu ac} \omega^c_{\mu b})] \tag{10.8}$$

is invariant under the coordinate transformations

$$\delta e^{\mu a} = -\eta^\lambda \, \partial_\lambda e^{\mu a} + e^{\lambda a} \, \partial_\lambda \eta^\mu;$$
$$\delta \omega_{\mu ab} = -\eta^\lambda \, \partial_\lambda \omega_{\mu ab} - \omega_{\lambda ab} \, \partial_\mu \eta^\lambda \tag{10.9}$$

and under the local Lorentz rotations

$$\delta e^{\mu a} = \eta^a_b e^{\mu b}, \qquad \delta \omega_{\mu ab} = -\eta^c_a \omega_{\mu cb} + \eta^c_b \omega_{\mu ac} + \partial_\mu \eta_{ab}. \tag{10.10}$$

Variations with respect to ω lead to equations expressing ω in terms of e. The solution can be written in the form

$$\omega_{\mu ab} = e^c_\mu \omega_{cab} \equiv \tfrac{1}{2} e^c_\mu (\Omega_{abc} + \Omega_{bca} - \Omega_{cab}), \tag{10.11}$$

where

$$\Omega_{abc} = e_{\mu a} \Omega^\mu_{bc} \equiv e_{\mu a} (e^\nu_b \, \partial_\nu e^\mu_c - e^\nu_c \, \partial_\nu e^\mu_b).$$

Supposing that this solution has been employed, then S is a functional of $e^{\mu a}$ only.

The formalism is said to be of the first order if the variables $g_{\mu\nu}$ and $\Gamma^\rho_{\mu\nu}$, or $e^{\mu a}$ and $\omega_{\mu,ab}$ are considered independent. If Γ are expressed via g and ω via e, then the formalism is said to be of the second order.

The descriptions of the pure gravitational field in terms of $g^{\mu\nu}$ or $e^{\mu a}$ are equivalent. The different number of components – 10 in the former case and 16 in the latter – is compensated by different gauge groups that are parametrized by four functions in the former case and 10 functions in the latter. The tetrad formalism is suitable for the description of the interaction with a spinor field.

The equivalence of the first and second orders fails when an interaction with other fields is switched on. The geometrical formula (10.11) defines the connexion without a torsion. The minimal interaction of the gravitational and spinorial fields induces the appearance of the torsion.

Further explanation in this section is restricted to an example of the tensor second-order formalism. The coordinate transformations of the metric tensor form a nonabelian gauge group parametrized by four functions (the infinitesimal transformations are given by (10.4)). Therefore, in agreement with the general scheme of gauge field quantization, when a system with action S is quantized, it is necessary to integrate the functional exp iS over a surface in the manifold of fields determined by four equations. It is suitable for the equations to choose the de Donder–Fock harmonic conditions

$$\partial_\nu(\sqrt{-g}g^{\mu\nu}) = l^\mu(x). \tag{10.12}$$

where $l^\mu(x)$ is a given vector field. The arbitrariness in the choice of $l^\mu(x)$ will be useful for formal transformations. Harmonic conditions (10.12) are an analogue of the Lorentz gauge in electrodynamics and the Yang–Mills theory.

Equations (10.12) are not generally covariant and that is why they can be used for the class parametrization. The equation $f(A^\Omega) = 0$ is represented by a complicated nonlinear equation for the parameters of the coordinate transformation that transforms the metric to a harmonic one. In the framework of the perturbation theory the equation has a unique solution.

The local gauge invariant measure is of the form*

$$\prod_x g^{5/2}(x) \prod_{\mu \leq \nu} dg^{\mu\nu}(x) = \prod_x h^{-5/2}(x) \prod_{\mu \leq \nu} dh^{\mu\nu}(x). \tag{10.13}$$

To prove the gauge invariance of measure (10.13), we shall regard Π_x

---

*In References [161–163] arguments are given supporting the use of the Leutwyler measure $d\mu = \Pi_x(g^{7/2}g^{00}\Pi_{\mu \leq \nu} dg^{\mu\nu})$ in gravity theory instead of measure (10.13) used here. In these works, however, it is shown that differences between the measures produce only components of a purely renormalization type, which serve for eliminating divergences from the perturbation theory. The structure of the perturbation theory is otherwise unchanged. Thus, using our convention, we do not change the physical consequences of the perturbation theory of the gravitational field.

as the product over the 'physical' points. The transformation of the metric tensor in one point is given by formulae (10.4) without the last term, so that

$$\delta g^{\mu\nu} = g^{\mu\nu} \partial_\lambda \eta^\nu + g^{\nu\lambda} \partial_\lambda \eta^\mu; \qquad \delta g = -2g \partial_\mu \eta^\mu. \tag{10.14}$$

Therefore

$$\prod_{\mu \leq \nu} d(g^{\mu\nu} + \delta g^{\mu\nu}) = (1 + 5\partial_\mu \eta^\mu) \prod_{\mu \leq \nu} dg^{\mu\nu};$$

$$g + \delta g = (1 - 2\partial_\mu \eta^\mu)g. \tag{10.15}$$

The invariance of the factor $g^{5/2} \prod_{\mu \leq \nu} dg^{\mu\nu}$, that implies the gauge invariance of measure (10.13), follows from (10.15).

From the knowledge of class parametrization (10.12) and measure (10.13) we obtain the continual integral

$$\int \exp{(iS)} \Delta_h[g] \prod_x \left(\prod_\mu \delta(\partial_\nu g^{\mu\nu} - l^\mu) \left(g^{5/2} \prod_{\mu \leq \nu} dg^{\mu\nu} \right) \right). \tag{10.16}$$

where, in correspondence with (7.8), the functional $\Delta_h[g]$ is determined by the equation

$$\Delta_h[g] \int \prod_x \left(\prod_\mu \delta(\partial_\nu (h^{\mu\nu})^{a(x)} - l^\mu(x)) \right) da(x) = 1 \tag{10.17}$$

and can be expressed as an integral of the δ functional over the gauge group.

Let us calculate integral (10.16). The expression $\Delta_h[g]$ contributes to the integral only at the surface defined by Equations (10.12). For such $g^{\mu\nu}$ only an infinitesimally-small neighbourhood of the unit element of the group contributes to integral (10.17). In this neighbourhood the group action on $h^{\mu\nu}$ and on the gauge group measure da can be parametrized by infinitesimal functions (10.5), which were introduced before. In this parametrization

$$\partial_\nu (h^{\mu\nu})^a - l^\mu = \partial_\nu (h^{\nu\lambda} \partial_\lambda \eta^\mu) - \partial_\lambda (\partial_\nu h^{\mu\nu} n^\lambda). \tag{10.18}$$

The measure da at the unit element has a simple form

$$da = \prod_x \prod_\mu d\eta^\mu(x). \tag{10.19}$$

implying that the integral can be written as

$$\int \prod_{x,\mu} \delta(\partial_\nu (h^{\nu\lambda} \partial_\lambda \eta^\mu) - \partial_\lambda (\partial_\nu h^{\mu\nu} \eta^\lambda)) \, d\eta^\mu(x). \tag{10.20}$$

Formally, this integral equals $(\det \hat{A})^{-1}$ where the operator $\hat{A}$ acts on the quadruplet of the functions η^μ according to the rule

$$(\hat{A}\eta)^\mu = \partial_\nu(h^{\nu\lambda}\partial_\lambda\eta^\mu) - \partial_\lambda(\partial_\nu h^{\mu\nu}\eta^\lambda). \qquad (10.21)$$

Thus, we have found that

$$\Delta_h[g] = \det \hat{A}. \qquad (10.22)$$

It is convenient for the perturbation theory formulation to represent $\det \hat{A}$ as a Gaussian integral over ghost fields. The fields should anticommute in order that the integral be equal to the first power of the determinant. This requirement satisfies the expression

$$\det \hat{A} = \int \exp\left(i \int \bar{\theta}^\mu(x) A_{\mu\nu} \theta^\nu(x) \, d^4x \right) \prod_{x,\mu} d\theta^\mu(x) \, d\bar{\theta}^\mu(x), \qquad (10.23)$$

where $\theta^\mu(x), \bar{\theta}^\mu(x)$ are anticommuting classical fields meeting

$$\theta^\mu(x)\theta^\nu(y) + \theta^\nu(y)\theta^\mu(x) = 0 \qquad (10.24)$$

and analogous relations for $(\theta, \bar{\theta})$, $(\bar{\theta}, \bar{\theta})$.

Now integral (10.16) can be written in the form

$$\int \exp\left\{ iS[g] + i \int \theta^\mu A_{\mu\nu}[g]\theta^\nu \, d^4x \right\} \prod_x \left(\prod_\mu \delta(\partial_\nu h^{\mu\nu} - l^\mu) \times \right.$$

$$\left. \times \left(g^{5/2} \prod_{\mu \le \nu} dg^{\mu\nu} \right) \prod_\mu d\theta^\mu \, d\bar{\theta}^\mu \right) \qquad (10.25)$$

and can be immediately used for the formulation of the perturbation theory. Nevertheless, we shall further transform (10.25) by exploiting the arbitrariness in the choice of l^μ. The transformation proposed by t'Hooft [18] has already been explained in Section 9 on the example of the Yang–Mills field. The integral (10.25) is independent of l^μ due to its construction. Therefore, it may be averaged over l^μ with respect to any weight. For the weight we take the exponential of the quadratic form of the fields

$$\exp\left(\frac{i\alpha}{4} \int l^\mu(x)\theta_{\mu\nu}l^\nu(x) \, d^4x \right), \qquad (10.26)$$

where $\eta_{\mu\nu}$ is the Minkowski tensor. The averaging is performed directly and yields

$$\int \exp\left\{ iS[g] + \frac{i\alpha}{4} \int \partial_\rho h^{\mu\rho} \eta_{\mu\nu} \partial_\sigma h^{\nu\sigma}\, d^4x + \right.$$

$$\left. + i \int \bar{\theta}^\mu A_{\mu\nu} \theta^\nu\, d^4x \right\} \prod_x g^{5/2} \left(\prod_{\mu \le \nu} dg^{\mu\nu} \right) \left(\prod_\mu d\theta^\mu\, d\bar{\theta}^\mu \right), \tag{10.27}$$

containing the quadratic form of the longitudinal parts of the field $h^{\mu\nu}$ multiplied by an arbitrary coefficient α. It follows from the previous considerations that the integral does not depend on α.

From (10.27) diagrams of the perturbation theory can be derived. The variables $h^{\mu\nu} = \sqrt{-g}\, g^{\mu\nu}$ and $\theta^\mu, \bar{\theta}^\mu$ in integral (10.27) will be considered independent. Let

$$h^{\mu\nu} = \eta^{\mu\nu} + \kappa u^{\mu\nu} \tag{10.28}$$

where $u^{\mu\nu}$ is a tensor field that is supposed to describe the gravitational field. The action functional acquires the form

$$S = S_2 + \sum_{n=1}^{\infty} \kappa^n S_{n+2}, \tag{10.29}$$

where S_2 is a quadratic form and S_n are forms of the nth degree in the variables $u^{\mu\nu}$ and their first derivatives.

The linearization of (10.28) may be unnatural in many respects. When $u^{\mu\nu}$ is not sufficiently small, it may change the signature of the metric tensor. There are parametrizations that do not exhibit this deficiency, e.g., the exponential one:

$$h^{\mu\nu} = \eta^{\mu\nu} [\exp(\kappa\Phi)]^{\lambda\nu}. \tag{10.30}$$

The decomposition (10.29) can be obtained in this parametrization, too. Notice that the quadratic form S_2 does not depend on the choice of parametrizations. Due to the invariance of the action under transformations (10.4), the quadratic form

$$S_2 = \tfrac{1}{4} \int (-\eta_{\nu\sigma} \delta^\alpha_\rho \delta^\beta_\mu + \tfrac{1}{2} \eta^{\alpha\beta} \eta_{\mu\rho} \eta_{\nu\sigma} +$$

$$+ \tfrac{1}{4} \eta_{\mu\nu} \eta_{\rho\sigma} \eta^{\alpha\beta}) \partial_\alpha u^{\mu\nu}\, \partial_\beta u^{\rho\sigma}\, d^4x \tag{10.31}$$

is degenerated. It does not contain the longitudinal components. Having introduced the ghost fields $\theta^\mu, \bar{\theta}^\mu$ we succeeded in removing the degeneracy of the quadratic form occurring in the exponent of the integrand of (10.27). The operators inverse to the operators of the forms quadratic in $h^{\mu\nu}, \theta^\mu, \bar\theta^\mu$, i.e., particle propagators corresponding to diagram lines, are determined in the same way.

The graviton propagators $\langle h^{\mu\nu}h^{\rho\sigma}\rangle$ will be depicted by full lines and the propagators of the ghost vector particles $\langle\theta^\mu\bar\theta^\nu\rangle$ by dashed lines. Vertices of diagrams are generated by the forms S_{n+2} from decomposition (10.29) as well as by the form $\int\theta^\mu u_{\mu\nu}\theta^\nu\,d^4x$, introducing the trilinear interaction $\sim\bar\theta\mu\theta$ of the graviton and the ghost vector particle.

The elements of the diagram method are

$$(10.32)$$

The expression for the graviton propagator is

$$G^{\mu\nu,\rho\sigma}(k) = \frac{2}{k^2}(\eta^{\mu\rho}\eta^{\nu\sigma} + \eta^{\mu\sigma}\eta^{\nu\rho} + (\alpha^{-1} - 2)\eta^{\mu\nu}\eta^{\rho\sigma}) +$$

$$+ \frac{2(1-\alpha^{-1})}{k^4}(2k^\mu k^\nu\eta^{\rho\sigma} + 2k^\rho k^\sigma\eta^{\mu\sigma} - k^\mu k^\rho\eta^{\nu\sigma} - \quad (10.33)$$

$$- k^\nu k^\rho\eta^{\mu\sigma} - k^\mu k^\sigma\eta^{\nu\rho} - k^\nu k^\sigma\eta^{\mu\rho}).$$

The formula contains a parameter α. The quantity α is analogous to the parameter d_l in quantum electrodynamics and in the Yang–Mills theory, and is the coefficient in the longitudinal part of the propagator. The physical results are independent of the choice of the constant α. The propagator of the ghost vector particle is

$$G^{\mu\nu} = -\eta^{\mu\nu}/k^2, \qquad (10.34)$$

and its interaction with the graviton is given by the vertex

$k_{3\rho\sigma} = \frac{\kappa}{2}[k_{1\nu}(\delta^\mu_\sigma k_{3\rho} + \delta^\mu_\rho k_{3\sigma}) - \delta^\mu_\nu(k_{1\rho}k_{2\sigma} + k_{1\sigma}k_{2\rho})], \quad (10.35)$

where $k_1 + k_2 + k_3 = 0$.

Let us also present the explicit formula for the third order vertex corresponding to the linearization (10.28)

$$k_1 \mu \nu$$
$$k_3 \sigma \tau$$
$$k_2 \lambda \rho$$

$$
= \frac{\kappa}{32} \Bigg\{ \frac{k_1^2}{2} (\eta_{\mu\nu}\eta_{\rho\sigma}\eta_{\tau\nu} + \eta_{\mu\tau}\eta_{\nu\lambda}\eta_{\rho\sigma} + \eta_{\mu\sigma}\eta_{\nu\lambda}\eta_{\rho\tau} +
$$

$$
+ \eta_{\nu\sigma}\eta_{\mu\nu}\eta_{\rho\tau} + \eta_{\mu\tau}\eta_{\lambda\rho}\eta_{\nu\sigma} + \eta_{\mu\sigma}\eta_{\nu\rho}\eta_{\lambda\tau} +
$$

$$
+ \eta_{\nu\tau}\eta_{\mu\rho}\eta_{\lambda\sigma} + \eta_{\nu\sigma}\eta_{\mu\rho}\eta_{\lambda\tau}) + k_1^2 \eta_{\mu\nu}(\eta_{\rho\sigma}\eta_{\lambda\tau} + \eta_{\rho\tau}\eta_{\lambda\sigma}) +
$$

$$
+ (k_{2\mu}k_{3\nu} + k_{2\nu}k_{3\mu})\eta_{\lambda\rho}\eta_{\sigma\tau} +
$$

$$
+ (k_{2\mu}k_{3\nu} + k_{2\nu}k_{3\mu})(\eta_{\lambda\sigma}\eta_{\rho\tau} + \eta_{\lambda\rho}\eta_{\sigma\tau}) - \qquad (10.36)
$$

$$
- k_1 k_{1\tau}(\eta_{\mu\lambda}\eta_{\rho\sigma} + \eta_{\mu\rho}\eta_{\lambda\sigma}) - k_{1\nu}k_{1\sigma}(\eta_{\mu\lambda}\eta_{\rho\tau} + \eta_{\mu\rho}\eta_{\lambda\tau}) -
$$

$$
- k_{1\mu}k_{1\tau}(\eta_{\nu\lambda}\eta_{\rho\sigma} + \eta_{\nu\rho}\eta_{\lambda\sigma}) - k_{1\mu}k_{1\sigma}(\eta_{\nu\lambda}\eta_{\rho\tau} + \eta_{\nu\rho}\eta_{\lambda\tau}) -
$$

$$
- k_{2\nu}k_{3\rho}\eta_{\nu\sigma}\eta_{\mu\tau} - k_{2\nu}k_{3\lambda}\eta_{\rho\sigma}\eta_{\mu\tau} - k_{2\mu}k_{3\rho}\eta_{\lambda\sigma}\eta_{\nu\tau} -
$$

$$
- k_{2\mu}k_{3\lambda}\eta_{\rho\sigma}\eta_{\nu\tau} - k_{2\nu}k_{3\rho}\eta_{\lambda\tau}\eta_{\mu\sigma} - k_{2\nu}k_{3\lambda}\eta_{\rho\tau}\eta_{\mu\sigma} -
$$

$$
- k_{2\mu}k_{3\rho}\eta_{\lambda\tau}\eta_{\nu\sigma} - k_{2\mu}k_{3\lambda}\eta_{\rho\tau}\eta_{\nu\sigma} +
$$

$$
+ \text{ the sum over the permutations of pairs } (\mu, \nu),
$$

$$
(\sigma, \tau), (\lambda, \rho) \Bigg\}.
$$

The contribution of a diagram is obtained when the product of terms (10.33)–(10.36), corresponding to the elements of the diagram, is integrated over internal momenta and the result is multiplied by

$$
r^{-1}(-1)^s \left(\frac{i}{(2\pi)^4} \right)^{l-v-1}, \qquad (10.37)
$$

where r is the order of the diagram symmetry group, l is the number of internal lines, v is the number of vertices and s is the number of ghost vector-particle loops.

Particles corresponding to the ghost vector fields are fermions so that the spin-statistic relations are broken. Their role reduces to subtraction of the contribution of the unphysical degrees of freedom.

Besides the described diagrams, the perturbation theory contains contributions from the renormalization procedure proportional to degrees of $\delta^{(4)}(0)$. The contributions are generated by the local factor $\Pi_x h^{-5/2}(x)$ that is present in the measure. Linearization (10.28) implies

$$\prod_{x,\mu,\nu} dh^{\mu\nu} = \prod_{x,\mu,\nu} du^{\mu\nu} \tag{10.38}$$

It is necessary to include the influence of the local factor into the perturbation theory. Formally its role reduces to the addition of the term

$$\Delta S = \tfrac{5}{2} i \delta^{(4)}(0) \int \ln h(x) \, d^4 x, \tag{10.39}$$

to the action. This term generates vertices proportional to $\delta^{(4)}(0)$. The occurrence of such renormalization terms is not considered in many works investigating nonlinear theories (cf. [46], [47]). Let us note that the terms do not appear in the exponential parametrizations (10.30). Measure (10.13) has the simple form

$$\prod_x \prod_{\mu \leq \nu} d\Phi^{\mu\nu} \tag{10.40}$$

up to a constant factor and there are no additional terms in this parametrization.

In such a way we have obtained the diagrams of the perturbation theory in the formalism of continual integration over the fields $g^{\mu\nu}$ (or $h^{\mu\nu}$). In many cases, especially when we pass to the Hamiltonian formulation, the first-order formalism proves to be more suitable. The variables $g^{\mu\nu}$ and $\Gamma^\rho_{\mu\nu}$ are considered independent in this case. The measure has (up to a degree of volume) the form

$$g^{15/2} \prod_{\mu \leq \nu} dg^{\mu\nu} \prod_{\substack{\mu \leq \nu \\ \rho}} d\Gamma^\rho_{\mu\nu} = h^{5/2} \prod_{\mu \leq \nu} dh^{\mu\nu} \prod_{\substack{\mu \leq \nu \\ \rho}} d\Gamma^\rho_{\mu\nu}. \tag{10.41}$$

The degree of the determinant g in the measure is such that after the Gaussian integration with respect to Γ, the measure coincides with (10.13).

We shall describe the diagrams occurring in the first-order formalism. The elements of diagrams corresponding to the ghost vector particles are unchanged. Besides the tensor propagator $\langle \mu, \mu \rangle$ (10.33), other propagators $\langle \mu, \gamma \rangle$ and $\langle \gamma, \gamma \rangle$ enter into the perturbation theory scheme. They are depicted by the following lines

$$\langle uu \rangle = \underline{\hspace{5cm}} \; ;$$

$$\langle u\gamma \rangle = \underline{\hspace{3cm}}\,\boxed{} \; ; \tag{10.42}$$

$$\langle \gamma\gamma \rangle = \boxed{} \; .$$

The propagators in the momentum representation are

$$G_{\sigma\tau}^{\mu\nu\rho}(k) = \frac{i}{2}(\eta_{\sigma\alpha}\delta_\beta^\rho k_\tau + \eta_{\tau\alpha}\delta_\beta^\rho k_\sigma - \eta_{\sigma\alpha}\eta_{\tau\beta}k^\rho)G^{\mu\nu,\alpha\beta}(k)$$

$$\equiv \Omega_{\sigma\tau,\alpha\beta}^\rho(k)G^{\alpha\beta,\mu\nu}(k);$$

$$G_{\sigma\tau,\mu\nu}^{\rho\lambda}(k) = \tfrac{1}{4}(\delta_\mu^\rho\delta_\sigma^\lambda\eta_{\nu\tau} + \delta_\nu^\rho\delta_\sigma^\lambda\eta_{\mu\tau} + \delta_\mu^\rho\delta_\tau^\lambda\eta_{\nu\sigma} + \tag{10.43}$$

$$+ \delta_\nu^\rho\delta_\tau^\lambda\delta_{\mu\sigma}) - \tfrac{1}{6}(\delta_\nu^\lambda\delta_\tau^\rho\eta_{\mu\sigma} + \delta_\mu^\lambda\delta_\tau^\rho\eta_{\nu\sigma} + \delta_\nu^\lambda\delta_\sigma^\rho\eta_{\mu\tau} +$$

$$+ \delta_\mu^\lambda\delta_\sigma^\rho\eta_{\nu\tau}) + \Omega_{\mu\nu,\alpha\rho}^\lambda(k)\Omega_{\rho\tau,\gamma\delta}^\rho(-k)G^{\alpha\beta,\gamma\delta}(k).$$

The only graviton vertex is generated by the trilinear form

$$\frac{\kappa}{2}\int u^{\mu\nu}(\gamma_{\mu\sigma}^\rho\delta_{\rho\nu}^\sigma - \gamma_{\mu\nu}^\rho\gamma_{\rho\sigma}^\sigma)\,\mathrm{d}^4x \tag{10.44}$$

and is expressed by

$$= \frac{\kappa}{8}\{2(\delta_\mu^\sigma\delta_\nu^\tau + \delta_\nu^\sigma\delta_\mu^\tau)(\delta_\rho^\beta\delta_\alpha^\gamma + \delta_\alpha^\beta\delta_\rho^\gamma) - \delta_\mu^\sigma\delta_\alpha^\tau\delta_\nu^\gamma\delta_\rho^\beta -$$

$$- \delta_\mu^\sigma\delta_\nu^\beta\delta_\rho^\gamma\delta_\alpha^\tau - \delta_\nu^\sigma\delta_\mu^\tau\delta_\alpha^\beta\delta_\rho^\gamma - \delta_\nu^\sigma\delta_\alpha^\tau\delta_\mu^\gamma\delta_\rho^\beta - \delta_\mu^\tau\delta_\alpha^\sigma\delta_\nu^\beta\delta_\rho^\gamma - \tag{10.45}$$

$$- \delta_\mu^\tau\delta_\alpha^\sigma\delta_\nu^\gamma\delta_\rho^\beta - \delta_\nu^\tau\delta_\alpha^\sigma\delta_\mu^\beta\delta_\rho^\gamma - \delta_\nu^\tau\delta_\alpha^\sigma\delta_\mu^\gamma\delta_\rho^\beta\}.$$

Renormalization terms proportional to $\delta^{(4)}(0)$ are generated by the local factor $\Pi_x h^{5/2}(x)$ of measure (10.41). Its contribution may be interpreted as an additional term to the action

$$\Delta S = -\tfrac{5}{2}i\delta^{(4)}(0)\int \ln h(x)\,\mathrm{d}^4x. \tag{10.46}$$

We have explored the case of the pure gravitational field in detail. Interactions with other fields do not essentially change the scheme of perturbation theory construction. There are no new ghost particles in cases of matter fields with nonsingular Lagrangians. The ghost particles and the corresponding diagrams appear only when fields with other gauge groups, e.g., electromagnetic or Yang–Mills fields, are introduced. Let us present the expression for the continual integral corresponding to interacting electromagnetic and gravitational fields without detailed investigation:

$$\int \exp\{iS[g^{\mu\nu}, A_\mu]\}\Delta[g] \prod_x \delta(\partial_\mu(h^{\mu\nu}A_\mu)) \prod_\mu \delta(\partial_\nu h^{\mu\nu}) \times$$
$$\times g^{5/2} \prod_{\mu \leq \nu} dg^{\mu\nu} \prod_\mu dA^\mu; \tag{10.47}$$

$$S[g^{\mu\nu}, A_{\mu\nu}] = S_g - \tfrac{1}{4} \int (\partial_\mu A_\nu - \partial_\nu A_\mu) \times$$
$$\times (\partial_\lambda A_\rho - \partial_\rho A_\lambda) g^{\mu\nu} g^{\lambda\rho} \sqrt{-g}\, d^4x, \tag{10.48}$$

where S_g is the gravitational field action and $\Delta[g]$ equals the product of the determinants

$$\det \hat{A} \det (\partial_\mu(h^{\mu\nu}\partial_\nu)), \tag{10.49}$$

in which $\hat{A}$ is the operator (10.21). The occurrence of the second factor in the product indicates that the ghost particle, which is inessential for the description of a pure electromagnetic field, interacts with a gravitational field, too. In this way, a ghost scalar neutral particle together with the previously-mentioned elements, appears in the covariant perturbation theory of electromagnetic and gravitational fields.

Let us investigate the Hamilton theory in continual integral formalism. The Hamilton formulation of the gravity theory was worked out for the first time by Dirac [48]. When the explicit Hamilton form of the Einstein equation is derived, a complicated problem arises – the solution of constraint equations. We shall consider a general Hamilton formulation of the gravity theory in which it is not necessary to solve the constraint equations but only check their commutation relations. This generalized formulation is a field theory analogue of the formulation developed for finite-dimensional mechanical systems in Section 3. We shall show that the action of the gravitational field can acquire form analogous to (3.1) for the finite-dimensional systems and that the corresponding constraints and the Hamiltonian satisfy conditions (3.2). We shall follow the method proposed by L. D. Fadeev and specially adapted to the gravitational field [53].

For that purpose it is convenient to employ the first-order formalism. We take the expression of the gravitational field action (10.3) and pick out terms of the Lagrangian containing time derivatives:

$$
\frac{1}{2\kappa^2}(\Gamma^0_{\mu\nu}\partial_0 h^{\mu\nu} - \Gamma^\rho_{\mu\rho}\partial_0 h^{\mu 0})
$$

$$
= \frac{1}{2\kappa^2}(\Gamma^0_{ik}\partial_0 h^{ik} + (\Gamma^0_{i0} - \Gamma^k_{ik})\partial_0 h^{i0} - \Gamma^i_{0i}\partial_0 h^{00}). \tag{10.50}
$$

This expression does not contain the variables Γ^μ_{00} that occur in $L(h,\Gamma)$ linearly and can be regarded as Lagrangian multiplicators. The factors A^{00}_μ at Γ^μ_{00} are the constraints. The constraint equations

$$
A^{00}_0 = h^{ik}\Gamma^0_{ik} + h^{00}\Gamma^i_{0i} + \partial_i h^{i0} = 0;
$$

$$
A^{00}_i = 2h^{k0}\Gamma^0_{ik} + h^{00}(\Gamma^0_{i0} - \Gamma^k_{ik}) + \partial_i h^{00} \tag{10.51}
$$

permit one to express the variables $\Gamma^i_{0i}, \Gamma^0_{i0} - \Gamma^k_{ik}$ via Γ^0_{ik} and $h^{\mu\nu}$. Meanwhile, the terms containing time derivatives acquire the form

$$
\frac{1}{2\kappa^2}\frac{\Gamma^0_{ik}}{h^{00}}\partial_0(h^{00}h^{ik} - h^{i0}h^{k0}), \tag{10.52}
$$

if the terms

$$
\frac{1}{2\kappa^2 h^{00}}(\partial_0 h^{00}\partial_i h^{i0} - \partial_i h^{00}\partial_0 h^{i0})
$$

$$
= \frac{1}{2\kappa^2}(\partial_0 \ln h^{00}\partial_i h^{i0} - \partial_i \ln h^{00}\partial_0 h^{i0}), \tag{10.53}
$$

which disappear after an integration by parts, are left out.

Formula (10.52) indicates that the natural dynamic variables are

$$
q^{ik} = h^{i0}h^{k0} - h^{00}h^{ik}; \qquad \pi_{ik} = -\frac{1}{h^{00}}\Gamma^0_{ik}. \tag{10.54}
$$

The variables $\Gamma^\rho_{\mu\nu}$ that are different from Γ^0_{ir} are not dynamical. They can be excluded by constraint equations

$$
\frac{\partial L(h,\Gamma)}{\partial \Gamma^\rho_{\mu\nu}} = 0; \qquad (\Gamma^\rho_{\mu\nu} \neq \Gamma^0_{ik}). \tag{10.55}
$$

System (10.55) contains Equations (10.51) together with

$$\partial_k h^{i0} + h^{is}\Gamma^0_{sk} + h^{00}\Gamma^i_{k0} + h^{0s}\Gamma^i_{sk} - h^{i0}\Gamma^s_{ks} = 0; \tag{10.56}$$
$$\partial_k h^{ij} + h^{i\sigma}\Gamma^j_{pk} + h^{j\sigma}\Gamma^i_{\sigma k} - h^{ij}\Gamma^\sigma_{k\sigma} = 0.$$

The solution of systems (10.51) and (10.56), which expresses the 'nondynamical' variables $\Gamma^0_{i0}, \Gamma^k_{i0}, \Gamma^k_{ij}$ via $h^{\mu\nu}, \Gamma^0_{ik}$, is

$$\Gamma^0_{i0} = \Gamma^s_{is} - \frac{\partial_i h^{00}}{h^{00}} - \frac{h^{0s}}{h^{00}}\Gamma^0_{is};$$

$$\Gamma^k_{i0} = -\frac{1}{h^{00}}(\partial_i h^{k0} + h^{0s}\Gamma^k_{is} - h^{k0}\Gamma^s_{is} + h^{ks}\Gamma^0_{is}); \tag{10.57}$$

$$\Gamma^k_{ij} = \overset{*}{\Gamma}{}^k_{ij} + \frac{h^{k0}}{h^{00}}\Gamma^0_{ij}.$$

where $\overset{*}{\Gamma}{}^k_{ij}$ are the three-dimensional connection coefficients generated by the three-dimensional metrics g_{ik} $(i, k = 1, 2, 3)$.

Let us insert the obtained expressions of $\Gamma^0_{i0}, \Gamma^k_{i0}, \Gamma^k_{ij}$ (10.57) into the Lagrangian $L(h, \Gamma)$ neglecting divergence-type terms which vanish due to asymptotical conditions (10.51) after the three-dimensional integration is performed. We obtain the result

$$\frac{1}{2\kappa^2}(\pi_{ik}(x)\,\partial_0 q^{ik}(x) - H(x) - \left(\frac{1}{h^{00}(x)} - 1\right)T_0(x) - \tag{10.58}$$

$$-\frac{h^{i0}(x)}{h^{00}(x)}T_i(x));$$

$$T_0(x) = q^{ij}q^{kl}(\pi_{ik}\pi_{jl} - \pi_{ij}\pi_{kl}) + g_3 R_3;$$
$$T_i(x) = 2(\nabla_i(q^{kl}\pi_{kl}) - \nabla_k(q^{kl}\pi_{il})); \tag{10.59}$$
$$H(x) = T_0(x) - \partial_i \partial_k q^{ik}(x).$$

where $g_3 = \det g_{ik}, R_3$ is a three-dimensional scalar curvature generated by the three-dimensional metrics g_{ik} $(i, k = 1, 2, 3)$. The symbol ∇_k in the expressions of the constraints T_i denotes the covariant differentiation with respect to g_{ik}.

Arnowitt *et al.* [49] showed that canonical variables in constraints have a simple geometrical interpretation. The functions q^{ik}, π_{ik} are coefficients of the first and second quadratic forms of the surface $x = \text{const}$ that is embedded in the fourth-dimensional space time with the metric $g_{\mu\nu}$ and connection $\Gamma^\rho_{\mu\nu}$. More exactly, q^{ik} is the covariant density of the weight $+2$ and π_{ik} is the covariant density of the weight -1. The constraints are the well-known Codazzi–Gauss relations of the surface theory (see, e.g., [54]).

Formula (10.58) solves the problem of transforming the gravitational field action to the generalized Hamiltonian form analogous to (3.1). For the finite-dimensional systems with constraints, formula (10.58) can be used. The constraints T_μ are related by an involution. To express the explicit relations it is convenient to introduce the quantities

$$T(\eta) = \int T_k(x)\eta^k(x)\,d^3x; \qquad T_0(\varphi) = \int T_0(x)\varphi(x)\,d^3x. \qquad (10.60)$$

Here η is a vector field, φ is a scalar field or, more exactly, the scalar density of the weight -1. It holds

$$\{T(\eta_1), T(\eta_2)\} = T([\eta_1, \eta_2]); \qquad \{T(\eta), T_0(\varphi)\} = T_0(\eta\varphi);$$
$$\{T_0(\varphi), T_0(\psi)\} = T(\varphi\eta_\psi - \psi\eta_\varphi) \qquad (10.61)$$

where $[\eta_1, \eta_2]$ is the commutator of vector fields, i.e., the vector field with components $\eta_1^j\,\partial_j\eta_2^k - \eta_2^j\,\partial_j\eta_1^k$; $\eta\varphi = \eta^k\partial_k\varphi - \partial_k\eta^k\varphi$; η_k, is a vector field with the components $q^{ik}\partial_k\varphi$. Relations (10.61) are field theory analogues of (3.2). The first line of (10.61) shows that the constraints $T_k(x)$ $(k = 1, 2, 3)$ play the role of coordinate transformation generators. The other relations have no simple group theory interpretation.

Let us note the divergence term $(-\partial_i\partial_k q^{ik})$ in the Hamiltonian density $H(x)$. If the constraint equations $T_\mu = 0$ are fulfilled, the Hamiltonian H is reduced to the three-dimensional integral of the divergence, i.e., an integral over a surface at infinity. That integral is determined by the asymptotical behaviour of the functions q^{ik} as $r = |x| \to \infty$ For the asymptotically-flat gravitational field we have

$$q^{ik} = \delta^{ik}\left(1 + \frac{\kappa^2 M}{2\pi r}\right) + O\left(\frac{1}{r^2}\right), \qquad (10.62)$$

where M is a total mass that can be obtained by integrating $H(x)$.

$$H = \int H(x)\,d^3x = -\frac{1}{2\kappa^2}\int \partial_i\partial_k q^{jk}\,d^3x$$
$$= -\frac{1}{2\kappa^2}\lim_{S\to\infty}\oint \partial_k q^{jk}\,dS_i = M. \qquad (10.63)$$

In this way $H = \int H(x)\,d^3x$ can be considered to be an energy. The integrand

$$H(x) = T_0(x) - \partial_i\partial_k q^{ik}(x), \qquad (10.64)$$

which is an energy density equals the sum of two quadratic forms – the quadratic form of the first derivatives of q^{ik} and the quadratic form of the 'moments' π_{ik}. It is similar to the energy of the gravitational field with two polarizations in agreement with the calculus

$$2 = 6(\text{coordinates}) - 4(\text{constraints}). \tag{10.65}$$

In the weak gravitational field approximation the Hamiltonian is represented by a quadratic form of entirely transverse components of the linearized field.

Especially this fact, as well as the above-mentioned equality (10.63) of the gravitational field energy and a mass, is a justification of the field action (10.3) where the derivatives act not on the Christoffel symbols $\Gamma^{p}_{\mu\nu}$, but on the metric tensor $g^{\mu\nu}$.

After the gravitational field action has been transformed to the generalized Hamilton form, the continual integral can be constructed. The conditions originally proposed by Dirac [48]

$$\partial_k(q^{-1/3}q^{ik}) = 0, \, i = 1, 2, 3, \qquad \pi = q^{ik}\pi_{ik} = 0, \tag{10.66}$$

where $q = \det q^{ik}$ are usually used. The conditions have a simple geometrical meaning: the surface $x^0 = \text{const}$ is minimal and its coordinates x^1, x^2, x^3 are harmonic (equations $\partial_k(q^{-1/3}q^{ik})$ are the 'three-dimensional harmonic conditions').

Other conditions turned out to be more suitable for the proof of equivalence of the canonical and relativistic continual integral forms, namely

$$\ln q = \Phi(x); \qquad q^{ik} = 0; \, i \neq k, \tag{10.67}$$

where Φ is a function that behaves like c/r at infinity. These conditions satisfy commutation conditions (3.2). The matrix of Poisson brackets for conditions (10.67) with constraints is determined by

$$
\begin{aligned}
(\hat{C}\eta)^0 &= \{T_\eta, \ln q - \Phi(x)\} = -\eta^s \partial_s \ln q - 4 \partial_s \eta^s + 4\pi \eta^0; \\
(\hat{C}\eta)^1 &= \{T_\eta, q^{23}\} = -\eta^s \partial_s q^{23} + q^{2s} \partial_s \eta^3 + q^{3s} \partial_s \eta^2 - \\
&\quad - 2q^{23} \partial_s \eta^s - 2(\pi^{23} - q^{23}\pi)\eta^0; \\
(\hat{C}\eta)^2 &= \{T_\eta, q^{31}\} = -\eta^s \partial_s q^{31} + q^{3s} \partial_s \eta^1 + q^{1s} \partial_s \eta^3 - \\
&\quad - 2q^{31} \partial_s \eta^s - 2(\pi^{31} - q^{31}\pi)\eta^0; \\
(\hat{C}\eta)^3 &= \{T_\eta, q^{12}\} = -\eta^s \partial_s q^{12} + q^{1s} \partial_s \eta^2 + q^{2s} \partial_s \eta^1 - \\
&\quad - 2q^{12} \partial_s \eta^s - 2(\pi^{12} - q^{12}\pi)\eta^0,
\end{aligned}
\tag{10.68}
$$

where

$$T_\eta = \int (T_0 \eta^0 + T_i \eta^i)\, d^3 x. \qquad (10.69)$$

The matrix C is regular if the curvature of the metric tensor g_{ik} is different from zero.

Let us denote

$$\ln q - \Phi = \chi_0; \qquad q^{23} = \chi_1; \qquad q^{31} = \chi_2; \qquad q^{12} = \chi_3. \qquad (10.70)$$

The continual integral in the Hamilton form for the gravitational field is

$$\int \exp\left\{ i \int \left(\pi_{ik}\, \partial_0 q^{ik} - \frac{h^{0i}}{h^{00}} T_i - \left(\frac{1}{h^{00}} - 1 \right) T_0 - H(x) \right) d^4 x \right\} \times$$

$$(10.71)$$

$$\times \det\{T_\mu, \chi_0\} \prod_x \left(\prod_{a=0}^{3} \delta(\chi_a) \prod_{i \leq k} d\pi_{ik}\, dq^{ik}\, d\frac{1}{h^{00}} \prod_{i=1}^{3} d\left(\frac{h^{0i}}{h^{00}} \right) \right).$$

We shall transform this expression to the form where the integration is performed solely over the field $g^{\mu\nu}$. It enables us to identify the invariant measure we are searching for. For that purpose it is necessary to integrate (10.71) over π_{ik} that occur not only in the functional $\exp iS$ but also in the determinant $\det\{T_\mu, \chi_a\}$. We write down the determinant as a continual integral over the anticommuting variables $\eta^\mu, \bar{\eta}^\mu$ ($\mu = 0, 1, 2, 3$)

$$\det\{T_\mu, \chi_a\} = \int \exp\left(i \int \bar{\eta}^\mu C_{\mu\nu}(\pi, q) \eta^\nu\, d^4 x \right) \prod_x d\bar{\eta}^\mu(x)\, d\eta^\mu(x). \qquad (10.72)$$

The functions π_{ik} occur only in the coefficients $C_{\mu 0}$ of the operator $\hat{C}$ and in addition, they appear in the linear way without derivatives. We perform the shift

$$\pi_{ik} \to \pi_{ik} + \pi_{ik}(g), \qquad (10.73)$$

in the integral, where $\pi_{ik} = -(1/h^{00})\Gamma_{ir}^0$ is the expression of $\pi_{ik}(g)$ via the metric tensor because of (10.1). This shift transforms the action functional $S[g^{\mu\nu}, \pi_{ik}]$, which is the integral of (10.58), to

$$S[g^{\mu\nu}] - \frac{1}{2\kappa^2} \int \frac{1}{h^{00}} q^{ij} q^{kl} (\pi_{ik}\pi_{jl} - \pi_{ij}\pi_{kl})\, d^4 x. \qquad (10.74)$$

where $S[g^{\mu\nu}]$ is the action (10.3) in which the Christoffel symbols $\Gamma_{\mu\nu}^\rho$ are expressed via the metric tensor. The quadratic form $\bar{\eta}^\mu C_{\mu\nu}(\pi_{ik}, q^{ik}) \eta^\nu$

transforms to

$$\bar{\eta}^{\mu} C_{\mu\nu}(\pi_{ik}(g), q^{ik})\eta^{\nu} + \bar{\eta}^{\mu} l_{\mu}(\pi_{ik})\eta^{0}, \tag{10.75}$$

where $l_{\mu}(\pi_{ik})$ are linear forms of π_{ik} whose explicit form is irrelevant at this stage. Let us perform another shift of π_{ik}. Instead of the former form we obtain another quadratic form of $\bar{\eta}^{\mu}\eta^{0}$ without derivatives. It vanishes identically because of $(\eta^{0})^{2} = 0$. Then we may perform the Gaussian integral over π_{ik}. The integral over $\eta^{\mu}, \bar{\eta}^{\mu}$ could be again written down as a determinant of an operator $\hat{C}_{1}$ that differs from $\hat{C}$ in substituting the symbols π_{ik} by their expressions via the metric tensor. The action of the operator $\hat{C}_{1}$ is defined:

$$(\hat{C}_{1}\eta)^{0} = -\eta^{\lambda} \partial_{\lambda} \ln q - 4 \partial_{s}\eta^{s} - \left(\frac{h^{0s}}{h_{00}} \partial_{s} \ln q + 4 \partial_{s}\left(\frac{h^{0s}}{h^{00}} \right) \right)\eta^{0};$$

$$(\hat{C}_{1}\eta)^{1} = -\eta^{\lambda} \partial_{\lambda}q^{23} + q^{2s} \partial_{s}\eta^{3} + q^{3s} \partial_{s}\eta^{2} - 2q^{23} \partial_{s}\eta^{s} +$$
$$+ \left(-\frac{h^{0s}}{h^{00}} \partial_{s}q^{23} + q^{2s} \partial_{s}\left(\frac{h^{03}}{h^{00}} \right) + \right.$$
$$\left. + q^{3s} \partial_{s}\left(\frac{h^{02}}{h_{00}} \right) - 2q^{23}\partial_{s}\left(\frac{h^{0s}}{h^{00}} \right) \right)\eta^{0};$$

$$(\hat{C}_{1}\eta)^{2} = -\eta^{\lambda} \partial_{\lambda}q^{31} + q^{3s} \partial_{s}\eta^{1} + q^{1s} \partial_{s}\eta^{3} - 2q^{31} \partial_{s}\eta^{s} + \tag{10.76}$$
$$+ \left(-\frac{h^{0s}}{h^{00}} \partial_{s}q^{31} + q^{3s} \partial_{s}\left(\frac{h^{01}}{h^{00}} \right) + q^{1s} \partial_{s}\left(\frac{h^{03}}{h^{00}} \right) - \right.$$
$$\left. - 2q^{31} \partial_{s}\left(\frac{h^{0s}}{h^{00}} \right) \right)\eta^{0};$$

$$(\hat{C}_{1}\eta)^{3} = -\eta^{\lambda} \partial_{\lambda}q^{12} + q^{1s} \partial_{s}\eta^{2} + q^{2s} \partial_{s}\eta^{1} - 2q^{12} \partial_{s}\eta^{s} +$$
$$+ \left(-\frac{h^{0s}}{h^{00}} \partial_{s}q^{12} + q^{1s} \partial_{s}\left(\frac{h^{02}}{h^{00}} \right) + \right.$$
$$\left. + q^{2s} \partial_{s}\left(\frac{h^{01}}{h^{00}} \right) - 2q^{12} \partial_{s}\left(\frac{h^{0s}}{h^{00}} \right) \right)\eta^{0}.$$

The local factors in the product of differentials as well as the local factor emerging from the integration over π_{ik} and differential yield.

$$\prod_{x} (h^{00})^{-4} q^{-2} \prod_{\mu \leq \nu} dh^{\mu\nu}. \tag{10.77}$$

The factor in front of the differentials may be transformed to the form
$$(h^{00})^{-1}h^{-5/2}q^{1/2}. \tag{10.78}$$
The last factor may be deleted as a result of the condition $q = \exp \Phi$. Eventually the continual integral acquires the form

$$\int \exp\left(iS[h]\right) \det \hat{B}_1 \prod_x \left(\left(\prod_a \delta(\chi_a)\right) h^{-5/2} \prod_{\mu \leq \nu} dh^{\mu\nu}\right), \tag{10.79}$$

where the operator $\hat{B}_1$ differs from $\hat{C}_1$ by the local factor $(h^{00})^{-1}$.

We shall prove that this integral is an integral over the gravitational field classes in the sense of Section 7. The classes are parametrized by conditions (10.67) and the invariant measure is of form (10.13). For that purpose it is sufficient to check that $\det \hat{B}_1$ coincides with the factor $\Delta_\chi[h]$ defined by

$$\Delta_\chi[h] \int \prod_x \left(\prod_a \delta(\chi_a)\right) da(x) = 1. \tag{10.80}$$

The integral in this formula can be evaluated identically with the integral in (10.17). Thus we obtain

$$\Delta_\chi[g] = \det \hat{B}, \tag{10.81}$$

where the operator $\hat{B}$ is defined by setting

$$(\hat{B}\zeta)^0 = -\zeta^\lambda \partial_\lambda \ln q - 4\partial_s\zeta^s + 4\frac{h^{0s}}{h^{00}}\partial_s\zeta^0;$$

$$(\hat{B}\zeta)^1 = -\zeta^\lambda \partial_\lambda q^{23} + q^{2s}\partial_s\zeta^3 + q^{3s}\partial_s\zeta^2 - 2q^{23}\partial_s\zeta^s -$$
$$-\left(\frac{h^{02}}{h_{00}}q^{3s} + \frac{h^{03}}{h^{00}}q^{2s} - \frac{2h^{0s}}{h_{00}}q^{23}\right)\partial_s\zeta^0;$$

$$(\hat{B}\zeta)^2 = -\zeta^\lambda \partial_\lambda q^{31} + q^{3s}\partial_s\zeta^1 + q^{1s}\partial_s\zeta^3 - 2q^{31}\partial_s\zeta^s - \tag{10.82}$$
$$-\left(\frac{h^{03}}{h^{00}}q^{1s} + \frac{h^{01}}{h_{00}}q^{3s} - \frac{2h^{0s}}{h^{00}}q^{31}\right)\partial_s\zeta^0;$$

$$(\hat{B}\zeta)^3 = -\zeta^\lambda \partial_\lambda q^{12} + q^{1s}\partial_s\zeta^2 + q^{2s}\partial_s\zeta^1 - 2q^{12}\partial_s\zeta^s -$$
$$-\left(\frac{h^{01}}{h^{00}}q^{2s} + \frac{h^{02}}{h^{00}}q^{1s} - 2\frac{h^{0s}}{h^{00}}q^{12}\right)\partial_s\zeta^0.$$

It is easy to verify that
$$\det \hat{B} = \det \hat{B}_1. \tag{10.83}$$

Indeed, the passage from the first operator to the second is mediated by the triangle substitution

$$\zeta^0 = \eta^0; \qquad \zeta^i = \zeta^i + \frac{h^{0i}}{h^{00}}\eta^0, \quad i = 1, 2, 3. \tag{10.84}$$

Hence, the explicitly unitary Hamiltonian form of the continual integral can be represented as an integral over classes of equivalent fields by formal interchanges of integration orders and by concrete choice of the class parametrization. The corresponding invariant measure is given by (10.13). In that way the Lorentz invariant form of continual integral (10.16), representing another realization of the same integral in another class parametrization, is justified.

Now we are confronted with a more demanding problem of realizing the renormalization procedure based on an invariant regularization. The difficulties stem from the complexity of the theory and from the fact that it is non-renormalizable from the formal point of view.

11. ATTEMPTS TO CONSTRUCT A GAUGE-INVARIANT THEORY OF ELECTROMAGNETIC AND WEAK INTERACTIONS

Examples of the gauge-invariant theories discussed above are constructed as part of a larger theoretical scheme that includes all existing interactions. The investigation of such a general model is a very difficult problem. A more special case is the unification of the electromagnetic and weak interactions ($EM + W$) via a multiplet of gauge fields. This approach has been attracting the attention of theoreticians for a long time. The realization of the idea has produced several models the most known of which is the *Weinberg model* [21]. In this paragraph we shall shortly deal with the Weinberg model and with a gauge-invariant model of the electromagnetic and weak interactions of leptons proposed by Faddeev [23]. Like any gauge field theory, these models are most naturally formulated in the language of continual integrals.

A fundamental idea of the Weinberg model is the spontaneous breaking of the original invariance with respect to the gauge transformation of massless vector fields of the Yang–Mills type. The gauge group of the model is group $U(2)$. This group is isomorphic to the group of 2×2 unitary matrices and equal to the product of the $U(1)$ group of phase transformation and the group of 2×2 unitary matrices with unit determinants.

The connection generated by group $U(2)$ consists of two types of vector fields – the Yang–Mills multiplet A_μ^a ($a = 1, 2, 3$) and the fields B_μ. Besides these fields, there are lepton fields and auxilliary scalar fields that induce the spontaneous symmetry breaking of the $U(2)$ gauge invariance. The lepton fields involved in the Weinberg model are the electron-type fields

$$L = \tfrac{1}{2}(1 + \gamma_5)\begin{pmatrix} v_e \\ \psi_e \end{pmatrix}, \qquad R = \tfrac{1}{2}(1 - \gamma_5)\psi_e, \tag{11.1}$$

where ψ_e is the electron field and v_e is the electron neutrino field. The scalar fields form the doublet

$$\varphi = \begin{pmatrix} \varphi_0 \\ \varphi_- \end{pmatrix}. \tag{11.2}$$

The Lagrangian of the model is

$$
\begin{aligned}
L = {}& -\tfrac{1}{4}(\partial_\mu A_\nu - \partial_\nu A_\mu + g[A_\mu, A_\nu])^2 - \tfrac{1}{4}(\partial_\mu B_\nu - \partial_\nu B_\mu)^2 - \\
& - \bar{R}\gamma^\mu(\partial_\mu - ig'B_\mu)R - \bar{L}\gamma^\mu\left(\partial_\mu + ig\frac{\tau^a}{2}A_\mu^a - \frac{i}{2}g'B_\mu\right)L - \\
& - \tfrac{1}{2}\left(\partial_\mu\varphi - ig\frac{\tau^a}{2}A_\mu^a + \frac{i}{2}g'B_\mu\varphi\right)^2 - \\
& - G_e(\bar{L}\varphi R + \bar{R}\bar{\varphi}L) - M_1^2\varphi^+\varphi - h(\varphi^+\varphi)^2.
\end{aligned}
\tag{11.3}
$$

where g, g' are the coupling constants of the multiplet A_μ and singlet B_μ, respectively.

The mechanism of spontaneous symmetry breaking and mass generation, which was first proposed by Higgs [56], is based on the appearance of the anomalous average value

$$\lambda = \langle \varphi^0 \rangle \tag{11.4}$$

of the zeroth component of the φ field. Such a mechanism was encountered in the superfluidity theory (see Chapter 6). Let us proceed from the original fields to the 'physical' ones subtracting from the φ fields their anomalous average values. For the physical fields we take the φ field and

$$\varphi_1 = (\varphi^0 + \bar{\varphi}^0 - 2\lambda)/\sqrt{2}; \qquad \varphi_2 = (\varphi^0 - \bar{\varphi}^0)/i\sqrt{2}. \tag{11.5}$$

In the first order of the perturbation theory the quantity λ is determined by the condition of the minimum of the expression $-M_1^2\varphi^+\varphi^- + h(\varphi^+\varphi)^2$ supposing that $\varphi^0 = \lambda, \varphi^- = 0$. This leads to

$$\lambda^2 = M_1^2/2h, \tag{11.6}$$

After these operations the φ_1 fields acquire the mass M_1 and the fields φ_2, φ^- remain massless. The appearance of massless excitations in models with spontaneous symmetry breaking was discovered by Goldstone [127]. In this case, however, the excitations have no immediate physical meaning and can be removed by a gauge $U(2)$ transformation.

The mass of the φ_1 meson appears to be too large (compared to the electron mass m_e) and that is why the coupling of φ to other fields may be neglected.

Eventually it became obvious that the effect of the appearance of anomalous average value (11.4) can be reduced in the first order to the substitution of the φ field by its vacuum expectation value

$$\langle \varphi \rangle = \lambda \begin{pmatrix} 1 \\ 0 \end{pmatrix}. \tag{11.7}$$

By this substitution Lagrangian (11.3) transforms to

$$-\tfrac{1}{4}(\partial_\mu A_\nu - \partial_\nu A_\mu + g[A_\mu, A_\nu])^2 - \tfrac{1}{4}(\partial_\mu B_\nu - \partial_\nu B_\mu)^2 -$$
$$- \bar{R}\gamma^\mu(\partial_\mu - ig'B_\mu)R - \bar{L}\gamma^\mu\left(\partial_\mu + ig\frac{\tau^a}{2}A_\mu^a - (\tfrac{1}{2})g'B_\mu\right)L - \tag{11.8}$$
$$- (\tfrac{1}{8})\lambda^2 g^2(A_\mu^1)^2 + (A_\mu^2)^2 - (\tfrac{1}{8})(gA_\mu^3 + g'B_\mu)^2 - \lambda G_e\bar{\psi}_e\psi_e.$$

The electron acquires the mass

$$m_e = \lambda G_e. \tag{11.9}$$

The charged vector field

$$W_\mu = 2^{-1/2}(A_\mu^1 + iA_\mu^2) \tag{11.10}$$

describes an intermediate boson with the mass

$$M_W = \tfrac{1}{2}\lambda g. \tag{11.11}$$

From the neutral fields A_μ^3, B_μ the combinations

$$Z_\mu = (g^2 + g'^2)^{-1/2}(gA_\mu^3 + g'B_\mu);$$
$$A_\mu = (g^2 + g'^2)^{-1/2}(-g'A_\mu^3 + gB_\mu) \tag{11.12}$$

with the masses

$$M_Z = \tfrac{1}{2}\lambda(g^2 + g'^2)^{1/2}; \qquad M_A = 0. \tag{11.13}$$

can be formed. In this way, one component of the vector field multiplet A_μ has zero mass and it is therefore considered to be the photon field.

The interaction term of lepton and vector fields can be written as

$$\frac{ig}{2\sqrt{2}}\bar{\psi}_e(1+\gamma_5)vW_\mu + \frac{igg'}{(g^2+g'^2)^{1/2}}\bar{\psi}_e\psi_\mu\psi_eA_\mu + \frac{i(g^2+g'^2)^{1/2}}{4}\times$$

$$\times\left[\frac{3(g'^2-g^2)}{g'^2+g^2}\bar{\psi}_e\gamma_\mu\psi_e - \bar{\psi}_e\gamma_\mu\gamma_5\psi_e + \bar{v}\gamma_\mu(1+\gamma_5)v\right]Z_\mu. \tag{11.14}$$

The second term in (11.14) implies that the electron charge e is

$$e = gg'(g^2+g'^2)^{-1/2} \tag{11.15}$$

and is therefore smaller than any of the two original charges g, g'. Supposing that W_μ is, as usual, coupled to hadrons and to a muon we obtain

$$G_W/\sqrt{2} = \frac{g^2}{8M_W^2} = \frac{1}{2\lambda^2}. \tag{11.16}$$

It follows from (11.12) and (11.16) that masses of intermediate bosons are very large

$$M_z > 80\,\text{GeV}, \quad M_W > 40\,\text{GeV} \tag{11.17}$$

compared not only to the electron mass but also to the hadron masses.

The Weinberg model predicts the interaction of neutral currents. To verify that we exclude the intermediate boson field by the transformation $W_\mu \to W_\mu + W_\mu^0$ excluding interaction terms linear in W_μ.

This produces the direct interaction of lepton currents in the form

$$\sum_{a,b}\int d^4x\, d^4y\, j^a(x)\, j^b(y)D(x-y), \tag{11.18}$$

where

$$D(x-y) = (2\pi)^{-4}\int(k^2 - M_W^2 - i0)^{-2}\exp\left[ik(x-y)\right]d^4k \tag{11.19}$$

is the propagator of the W field with the mass M_W. For large M_W we can neglect k^2 with respect to M_W^2 in the denominator of the integrand and replace (11.19) by

$$-M_W^2\delta(x-y). \tag{11.20}$$

After this replacement the lepton interaction terms acquire the form

$$\sum_{a,b}\int d^4x\, j^a(x)\, j^b(x). \tag{11.20a}$$

The terms

$$\int d^4x\, j^a(x) j^b(x) \tag{11.21}$$

correspond to the neutral current interaction. This interaction is characteristic of the Weinberg model and does not exist in the $(V - A)$ variant of the weak interaction model.

The existence of the neutral current interactions was confirmed experimentally [133], which provides a serious argument in favour of the Weinberg model.

The methods of gauge field theory [12–18] enable us to justify Weinberg's conjecture concerning the renormalizability of his model. 't Hooft proved this conjecture in simpler examples [57]. In this book we shall not deal with the proof of renormalizability of the Weinberg model. Let us remark only that the construction of the correct perturbation theory follows the general scheme of the gauge field quantization explained in this chapter.

Regardless of the quantitative agreements of the Weinberg model with the experiments [133, 134] this model cannot be considered the only acceptable variant of unified gauge theory of the weak and electromagnetic interactions. This model has many aesthetical shortcomings. Let us present two of them: (1) the choice of the nonsimple group $U(2)$ as a gauge group contradicts the idea of universality of interactions.

(2) The introduction of a linear multiplet of scalar fields with consequent spontaneous symmetry breaking (Higgs mechanism) has no natural explanation in the framework of the gauge invariance idea.

A gauge invariant model of the electromagnetic and weak interaction of leptons free from these problems was proposed by Faddeev [23]. In the remaining part of this section we shall explain Faddeev's model by following his work [23].

The model is based on the simplest nontrivial gauge group $O(3)$ (three-dimensional rotation group) and contains only a triplet of vector fields. Instead of the scalar Higgs field, the model introduces a field of directions $n(x)$ in the charge space; determining there a neutral subspace. The family of physical particles includes the electron, muon and its neutrinos, photon and charged intermediate bosons. Let us present the geometrical ideas that the model is founded on. Classical fields are divided into two classes:

(1) Sections of a fibre bundle over the space-time manifold invariant under the local action of the gauge group.
(2) Connections given on this fibre bundle defining a parallel transport of fields from the first class.

The fibre is usually the product of a linear space carrying a Lorentz group representation corresponding to the spin of the first class fields and an internal (charge) space in which the internal symmetry group acts. When the gravity effects are neglected, a nontrivial connection is generated only by the latter group and is given by the vector Yang–Mills fields [41]. Their number is given by the dimension of the gauge group. In the investigated model the fields of the first class are the spinorial lepton fields and the gauge group is $O(3)$. The internal space is the product of a linear space carrying an $O(3)$ representation and a manifold that is the action space of $O(3)$. The linear component of the fibre is assigned to the multiplet of spinorial lepton fields. The simplest possibility is to take the space of the vector representation R^3. The leptons $\bar{\mu}$, e, $v = v_e + v_\mu$ (antimuon, electron and neutrino) are described by a three-dimensional vector making the R^3 space suitable from the physical point of view. The neutral lepton can be distinguished from the charged one by determining which of the $O(3)$ generators represents the charge. Accordingly, the definition of the charge is equivalent to the introduction of a direction in R^3 that will be called the neutral one. The directions generate the manifold S^2 that is supposed to be a direct factor of the internal space.

Hence, there are three fermions ψ_1, ψ_2, ψ_3 in the model united into an isovector $\psi \in R^3$, a family of scalar fields n_1, n_2, n_3 satisfying

$$n_1^2 + n_2^2 + n_3^2 = 1 \tag{11.22}$$

and generating a unit vector $n \in S^2$, and three vector fields $z_\mu^1, z_\mu^2, z_\mu^3$ generating an isovector $Z_\mu \in R^3$.

It is convenient to introduce three skew-symmetric matrices

$$V_1 = \begin{pmatrix} 0 & 0 & 0 \\ 0 & 0 & 1 \\ 0 & -1 & 0 \end{pmatrix}; \quad V_2 = \begin{pmatrix} 0 & 0 & -1 \\ 0 & 0 & 0 \\ 1 & 0 & 0 \end{pmatrix}; \quad V_3 = \begin{pmatrix} 0 & 1 & 0 \\ -1 & 0 & 0 \\ 0 & 0 & 0 \end{pmatrix}$$

$$\tag{11.23}$$

representing the Lie algebra of the group $O(3)$ and regard the family V_1, V_2, V_3 as a three-dimensional vector V.

Let us write down an infinitesimal gauge group action

$$\delta\psi = \psi \wedge \varepsilon; \qquad \delta n = n \wedge \varepsilon; \qquad \delta Z_\mu = \partial_\mu \varepsilon + Z_\mu \wedge \varepsilon, \qquad (11.24)$$

where $\varepsilon = \varepsilon(x)$ is the vector of an infinitesimal local rotation. The notation $(,)$, $\wedge$ for the scalar and vector products in R^3 is used, e.g., $\psi \wedge \varepsilon = (V, \varepsilon)\psi$.

From the fields Z_μ, n another two triplets of vector fields can be constructed

$$Y_\mu = \partial_\mu n + Z_\mu \wedge n; \qquad X_\mu = n \wedge Y_\mu, \qquad (11.25)$$

They transform under the gauge transformation like vectors

$$\delta Y_\mu = Y_\mu \wedge \varepsilon; \qquad \delta X_\mu = X_\mu \wedge \varepsilon. \qquad (11.26)$$

These fields together with Z_μ can be exploited for the definition of covariant derivatives of the field ψ, where matrices are obtained by the tensor multiplication X_μ, Y_μ to n.

Let us investigate a concrete combination

$$\nabla_\mu \psi = \partial_\mu \psi + (Z_\mu, V)\psi + (X_\mu \otimes n + n \otimes X_\mu)\gamma_5 \psi, \qquad (11.27)$$

constructed by means of the matrix $\gamma_5(\gamma_5^+ = -\gamma_5, \gamma_5^2 = -1)$. It is distinguished by a supplementary condition of the commutation ∇_μ with the transformation

$$\delta\psi = \alpha n(n, \gamma_5 \psi); \qquad \delta n = 0; \qquad \delta Z_\mu = \alpha Y_\mu; \qquad \delta X_\mu = \alpha Y_\mu, \qquad (11.28)$$

where α is an infinitesimally small constant. Formula (11.27) defines an infinitesimal connection associated with the group $SU(3)$.

The lepton number L generated by the ordinary phase transformation

$$\delta\psi = i\beta\psi; \qquad \delta n = 0; \qquad \delta Z_\mu = 0, \qquad (11.29)$$

where β is a constant (independent of x), together with the charge Q generated by (V, n), forms the full system of quantum numbers for lepton classification.

Transformation (11.28) produces a supplementary classification of neutral leptons absolutizing the difference their chiralities.

The Lagrangian invariant under the described transformations is

$$L = \frac{1}{2i}(\bar\psi\gamma_\mu\nabla_\mu\psi - (\gamma_\mu\nabla_\mu\bar\psi)\psi)\frac{1}{e^2}L_{YM} + \frac{m^2}{2e^2}(Y_\mu, Y_\mu), \qquad (11.30)$$

where L_{YM} is the Yang–Mills Lagrangian of the field Z_μ:

$$L_{YM} = -\tfrac{1}{4}(Z_{\mu\nu}, Z_{\mu\nu}); \qquad Z_{\mu\nu} = \partial_\mu Z_\nu - \partial_\nu Z_\mu + Z_\mu \wedge Z_\nu. \qquad (11.31)$$

The constant e is dimensionless, m has the dimension of mass.

The terms of Lagrangian (11.31) describing the interaction of the fermions and vector fields are

$$L_1 = i(Z_m, v_\mu) + i(X_\mu a_\mu), \tag{11.32}$$

where

$$v_\mu = \bar{\psi}\gamma_\mu v\psi; \qquad a_\mu = (\bar{\psi}n)\gamma_\mu\gamma_5\psi + \bar{\psi}\gamma_\mu\gamma_5(n\psi). \tag{11.33}$$

Expression (11.33) is analogous to the standard $(V - A)$ structure in the ordinary weak interaction theory. This analogy is confirmed by the behaviour of L_1 under the parity P transformation and charge conjugation R_Q. Schwinger [58] defined the charge conjugation R_Q geometrically as the transition to antiparticles which is accompanied by the change of the sign of the neutral direction in the charge space R^3. In the explicit form, R_Q is defined as:

$$\psi'' \to \psi^{*''}; \qquad \psi^\perp \to \psi^{*\perp}; \qquad n \to -n;$$

$$Z_\mu'' \to -Z_\mu''; \qquad Z_\mu^\perp \to Z_\mu^\perp; \qquad Y_\mu \to Y_\mu; \qquad X_\mu \to X_\mu, \tag{11.34}$$

where

$$Z_\mu'' \to (Z_\mu, n); \qquad Z_\mu^\perp = Z_\mu - (Z_\mu, n)n. \tag{11.35}$$

Note that $X_\mu'' = Y_\mu'' = 0$. Bilinear forms v_μ, a_μ transform

$$v_\mu'' \to -v_\mu''; \qquad v_\mu^\perp \to v_\mu^\perp; \qquad a_\mu^\perp \to a_\mu''; \qquad a_\mu^\perp \to -a_\mu^\perp, \tag{11.36}$$

so that the charge conjugation changes the sign of the second summand of (11.32). The space inversion changes the sign too, so that the interaction L_1 is invariant under the combined $R_Q P$ parity.

We shall investigate a possible structure of fermion mass terms, having in mind that they may be added to the original Lagrangian (11.30) or calculated dynamically. There are three suitable matrices $I, i(Vn), n \otimes n$ from which the mass term matrix M can be constructed by linear combination. Invariance under transformation (11.29) provides the mass matrix

$$M = a(I - n \otimes n) + ib(Vn), \tag{11.37}$$

where the coefficients a and b have the dimension of mass. For the physical interpretation of the model we shall use a particular gauge $n = n_0$ where n_0 is the constant vector $(0, 0, 1)$. This condition implies that the charge is connected with the matrix $-iV_3$ and that the field ψ_3 is neutral. Masses of the fields $\psi_1 + i\psi_2$, $\psi_1 + i\psi_2$ are $a + b$, $a - b$ and the charges $+1, -1$, respectively. The neutral field ψ_3 has the zero mass. The components of X_μ and Y_μ are

$$X_\mu^1 = Z_\mu^1; \qquad X_\mu^1 = Z_\mu^2; \qquad X_\mu^3 = 0;$$
$$Y_\mu^1 = Z_\mu; \qquad Y_\mu^2 = -Z_\mu; \qquad Y_\mu^3 = 0, \tag{11.38}$$

so that the interaction acquires the form

$$L_1 = Z_\mu(v_\mu + a_\mu) + Z_\mu^2(v_\mu^2 + a_\mu^2) + Z_\mu^3 v_\mu^3. \tag{11.39}$$

The last term in the Lagrangian (11.30) provides the mass term of vector mesons Z_μ^1, Z_μ^2 In the result the Lagrangian L_1 describes the weak $(V - A)$ interactions of electrons, muons and neutrinos mediated by the massive vector bosons $W_\mu^\pm$ and the interaction of charged leptons with electromagnetic field in the standard form

$$e\{A_\lambda(\bar{\psi}_e \gamma_\lambda \psi_e + \bar{\psi}_\mu \gamma_\lambda \psi_\mu) + W_\lambda[\bar{\psi}_e \gamma_\lambda(1 + i\gamma_5)\bar{\psi}_e + \psi_\mu \gamma_\lambda \times$$
$$\times (1 + i\gamma_5)\psi_\mu] + \text{c.c.}\}, \tag{11.40}$$

where the identification

$$Z_\mu^3 = eA_\mu; \qquad Z_\mu^1 \pm Z_\mu^2 = eW_\mu^\pm;$$
$$\psi_1 + i\psi_2 = \psi_\mu^*; \qquad \psi_1 - i\psi_2 = \psi_e; \qquad \psi_3 = \psi_{\nu_e} + \psi_{\nu_\mu}^*. \tag{11.41}$$

was made.

The constant e has the meaning of the electrical charge, e/m is the weak coupling constant and m is the mass of intermediate bosons. To quantize the described model it is suitable to use the transversality of the vector fields

$$\partial_\mu Z_\mu = 0. \tag{11.42}$$

as a gauge condition. Meanwhile, in correspondence with general methods of gauge field quantization, it is necessary to add a compensating term containing the interaction of vector particles with ghost scalar fields to Lagrangian (11.30). In the transverse gauge the field $n(x)$ does not disappear from the Lagrangian but is present in a nonlinear way. The question of correct quantization of the field $n(x)$ is very important. It is hoped that the continual integration method will facilitate the obtaining of the right answer.

CHAPTER 4

INFRARED ASYMPTOTICS OF
GREEN'S FUNCTIONS

12. METHOD OF SUCCESSIVE INTEGRATION OVER 'RAPID' AND 'SLOW' FIELDS

This section is devoted to the possibilities of applying the continual integration method for the evaluation of the asymptotic behaviour of Green's functions at small energies and momenta (*the infrared asymptotics*).

If quanta with arbitrarily small energies are present, the standard perturbation theory leads to difficulties. Such situations are met in quantum electrodynamics and in some examples of statistical physics – superfluidity theory, plasma theory and the general theory of phase transitions. The difficulties mentioned stem from the fact that every graph of the ordinary perturbation theory has a so-called *infrared singularity*. It means that the expression assigned to a graph that is an integral over momentum variables is singular when the external momenta tend to zero. In such cases, a reformulation of the perturbation theory is desirable. One possible variant of the reformulation may be called the method of subsequent integration, first over '*rapid*' and then over '*slow*' fields.

We shall represent every fields ψ that is integrated over as a sum of two terms. One of them will be called the slowly oscillating field ψ_0 and the other the *rapidly oscillating field* ψ_1,

$$\psi = \psi_0 + \psi_1 \qquad (12.1)$$

To be explicit, we shall investigate the example of a complex scalar field ψ. In the relativistic theory the functions ψ, ψ_0, ψ_1 depend on the space time points (x_0, x_1, x_2, x_3). In the statistical physics $\psi, \bar{\psi}$ are functions of $(\mathbf{x}, t)$ where $\mathbf{x} \in V$, $\tau \in [0, \beta]$ (see Section 6). We shall start with determining the meaning of the rapidly and slowly oscillating parts of the field in the statistical physics. As has already been stated in Section 6, the function $\psi(\mathbf{x}, t)$ is supposed to be periodic in its arguments and it is decomposed into the Fourier series

$$\psi(\mathbf{x}, \tau) = \frac{1}{\sqrt{\beta V}} \sum_{\mathbf{k}, \omega} \exp\left(i(\omega\tau - \mathbf{k} \cdot \mathbf{x})\right) a(\mathbf{k}, \omega). \qquad (12.2)$$

$$|\mathbf{k}| < k_0 \, ; |\omega| < \omega_0. \tag{12.3}$$

The sum in (12.2) with (12.3) will be called the slowly oscillating part $\psi_0(\mathbf{x}, t)$ of $\psi(\mathbf{x}, t)$. The difference $\psi(\mathbf{x}, t) - \psi_0(\mathbf{x}, t)$ will be called the rapidly oscillating part $\psi_1(\mathbf{x}, t)$. This difference is, of course, the sum of terms in (12.2) with $|\mathbf{k}| > k_0$ or $|\omega| > \omega_0$.

In the relativistic theory the analogue of decomposition (12.2) is

$$\psi(x) = \int \exp{(ikx)} \tilde{\psi}(k) \, d^4 k, \tag{12.4}$$

where k is a four-vector (k_0, k_1, k_2, k_3), $kx = k_0 x_0 - k_1 x_1 - k_2 x_2 - k_3 x_3$. One of the possibilities of decomposing the function into rapidly and slowly oscillating parts is to call the slowly oscillating part the integral over the region $|k^2| \equiv |k_0^2 - \mathbf{k}^2| \leq m_0^2$ and the rapidly oscillating part the integral over the region $|k^2| > m_0^2$.

Another possibility is to pass to the Euclidean field theory in which the space and time coordinates are treated equally. The scalar product is $kx = \Sigma_{i=1}^4 k_i \cdot x_i$. The slowly oscillating part of a function $\psi(x)$ is defined as integral (12.4) over the region

$$k^2 = \sum_{i=1}^{4} k_i^2 \leq m_0^2, \tag{12.5}$$

and the rapidly oscillating part as the integral over

$$k^2 = \sum_{i=1}^{4} k_i^2 > m_0^2. \tag{12.6}$$

The transition to the physical case of the pseudo-Euclidean metric is realized by the analytic continuation of results obtained by the continual integral methods in the Euclidean field theory.

It is worth noting that the boundary between the 'slow' and 'rapid' fields in the concrete problems of quantum field theory and statistical physics is to some extent conditional. It reflects in the fact that k_0, ω_0, m_0 distinguishing the 'slow' and the 'rapid' variables, are not determined exactly but only in order.

In statistical physics, as was proved in Section 6, the continual integral is the integral over the measure

$$\prod_{\mathbf{k}, \omega} da^+(\mathbf{k}, \omega) \, da(\mathbf{k}, \omega). \tag{12.7}$$

In Euclidean field theory, the measure can be represented by the form

$$\prod_k d\bar{\psi}(k)\, d\tilde{\psi}(k), \tag{12.8}$$

i.e., as a limit of the lattice formulation wherein the distance between the nearest points tends to zero and the volume of the k-space covered by the lattice tends to infinity.

The fundamental idea of the reformulation of the perturbation theory consists in the successive integration, at first over the rapidly oscillating field and afterwards over the slowly oscillating one, using different schemes of perturbation theory at the two different stages of integration. At the first stage we integrate over the Fourier coefficients $a(\mathbf{k}, \omega)$, $a^*(\mathbf{k}, \omega)$ or alternatively $\tilde{\psi}(k)$, $\bar{\psi}(k)$, whose indices $(\mathbf{k}, \omega)$ and k, respectively, lie in the region of rapid variables. At the second stage, integration over the region of slow variables is carried out.

When integration over the rapid fields we can use the perturbation theory and corresponding graphs that differ from the standard one explained in Sections 5 and 6 in two points.

(1) The integrals (sums) over the four-momenta are cut off at a low limit.
(2) Suplementary vertices will emerge describing the interaction of the field ψ_1 with slowly oscillating ψ_0.

The first point is obvious because the variables $(\mathbf{k}, \omega)$ (or k) of the rapidly oscillating field lie in the 'rapid region'.

The rationale of the supplementary vertices appearance is that the original action S is expressed via $\psi = \psi_0 + \psi_1$ so that the yield-crossing terms that may not appear in the quadratic form do contribute to the terms of the third and higher degrees.

The cutting off integrals at a low limit prevents the emergence of infrared divergencies at the first stage of the integration. At the second stage (integration over the slowly oscillating fields) we may achieve the vanishing of infrared divergencies if a nontrivial theory scheme exploiting the specificity of the system is adopted. In the superfluidity theory, for example, when integrating over the slowly oscillating Bose fields $\psi_0(\mathbf{x}, t)$, $\bar{\psi}_0(\mathbf{x}, t)$, it proves convenient to pass to the polar coordinates

$$\psi_0(\mathbf{x}, \tau) = \sqrt{\rho(\mathbf{x}, \tau)} \exp\left[i\varphi(\mathbf{x}, \tau)\right];$$
$$\bar{\psi}_0(\mathbf{x}, \tau) = \sqrt{\rho(\mathbf{x}, \tau)} \exp\left[-i\varphi(\mathbf{x}, \tau)\right] \tag{12.9}$$

and integrate over fields $\rho(\mathbf{x}, t)$, $\varphi(\mathbf{x}, t)$. The perturbation expressions of the integral over slowly oscillating fields will be formulated in terms of Green's functions of the fields $\rho(\mathbf{x}, t)$, $\varphi(\mathbf{x}, t)$. As will be shown in the next chapter, such a perturbation theory is free from infrared divergencies. The evaluation of infrared singularities of the Green's functions in quantum electrodynamics will be performed in the next section.

In phase transition theory, the method of multi-sequential integration is used. Far as it is from any possibility of theoretical foundation, that approach is nevertheless useful because it leads to a qualitative picture of critical phenomena and provides an approximative calculation method of critical indices.

13. INFRARED ASYMPTOTIC BEHAVIOUR OF GREEN'S FUNCTIONS IN QUANTUM ELECTRODYNAMICS

The physical reason for the infrared singularities in quantum electrodynamics is the existence of photons with arbitrary small energy. The singularities occur in the electron Green's function yielding power singularity instead of the pole $p^2 = m^2$. It is connected with the fact that the real electron is surrounded by a cloud of virtual photons among which the photons with arbitrarily small energy (arbitrarily large length of the wave) exist.

The power singularity of the electron Green's function at $p^2 \approx m^2$ was mentioned for the first time in the paper of Abrikosov et al. [59] who summed up the sequence of the main graph of perturbation theory. Functional methods were applied to this problem by Fradkin in [11] and by other authors [67].

The characteristic infrared singularities appear not only in Green's functions but also in the S matrix elements, if they are determined by standard methods of the quantum field theory. Related to that is the problem of redefining the S matrix and cross-sections so that the redefined expressions will not contain the infrared singularities. This problem is not treated in this book and one can learn about it in [61–65].

We shall calculate the infrared singularities of the Green's function of quantum electrodynamics. First we shall find the Green's function of the Euclidean quantum electrodynamics and proceed to the pseudo-euclidean metrices by an analytic continuation. Unified treatment of space and time coordinates in Euclidean theory simplifies the explanations. The functional $\exp iS$ of the pseudoeuclidean theory transforms to a functional that decreases with large values of the fields.

The one-particle electron Green's function of the Euclidean quantum electrodynamics is defined

$$G(x, y) = - \langle \psi(x)\bar{\psi}(y) \rangle$$

$$\equiv \frac{- \int \psi(x)\bar{\psi}(y)(\exp(S))(\Pi_x \delta(\partial_\mu A_\mu)) \, d\bar{\psi} \, d\psi \, dA}{\int (\exp(S))(\Pi_x \delta(\partial_\mu A_\mu)) \, d\bar{\psi} \, d\psi \, dA}, \qquad (13.1)$$

where the functional

$$S = -\tfrac{1}{4} \int (\partial_\mu A_\nu - \partial_\nu A_\mu)^2 \, d^4x - \int \bar{\psi}(\gamma_\mu(\partial_\mu + ieA_\mu) - m)\psi \, d^4x \quad (13.2)$$

plays the role of action. The Dirac matrices satisfy the relations

$$\gamma_\mu\gamma_\nu + \gamma_\nu\gamma_\mu = 2\delta_{\mu\nu}. \qquad (13.3)$$

The continual integrals in (13.1) are the integrals in the Lorentz gauge respecting requirements of the gauge field quantum theory (see Chapter 3). We shall determine Green's function (13.1) by calculating:

(1) Integrals over Fermi fields in the numerator and the denominator of (13.1);
(2) Integrals over rapidly oscillating electromagnetic fields;
(3) Integrals over slowly oscillating electromagnetic fields.

In correspondence with Section 12, we shall define the slowly and rapidly oscillating parts $A_\mu^{(0)}(x)$ and $A_\mu^{(1)}(x)$ of the field $A_\mu(x)$ as integrals over the region $k \le k_0$ and $k > k_0$ respectively in

$$A_\mu(x) = \int \exp(ikx))\tilde{A}_\mu(k) \, d^4k. \qquad (13.4)$$

The integrals over Fermi fields can be computed explicitly using the formulae

$$\int (\exp S) \, d\bar{\psi} \, d\psi = \exp S_0[A] \det(\gamma_\mu(\partial_\mu + ieA_\mu) - m);$$

$$\int (\exp S)\psi(x)\bar{\psi}(y) \, d\bar{\psi} \, d\psi \qquad (13.5)$$

$$= \exp S_0[A]G(x, y; A) \det(\gamma_\mu(\partial_\mu + ieA_\mu) - m).$$

Here, $S_0[A]$ is the action of the free electromagnetic field, the functional $\exp S_0[A]$ is constant with respect to the integration over $\psi, \bar{\psi}$; $G(x, y; A)$ is the Green's function of an electron in the electromagnetic field A_μ; $\det (\gamma_\mu(\partial_\mu + ieA_\mu) - m)$ is the determinant of the Dirac's operator in the external field. This determinant regularizes when divided by the free Dirac's operator:

$$\frac{\det [\gamma_\mu(\partial_\mu + ieA_\mu) - m]}{\det (\gamma_\mu \partial_\mu - m)} = \det \left[1 + \frac{1}{\hat{\partial} - m} ie \hat{A} \right]$$

$$= \exp \left\{ \operatorname{Sp} \ln \left(1 + \frac{1}{\hat{\partial} - m} ie \hat{A} \right) \right\} \tag{13.6}$$

$$= \left\{ - \sum_{n=1}^{\infty} \frac{(-1)^n e^{2n}}{2n} \int \operatorname{tr} (\hat{G}(x_1 - x_2)\hat{A}(x_2), \ldots \right.$$

$$\left. \ldots, \hat{G}(x_n - x_1)\hat{A}(x_1)) \, d^4x_1, \ldots, d^4x_n \right\}.$$

Here, $\hat{G}(x)$ is the Green's function of the operator $(m - \hat{\partial})$

$$\hat{G}(x) = \frac{1}{(2\pi)^4} \int \frac{\exp (ipx)}{i\hat{p} + m} \, d^4p. \tag{13.7}$$

The symbol tr denotes the trace in the matrix sense. Restriction to the even powers e^{2m} in the sum over n is due to the fact that the functional (13.6) is even with respect to $A(x)$. The evenness of the determinant of Dirac's operator is usually called *Furry's theorem*. For proof of the theorem we choose the realization of the Dirac matrices satisfying (13.3)

$$\gamma_i = \begin{pmatrix} 0 & \sigma_i \\ \sigma_i & 0 \end{pmatrix}, \quad i = 1, 2, 3; \quad \gamma_4 = \begin{pmatrix} I & 0 \\ 0 & -I \end{pmatrix}. \tag{13.8}$$

Here, σ_i are the Pauli matrices

$$\sigma_1 = \begin{pmatrix} 0 & 1 \\ 1 & 0 \end{pmatrix}; \quad \sigma_2 = \begin{pmatrix} 0 & -i \\ i & 0 \end{pmatrix}; \quad \sigma_3 = \begin{pmatrix} 1 & 0 \\ 0 & -1 \end{pmatrix}; \tag{13.9}$$

I is the 2-dimensional unit matrix. It is easy to see that a transposition operation acts on γ_i in the following way:

$$\gamma_i^T = \gamma_i, \quad i = 1, 3, 4; \quad \gamma_2^T = -\gamma_2. \tag{13.10}$$

That functional (13.6) is even follows from

$$
\begin{aligned}
\det\left(1 + \frac{1}{\hat{\partial} - m} ie\hat{A}\right) &= \det\left(1 + ie\hat{A}\frac{1}{\hat{\partial} - m}\right) \\
&= \det\left(1 + \left(ie\hat{A}\frac{1}{\hat{\partial} - m}\right)^T\right) = \det\left(1 + \frac{1}{\hat{\partial}^T - m} ie\hat{A}^T\right) \\
&= \det\left(1 + \left(\gamma_2\frac{1}{\hat{\partial}^T - m}\gamma_2^{-1}\right)(\gamma_2 ie\hat{A}^T\gamma_2^{-1})\right) \quad (13.11) \\
&= \det\left(1 + \frac{1}{\gamma_2\hat{\partial}^T\gamma_2^{-1} - m} ie\gamma_2\hat{A}^T\gamma_2^{-1}\right) \\
&= \det\left(1 - \frac{1}{\hat{\partial} - m} ie\hat{A}\right).
\end{aligned}
$$

The last equality is the consequence of

$$\gamma_2\hat{\partial}^T\gamma_2^{-1} = \hat{\partial}; \qquad \gamma_2\hat{A}\gamma_2^{-1} = -\hat{A}, \qquad (13.12)$$

that holds because of (13.10) and the obvious relation

$$\partial_i^T = -\partial_i, \quad i = 1, 2, 3, 4. \qquad (13.13)$$

The integration over the electromagnetic field A does not result in explicit relations like (13.5) and, therefore, the approximative methods are applied. The first nontrivial approximation can be obtained in the following way:

(1) In the first approximation we put functional (13.6) equal to one. That means that the vacuum polarization effects are neglected.
(2) For the Green's function of electron in an electromagnetic field we use the approximative formula

$$G(x, y; A^{(0)} + A^{(1)}) \approx G(x, y; A^{(1)})\exp\left(ie\int_x^y A_\mu^{(0)}(z)\, dz_\mu\right), \quad (13.14)$$

according to which the influence of the slowly oscillating field $A_\mu^{(0)}(x)$ is limited to the multiplication of the Green's function by a phase factor. The integration in the formula is performed over the line connecting x and y.

Formula (13.14) is exact when $A_\mu^{(0)}(z)$ is the gradient $\partial_\mu c$ of a scalar function c. It can be proved that for the fixed x, y, formula (13.14) is asymptotically exact in the limit $k_0 \to 0$.

Accepting the approximations we can represent Green's function as a product of two factors:

$$G(x, y) = -\frac{\int \exp(S_0[A^{(1)}])G(x, y; A^{(1)})(\Pi_x \delta(\partial_\mu A_\mu^{(1)}))\, dA^{(1)}}{\int \exp(S_0[A^{(1)}])(\Pi_x \delta(\partial_\mu A_\mu^{(1)}))\, dA^{(1)}} \times$$

$$\times \frac{\int \exp(S_0[A^{(0)}] + ie\int_x^y A_\mu(z)\, dz^\mu)(\Pi_x \delta(\partial_\mu A_\mu^{(0)}))\, dA^{(0)}}{\int \exp(S_0[A^{(0)}])[(\Pi_x \delta(\partial_\mu A_\mu^{(0)}))\, dA^{(0)}]}.$$

(13.15)

The first factor is the quantum Green's function obtained when the rapidly oscillating fields are taken into account. It is given by the formula

$$\frac{1}{(2\pi)^4}\int \exp(ipx)(G(\hat{p})\, d^4p),$$

(13.16)

where

$$G(\hat{p}) = (i\hat{p} + m - \sum(\hat{p}))^{-1}$$

(13.17)

is expressed over the self-energy part $\Sigma(\hat{p})$ that in the first approximation is given by the graph

(13.18)

The momentum of the internal photon line in this graph is cut off at k_0. This implies that the first factor of (13.15) depends on k_0. This dependence will cancel the contribution of the second factor of (13.15). The second factor, which can be called infrared, will be investigated first. It is an integral of a Gaussian type that can be calculated in the standard way using a shift that cancels the linear form of $A_\mu^{(0)}$ in the exponent. Representing the linear form as

$$ie\int_x^y A_\mu^{(0)}(z)\, dz_\mu = ie\int l_\mu(z)A_\mu^{(0)}(z)\, d^4z,$$

(13.19)

it is easy to obtain

$$\exp\left\{-\frac{l^2}{2}\int l_\mu(z_1)D_{\mu\nu}^{(0)}(z_1 - z_2)l_\nu(z_2)\, d^4z_1\, d^4z_2\right\}$$

$$= \exp\left\{-\frac{e^2}{2}\int_x^y\int_x^y dz_{1\mu}\, dz_{2\nu}D_{\mu\nu}^{(0)}(z_1 - z_2)\right\}.$$

(13.20)

where $D_{\mu\nu}^{(0)}(z_1 - z_2)$ is the transverse Green's function of the slowly oscillating field

$$D^{(0)}_{\mu\nu}(z) = \frac{1}{(2\pi)^4} \int_{k<k_0} \exp{(ikz)} \left(\frac{k^2 \delta_{\mu\nu} - k_\mu k_\nu}{k^4} \right) d^4k. \qquad (13.21)$$

Inserting (13.21) into (13.20) we obtain

$$-\frac{e^2}{2(2\pi)^4} \int_{k<k_0} \frac{d^4k}{k^4} (k^2 \delta_{\mu\nu} - k_\mu k_\nu) \int_x^y \exp{(ikz_1)} dz_{1\mu} \int_x^y \exp{(-ikz_2)} dz_{2\mu}. \qquad (13.22)$$

in the exponent. The linear integrals are calculated according to the formula

$$\int_x^y \exp{(ikz)} dz_\mu = \frac{\exp{(iky)} - \exp{(ikx)}}{i(k, x-y)} (y-x)_\mu, \qquad (13.23)$$

obtained when the variable t

$$z_\mu = x_\mu + (y-x)_\mu t \qquad (13.24)$$

ranging from 0 to 1 is introduced. By means of (13.23), expression acquires the form

$$-\frac{e^2}{(2\pi)^4} \int_{k<k_0} \left(\frac{k^2(x-y)^2 - (k, x-y)^2}{(k, x-y)^2} \right) (1 - \cos{(k, x-y)}) \frac{d^4k}{k^4}. \qquad (13.25)$$

Let us denote $|x - y| = r$ and pass to the polar coordinates. Then $(x-y, k) = rk\cos\theta$ and (13.25) transforms to

$$-\frac{e^2}{4\pi^3} \int_0^\pi \frac{\sin^4\theta}{\cos^2\theta} d\theta \int_0^{k_0 r |\cos\theta|} \frac{1 - \cos x}{x} dx, \qquad (13.26)$$

that is independent of the parameter $-k_0 r$. Let us investigate the asymptotics for $k_0 r \gg 1$. We can use the formula

$$\frac{\sin^4\theta}{\cos^2\theta} = \frac{1}{\cos^2\theta} + \frac{\cos^2\theta}{2} - \frac{3}{2}$$

First we shall find the contribution of the last constant term to integral (13.26)
Since for $k_0 r \gg 1$

$$\int_0^{k_0 r |\cos\theta|} \frac{1 - \cos x}{x} dx \approx \ln k_0 r |\cos\theta| + C,$$

where C is the Euler constant, then

$$-\frac{3}{2}\int_0^\pi (\ln k_0 r |\cos\theta| + C)\,d\theta = -\tfrac{3}{2}\pi\left(\ln\frac{k_0 r}{2} + C\right).$$

To calculate contributions of the other terms we integrate by parts:

$$\int_0^\pi \left(\frac{1}{\cos^2\theta} + \frac{\cos 2\theta}{2}\right)d\theta \int_0^{k_0 r|\cos\theta|} \frac{1-\cos x}{x}\,dx$$

$$= \int_0^\pi \left(\tan\theta + \frac{\sin 2\theta}{4}\right)(1 - \cos k_0 r|\cos\theta|)\tan\theta\,d\theta$$

$$= \int_0^\pi \left(\frac{\sin^2\theta}{2} - 1\right)(1 - \cos k_0 r\cos\theta)\,d\theta +$$

$$+ \int_0^\pi \frac{1}{\cos^2\theta}(1 - \cos k_0 r\cos\theta)\,d\theta.$$

The first integral tends to the constant $-3r/4$ for $k_0 r \gg 1$ and the second is equal to the integral

$$\int_{-\infty}^\infty \frac{1 - \cos k_0 rx}{x^2}\,dx = \pi k_0 r \tag{13.27}$$

up to quantities tending to zero as $k_0 r \to \infty$.

Thus, for $k_0 r \to \infty$ the infrared multiplier shows to be

$$\exp\left(-\frac{e^2}{4\pi^2}k_0 r + \frac{3e^2}{8\pi^2}\ln k_0 r + \frac{3e^2}{16\pi^2}(1 + 2C - 2\ln 2)\right). \tag{13.28}$$

The first factor in (13.15) is determined by the self-energy term Σ and for its calculation we shall restrict ourselves to the simplest diagram (13.18) of the perturbation theory. The expression corresponding to the diagram is

$$-\frac{e^2}{(2\pi)^4}\int_{k > k_0} d^4k \frac{k^2\delta_{\mu\nu} - k_\mu k_\nu}{k^4}\gamma_\mu\frac{1}{i(\hat{p} - \hat{k}) + m}\gamma_\nu. \tag{13.29}$$

This integral diverges and must be cut off at an upper bound $|k| = \Lambda \gg m$. The asymptotics in the x space of a Green's function containing the self-energy part (13.29) is determined by the behaviour of the function in the p space in the neighbourhood $p^2 \approx -m^2$ (in the analytic condition). By computing (13.29) we obtain the following formula (for $p^2 \approx -m^2$)

$$G_{k_0} \approx a_{k_0} \frac{\hat{m}_{k_0} - i\hat{p}}{m_{k_0}^2 + p^2}, \qquad (13.30)$$

where

$$a_{k_0} = 1 - \frac{7e^2}{16\pi^2} - \frac{3e^2}{8\pi^2} \ln \frac{k_0}{m},$$

$$m_{k_0} = m_0 \left(1 + \frac{3e^2}{16\pi^2} \left(\ln \frac{\Lambda^2}{m^2} + 1 \right) \right) - \frac{e^2}{4\pi^2} k_0. \qquad (13.31)$$

Here, a_{k_0} is the renormalization multiplicator, m_{k_0} is the electron mass computed to the second order of the perturbation theory considering only the interaction with the rapidly oscillating field. The Green's function $G(x)$ is

$$G_{k_0}(x) = a_{k_0}(m_{k_0} - \hat{\partial}) \frac{1}{(2\pi)^4} \int \frac{\exp(ipx)}{p^2 + m_{k_0}^2} d^4p. \qquad (13.32)$$

Denoting $|x| = r$ we obtain for $rm \gg 1$

$$\frac{1}{(2\pi)^4} \int \frac{\exp(ipx)}{p^2 + m^2} d^4x = \frac{1}{4\pi^2 r} \int_0^\infty \exp\left(-r(p^2 + m^2)^{1/2}\right) dp$$

$$\approx \frac{1}{4\pi^2 r} \exp(-mr) \int_0^\infty \exp\left(-\frac{rp^2}{2m}\right) dp = \frac{m^{1/2}}{2(2\pi r)^{3/2}} \exp(-mr) \qquad (13.33)$$

and after inserting this expression into (13.32) we obtain

$$\hat{G}_{k_0}(x) \approx \frac{a_{k_0}}{2} \left(\frac{m_{k_0}}{2\pi r} \right)^{3/2} (1 + \hat{n}) \exp(-m_{k_0} r), \qquad (13.34)$$

where $\hat{n} = n_\mu \gamma_\mu = (x_\mu/r)\gamma_\mu$. Multiplying this function by the infrared factor (13.28) we obtain an expression that does not contain the arbitrary parameter k_0

$$G(x) = \frac{a}{2} \frac{m^2}{(2\pi)^{3/2}} (mr)^{-(3/2) + 3e^2/8\pi^2} (1 + \hat{n}) \exp(-mr). \qquad (13.35)$$

Here, a is a normalizing factor and m is the physical electron mass

$$a = 1 + \frac{3e^2}{8\pi^2}(C - 2 - \ln 2); \qquad (13.36)$$

$$m = m_0 \left(1 - \frac{3e^2}{16\pi^2} \left(\ln \frac{\Lambda^2}{m^2} + 1 \right) \right). \qquad (13.37)$$

Proceeding from function $G(x)$ (13.35) to its Fourier transform we obtain

$$G(\hat{p}) \approx \frac{\hat{m} - i\hat{p}}{m^2\left(1 + \dfrac{p^2}{m^2}\right)^{1 + 3e^2/8\pi^2}}\left(1 - \frac{3e^2}{4\pi^2}\right). \tag{13.38}$$

in the neighbourhood of the singularity $p^2 \approx -m^2$. The parameters contained in this formula, i.e., the coefficient in front of the Green's function $1 - (3e^2/4\pi^2)$, the physical electron mass and the exponent $1 + (3e^2/8\pi^2)$ are computed to the second order of the perturbation theory.

We shall prove that the asymptotic formula

$$G(\hat{p}) = a\frac{m - i\hat{p}}{(m^2 + p^2)^{1 + 3e^2/8\pi^2}} \tag{13.39}$$

holds even if the above-mentioned approximations 1, 2 are not used. In that case the exponent e has the meaning of the physical (renormalized) charge. We choose an arbitrarily small k_0 and use formula (13.14), the exactness of which grows when k_0 decreases. After integration over Fermi fields it is necessary to perform the integration of product (13.14) over the rapidly oscillating fields. Let us note that the parameter e in this formula is the bare (unrenormalized) charge. The exponential factor in (13.14) can be put before the integral over $A^{(1)}$. The integral of the Green's function $G(x, y; A)$ is calculated according to

$$\int G(x, y; A^{(1)})\,\exp(S_0[A])\,\det[\hat{\partial} - m - ie_0(A^{(0)} + A^{(1)})] \times$$
$$\times \prod_x \delta(\partial_\mu A_\mu)\,dA = G_{k_0}(x, y)\,\exp(S[A^{(0)}]), \tag{13.40}$$

where

$$G_{k_0}(x, y) \tag{13.41}$$

$$= \frac{\int G(x, y; A^{(1)})\exp(S_0[A^{(1)}])\det[\hat{\partial} - m - ie_0 A^{(1)}]\prod_x \delta(\partial_\mu A_\mu)\,dA}{\int \exp(S_0[A])\det[\hat{\partial} - m - ie_0\hat{A}]\prod_x \delta(\partial_\mu A_\mu)\,dA}$$

is the quantum Green's function in the rapidly oscillating field and the functional

$$\exp(S[A^{(0)}]) \tag{13.42}$$

$$= \int \exp(S_0[A])\,\det[\hat{\partial} - m - ie_0(A^{(0)} + A^{(1)})]\prod_x \delta(\partial_\mu A_\mu)\,dA$$

is obtained as a result of the integration of the functional $\exp S$ over the Fermi fields and rapidly oscillating field $A_\mu^{(1)}$. The expression $S[A^{(0)}]$, which has the meaning of the action of the slowly oscillating field, can be written as

$$-\frac{1}{4Z}\int(\partial_\mu A_\nu^{(0)} - \partial_\nu A_\mu^{(0)})^2\, d^4x + \cdots, \tag{13.43}$$

where the dots represent higher derivative terms and functionals of a degree higher than two. The constant $Z = Z_3$ is the ratio of the squared bare and physical charges:

$$Z = Z_3 = e^2/e_0^2 \tag{13.44}$$

This is clear from the following: if the photon Green's function for small k is calculated, then in the first approximation it equals the Green's function of the quadratic operator (13.43) and is of the form

$$ZD_{\mu\nu}(k) = \frac{Z}{k^4}(k^2\delta_{\mu\nu} - k_\mu k_\nu). \tag{13.45}$$

Thus, Z is the renormalization constant Z_3 of the photon Green's function equal, as is known, to e^2/e_0^2.

Returning to the electron Green's function we can see that the task that should still be solved is finding the infrared multiplier according to the formulae above (13.20). Moreover, the parameter Z in (13.43) transforms e_0 into e. The function $G_{k_0}(x, y)$ (13.41), determined only by the rapidly oscillating part, has the pole $\sim(p^2 + m_k^2)$ in the p representation and asymptotics (13.34) in the x representations. In the result we obtain asymptotics (13.35) and (13.39) for the functions $G(x)$ and $G(p)$, respectively, where e^2 is the square of the renormalized charge.

We shall apply the investigated method of determining the infrared asymptotics to the arbitrary theory of the electromagnetic interaction in which the electromagnetic field is introduced by replacing the ordinary derivatives ∂_μ by the covariant ones $\nabla_\mu = \partial_\mu + iA_\mu$. For example, the Green's function of the scalar field in the Euclidean scalar electrodynamics with the action

$$-\int(|(\partial_\mu + ieA_\mu)\psi|^2 + m^2|\psi|^2)\,d^4x \tag{13.46}$$

in the p representation for $p^2 \approx -m^2$ is

$$-\frac{Z_2}{m}\left(1 + \frac{p^2}{m^2}\right)^{-1-3e^2/8\pi^2}, \tag{13.47}$$

where Z_2 is a renormalization constant. The point is that formula (13.14) of the Green's function of a particle with the charge e in the slowly oscillating field is universal. It leads to the universal infrared factor (13.47) defining the exponent of the asymptotics for $p^2 \approx -m^2$.

Let us investigate a general case of Green's function

$$\langle \psi(x_1), \ldots, \psi(x_n)\bar{\psi}(y_1), \ldots, \bar{\psi}(y_n)A_{\mu_1}(z_1), \ldots, A_{\mu_k}(z_k) \rangle. \qquad (13.48)$$

We shall find the asymptotics of this function for the case when its arguments $x_1 \ldots x_n, y_1 \ldots y_n, z_1 \ldots z_n$ are far from each other. This asymptotics actually determines the asymptotics in the p space in the vicinity of the mass shell.

Function (13.48) can be written as a continual integral and integrated first over the spinor fields $\psi, \bar{\psi}$ and $\lambda_\mu^{(1)}$ afterwards over the rapidly oscillating fields $A_\mu^{(1)}$. At this stage we can use the diagram technique with modifications mentioned in Section 12. The nontrivial part of function (13.48) is given by the sum of connected diagrams of the form

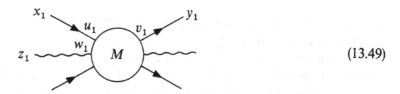

$$(13.49)$$

where M is an irreducible connected part.

The expression corresponding to (13.49) is

$$\prod_{i=1}^{n} G_{k_0}(x_i, u_i; A^{(0)})G_{k_0}(v_i, y_i; A^{(0)}) \times$$

$$\times \prod_{j=1}^{k} D_{\mu_j \nu_j}(z_j, w_j; A^{(0)})M_{k_0, \nu_1, \ldots, \nu_k}(u_i, v_i, w_j). \qquad (13.50)$$

Here, $G_{k_0}(x, u; A^{(0)})$ is the quantum Green's function of the electron in the slowly oscillating field $A^{(0)}$. It is given by the approximated formula

$$G_{k_0}(x, u; A^{(0)}) \approx G_{k_0}(x - u) \exp\left(ie \int_u^x A_\mu^{(0)}(z) \, dz_\mu \right), \qquad (13.51)$$

where the first factor is the quantum electron Green's function obtained by averaging over the rapidly oscillating field $A_\mu^{(1)}$. For the functions D and M we can use the quantum functions calculated from the interaction of electrons with the rapidly oscillating field only.

The integral over the slowly oscillating field is the Gaussian integral from

$$\exp K = \exp\left\{ -\frac{1}{4Z_3} \int (\partial_\mu A_\nu^{(0)} - \partial_\nu A_\mu^{(0)})^2 \, d^4x + \right.$$
$$\left. + i \sum_{i=1}^{2n} e_i \int_{u_i}^{x_i} A_\mu(z) \, dz_\mu \right\}, \tag{13.52}$$

where

$$x_i = x_i; \quad u_i = u_i; \quad e_i = e; \quad 1 \le i \le n;$$
$$x_i = y_i; \quad u_i = v_i; \quad e_i = -e; \quad n+1 \le i \le 2n. \tag{13.53}$$

The Gaussian integral from $\exp K$ will be called the infrared factor corresponding to Green's function (13.48). It has the form

$$\langle \exp K \rangle_{A^{(0)}} = \exp(-X), \tag{13.54}$$

where

$$X = \frac{1}{2(2\pi)^4} \int_{k<k_0} \frac{d^4k}{k^4} \sum_{i,j=1}^{2n} e_i e_j \left(\frac{k^2(x_i - u_i, x_j - u_j)}{(k, x_i - u_i)(k, x_j - u_j)} - 1 \right) \times \tag{13.55}$$
$$\times (\exp(ix_i k) - \exp(iu_i k))(\exp(-ix_j k) - \exp(-iu_j k)).$$

Thus we obtain

$$\exp(-X) \prod_{i=1}^{n} G_{k_0}(x_i - u_i) G_{k_0}(v_i - y_i) \times$$
$$\times \prod_{i=1}^{u} D_{\mu_i \nu_i}(z_i - w_j) M_{k_0, \nu_1, \ldots, \nu_k}(u_i, v_i, w_j). \tag{13.56}$$

for the expression corresponding to (13.49). The dependence of the functions G, D, M on k_0 must vanish when the functions are multiplied by the infrared factor $\exp(-X)$.

We shall find the momentum representation of (13.56). Note that in the expression that occurs in $\exp(-X)$ the factor $\exp[ik(u_i - u_j)]$ in

$$(\exp(ikx_i) - \exp(iku_i))(\exp(-ikx_j) - \exp(-iku_j))$$
$$= \exp[ik(u_i - u_j)] \times$$
$$\times (\exp[ik(x_i - u_i)] - 1)(\exp[-ik(x_j - u_j)] - 1)$$

may be replaced by one because the main contribution to the integral over u_i, v_j comes from the region where u_i and v_j are sufficiently close to each other ($k_0 |u_i - u_j| \ll 1$). We can therefore suppose that

$$\exp(-X) \prod_{i=1}^{n} G_{k_0}(x_i - u_i) G_{k_0}(v_i - y_i) \tag{13.57}$$

depends on the differences $x_i - u_i$, $y_i - v_i$ because M depends on u_i, v_i, w_i and $D_{\mu\nu}$ on $z_i - w_i$. Performing the Fourier transformation over the arguments $x_i - u_i, y_i - v_i, z_i - w_i, u_i, v_i, w_i$ we obtain

$$\prod_{i=1}^{n} (-i\hat{p}_i + m) \prod_{j=1}^{n} (-i\hat{p}_j - m) \prod_{s=1}^{k} D_{\mu_s \nu_s}(k_s) \times$$

$$\times M_{\nu_1 \ldots \nu_s}(p_1, \ldots, p_{2n}, q_1, \ldots, q_k) \delta\left(\sum_{i=1}^{2n} p_i + \sum_{s=1}^{k} q_s \right) \times \tag{13.58}$$

$$\times \int_0^\infty \ldots \int_0^\infty \prod_i d\sigma_i \exp\left(i\sum_i (p_i^2 + m_i^2)\sigma_i - Y \right),$$

where

$$Y = \frac{im^2 e^2}{2\pi^2} \sum_i \sigma_i + \sum_{i,j=1}^{2n} \frac{e_i e_j}{2(2\pi)^4} \int_{k < k_0} \frac{d^4 k}{k^4} (\exp[2i\sigma_i(p_i k)] - 1) \times$$

$$\times (\exp[-2i\sigma_j(p_j k)] - 1)\left(\frac{k^2(p_i p_j)}{(kp_i)(kp_j)} - 1 \right). \tag{13.59}$$

The integral in (13.58) has the power singularities $\sim (1 + p_i^2/m^2)^{-1-(3e^2/8\pi^2)}$ for all momentum variables in the vicinity of the mass shell $p_i^2 \approx -m^2$.

CHAPTER 5

SCATTERING OF HIGH-ENERGY PARTICLES

14. DOUBLE LOGARITHMIC ASYMPTOTICS IN QUANTUM ELECTRODYNAMICS

We shall apply the functional integration method to some problems of high-energy particle scattering. In this section we shall obtain the double-logarithmic asymptotics of the vertex Green's function. In the following section we shall develop the eikonal approximation for the two-particle scattering amplitude in the model of a scalar field u with the mass M interacting with another scalar field u with the mass $m \ll M$ via the term $gu\varphi^2$.

The concept of *double-logarithmic asymptotics* has appeared in the quantum electrodynamics during the investigation of the electron in an external field with a large transfer of momentum

$$(p_1 - p_2)^2 \equiv p_{12}^2 \gg m^2, \tag{14.1}$$

where p_1 and p_2 are the momenta of the initial and final states. It was shown in [137–138] that to obtain the correct result it is necessary to sum up the infinite number of 'main' diagrams of the perturbation theory. Afterwards the results were obtained by the more natural method of functional integration.

We shall derive only the main asymptotic term of the vertex part here. The method is analogous to that used for obtaining the infrared asymptotics in Section 13. We shall prove that the result we are seeking is essentially given by formulae (13.58), (13.59) for amplitudes of QED and can be extracted from them by a suitable modification. It is not necessary to use the formulae on the mass shell in the limit $k_i^2 \to 0, p_i^2 \to m^2$. Out of the mass shell the formulae describing virtual processes take exactly into account the interaction with long-wave photons. Namely this interaction induces the double-logarithmic asymptotics of the vertex part in the region

$$(p_1 - p_2)^2 \gg p_1^2, p_2^2 \gg m^2. \tag{14.2}$$

In the following we shall neglect terms of the order $\alpha = e^2/4\pi \approx \frac{1}{137}$ in comparison with one. We take for granted that the inequality

$$\alpha \ln \frac{p^2}{m^2} \ll 1 \tag{14.3}$$

is fulfilled for all momenta occurring below and we shall neglect terms of the order $\alpha \ln p^2/m^2$ in comparison with unity. The region where the quantity $\alpha \ln p^2/m^2$ is of the order of unity is that of very high energies and is experimentally inaccessible at present. Meanwhile we suppose that

$$\alpha \ln^2 \frac{p_{1,2}^2}{p_1^2} \gtrsim 1. \tag{14.4}$$

In other words, the logarithms of relative momenta that are small compared to α^{-1} are supposed to be comparable to $\alpha^{-1/2}$.

We obtain the vertex part putting $n = 1, k = 1$ in (13.58). Considering the momenta $p_1, p_2, p_{1,2}$, the main contribution to the integrals over σ_i comes from $\sigma_i \sim p_i^{-2}$; we impose the condition on the upper bound k_0 in the integrals over k in (13.59),

$$k_0^2 \lesssim p_{1,2}^2, \tag{14.5}$$

that means that for long-wave photons we take those with momenta smaller than the transferred momentum $p_{1,2}$. The components of the electron momenta p_1, p_2 must be of the order $p_{1,2}$ due to the conservation law of the 4-momentum $p_1 - p_2 = p_{1,2}$. That is why the single interaction with a photon having the momentum $k < k_0$ does not essentially influence the electron movement.

Let us explore the expression Y (13.59). In the region $\sigma_i \sim p_i^{-2}$ the first term $(ime^2/2\pi^2)\Sigma_i\sigma_i$ can be neglected in comparison with one. The sum of four integrals.

$$Y_{11} + Y_{22} + Y_{12} + Y_{21}. \tag{14.6}$$

remains. In the beginning we compute

$$Y_{12} = \frac{-e^2}{2(2\pi)^4} \int_{k<k_0} \frac{d^4k}{k^4} (\exp[2i\sigma_1(p_1 k)] - 1) \times$$

$$\times (\exp[-2i\sigma_2(p_2 k)] - 1) \left(\frac{k^2(p_1 p_2)}{(kp_1)(kp_2)} - 1 \right)$$

$$= -\frac{e^2(p_1 p_2)}{2(2\pi)^4} \int_{k<k_0} d^4k (\exp[2i\sigma_1(p_1 k)] - 1) \times$$

$$\times (\exp[-2i\sigma_2(p_2 k)] - 1)[k^2(kp_1)(kp_2)]^{-1} +$$

$$+ \frac{2}{2(2\pi)^4} \int_{k<k_0} \frac{d^4k}{k^4} (\exp[2i\sigma_1(p_1 k)] - 1) \times$$

$$\times (\exp[-2i\sigma_2(p_2 k)] - 1). \tag{14.7}$$

The second term on the right-hand side of (14.7) is the quantity of the order

$$\frac{\alpha}{2\pi} \ln \frac{(\sigma_1 p_1 - \sigma_2 p_2)^2}{\sigma_1^2 p_1^2 \sigma_2^2 p_2^2 k_0^2}, \tag{14.8}$$

that can be neglected when compared to unity. We calculate the first terms on the right-hand side of (14.7) in momentum region (14.2). Passing to pseudoeuclidean metrics for momenta p_1, p_2 and integration momenta k we obtain the expression

$$Y_{12} \approx -\frac{ie^2(p_1 p_2)}{2(2\pi)^4} \int d^4k (\exp[2i\sigma_1(p_1 k)] - 1) \times$$

$$\times (\exp[-2i\sigma_2(p_2 k)] - 1)[(p_1 k)(p_2 k)(k^2 - i0)]^{-1}, \tag{14.9}$$

in which the scalar product and squares of vectors are understood in the pseudoeuclidean metric $(ab) = a_0 b_0 - a_1 b_1 - a_2 b_2 - a_3 b_3$.

For calculation of integral (14.9) it is convenient to pass to new integration variables. For that reason we represent the vector k in the form

$$k = k_\perp + u(kp_1) + v(kp_2), \tag{14.10}$$

where u, v are two four-vectors in the subspace generated by the vectors p_1, p_2 and satisfying the conditions

$$u^2 = v^2 = 0; \tag{14.11}$$

$$(uk_\perp) = (vk_\perp) = 0. \tag{14.12}$$

The solution of Equations (14.11) and (14.12) is

$$u = a(p_2 + \alpha p_1); \quad v = b(p_1 + \beta p_2) \tag{14.13}$$

with $|\alpha|, |\beta| \ll 1$. In region (14.2) condition (14.11) leads to

$$\alpha \approx -\frac{p_2^2}{2p_1 p_2} \approx \frac{p_2^2}{p_{1,2}^2}; \quad \beta \approx -\frac{p_1^2}{2(p_1 p_2)} \approx \frac{p_1^2}{p_{1,2}^2}. \tag{14.14}$$

Condition (14.12) reduces to

$$(k_\perp p_1) = (k_\perp p_2) = 0, \tag{14.15}$$

and we obtain

$$a \approx b \approx \frac{1}{(p_1 p_2)} \approx -\frac{2}{p_{1,2}^2}. \tag{14.16}$$

Decomposition (14.10) results in

$$k^2 = 2(uv)(kp_1)(kp_2) + k_\perp^2. \tag{14.17}$$

The vector $k_\perp$ is orthogonal to the time-like vectors p_1, p_2, it is spacelike and its square

$$k_\perp^2 = -\mathbf{r}^2 \tag{14.18}$$

is the square of a Euclidean two-dimensional vector $\mathbf{r}$ with a negative sign. Components of this vector together with $(kp_1), (kp_2)$ are taken for new integration variables.
We get

$$d^4k = |J| \, d(kp_1) \, d(kp_2) d^2r, \tag{14.19}$$

where J is the Jacobian of the transition to the variables $(kp_1), (kp_2)$ in the plane generated by p_1, p_2 and defined by

$$|J| = \det \begin{pmatrix} u^2 & uv \\ uv & v^2 \end{pmatrix}^{1/2} = |(uv)| = (p_1 p_2)|^{-1}. \tag{14.20}$$

In the new variables (14.9) acquires the form

$$Y_2 \approx -\frac{ie^2}{2(2\pi)^4} \int d(kp_1) \, d(kp_2) d^2r (\exp\left[2i\sigma_1(kp_1)\right] - 1) \times$$

$$\times (\exp\left[-2i\sigma_2(kp_2)\right] - 1)(kp_1)^{-1}(kp_2)^{-1} \times \tag{14.21}$$

$$\left[\mathbf{r}^2 + \frac{4}{p_{1,2}^2}(kp_1)(kp_2) + i0 \right]^{-1}.$$

The integrals in (14.21) are supposed to be cut off at the boundaries $r \sim k_0, (kp_1), (kp_2) \sim k_0|p_{1,2}|$. The exact values of these boundaries are not important when calculations up to logarithmic terms are done. The integral over $\mathbf{r}$ in (14.21) is

$$\int \frac{d^2r}{r^2 + \frac{4}{p_{1,2}^2}(kp_1)(kp_2) + i0} = \pi \int_0^{k_0^2} \frac{d^2r}{r^2 + \frac{4}{p_{1,2}^2}(kp_1)(kp_2) + i0}$$

(14.22)

$$= \pi \ln \left| \frac{k_0^2 p_{1,2}^2}{4(kp_1)(kp_2)} \right| - i\pi^2 \theta(-(kp_1)(kp_2)).$$

After integration of the first term over (kp_1), (kp_2) we obtain the expression

$$\frac{ie^2\pi}{2(2\pi)^4} \left[\ln(k_0^2 p_{1,2}^2 \sigma_1 \sigma_2) \left(\int_{-\infty}^{\infty} \frac{\sin z}{z} dz \right)^2 + \right.$$

$$\left. + \int_{-\infty}^{\infty} \frac{\sin z}{z} \ln \frac{1}{|z|} dz \int_{-\infty}^{\infty} \frac{\sin z}{z} dz \right] \approx \frac{i\alpha}{8\pi} \ln(k_0^2 p_{1,2}^2 \sigma_1 \sigma_2), \quad (14.23)$$

which for $k_0^2 \lesssim p_{1,2}^2, \sigma_1 \sim p_1^{-2}, \sigma_2 \sim p_2^{-2}$ is small in comparison with one.

The main contribution to the integral over (kp_1), (kp_2) originates in the second term of (14.22). The integration yields

$$Y_{12} \approx -\frac{e^2}{16\pi^2} \int_0^{k_0|p_{1,2}|} \frac{d(kp_1)}{(kp_1)} (\exp[2i\sigma_1(kp_1)] - 1) \times$$

$$\times \int_{-k_0|p_{1,2}|}^0 \frac{d(kp_2)}{(kp_2)} (\exp[-2i\sigma_2(kp_2)] - 1) \quad (14.24)$$

$$= \frac{e^2}{16\pi^2} \ln(k_0|p_{1,2}|\sigma_1) \ln(k_0|p_{1,2}|\sigma_2).$$

up to the logarithmic terms. It is evident that $Y_{21} = Y_{12}$. Investigating analogously the integrals Y_{11}, Y_{22} we find that they are small compared with one (their order is $\alpha/2\pi \ln \sigma_i |p_i|k_0$) and therefore can be neglected. Within logarithmic accuracy we can replace $\sigma_1 \sim p_1^{-2}, \sigma_2 \sim p_2^{-2}, k_0^2 \sim p_{1,2}^2$ in (14.14) and obtain the approximate formula for Y

$$Y \approx Y_{12} + Y_{21} = \frac{e^2}{8\pi} \ln \frac{p_{1,2}^2}{p_1^2} \ln \frac{p_{1,2}^2}{p_2^2}$$

(14.25)

$$= \frac{\alpha}{2\pi} \ln \frac{p_{1,2}^2}{p_1^2} \ln \frac{p_{1,2}^2}{p_2^2},$$

which does not contain σ_i. It enables us to take $\exp(-Y)$ out of the integral over σ_i. Performing the integration over σ_i we obtain

$$\delta(p_1 - p_2 + k)\frac{(-i\hat{p}_1 + m)(-i\hat{p}_2 - m)}{p_1^2 + m^2}\frac{}{p_2^2 + m^2}$$

$$\times D_{\mu\nu}(k)M_\nu \exp(-Y). \tag{14.26}$$

Using the first approximation $M_\nu = e\gamma_\nu$, we obtain the irreducible vertex part Γ_μ in the form

$$\Gamma_\mu \approx e\gamma_\mu \exp\left(-\frac{\alpha}{2\pi}\ln\frac{p_{1,2}^2}{p_1^2}\ln\frac{p_{1,2}^2}{p_2^2}\right), \tag{14.27}$$

with the double logarithmic asymptotics.

Deriving formulae of this type by the perturbation theory method, it is necessary to sum up the series having the first terms of larger order than the result of summation. It implies that the series terms cancel each other. That is why the derivation of the double logarithmic asymptotics by the functional integration method is preferable.

Allow us to stress that the starting point for the electron Green's function in the slowly oscillating field was the approximate formula (13.11).

15. EIKONAL APPROXIMATION

We shall apply the functional integration method to the derivation of the so-called *eikonal approximation* of the scattering amplitude for high-energy particles. To illustrate the method, we have chosen a model of the relativistic quantum field theory the action of which (in the Euclidean metrics) is

$$S = -\tfrac{1}{2}\int[(\nabla\varphi)^2 + m^2\varphi^2 + (\nabla u)^2 + \mu^2 u^2 + g\varphi^2 u]\,\mathrm{d}^4x. \tag{15.1}$$

The action (15.1) describes a system of two real scalar fields φ and u interacting via $\varphi^2 u$. In the following we shall suppose that

$$m \gg \mu. \tag{15.2}$$

We shall investigate the scattering of two heavy particles (of the mass m) with high energy so that

$$s \gg m, \tag{15.3}$$

where s is the energy of the colliding particles in the centre of mass system. The scattering amplitude of this process may be obtained from the Green's function of the Euclidean theory

$$\langle \varphi(x_1)\varphi(x_2)\varphi(x_3)\varphi(x_4) \rangle \tag{15.4}$$

by the Fourier transformation followed by analytic continuation of the mass shell. The sign$\langle \ldots \rangle$ in (15.4) means the averaging over fields with the weight $\exp S$

$$\langle X \rangle = \frac{\int X \exp S \Pi_x \, \mathrm{d}\varphi(x) \mathrm{d}u(x)}{\int \exp S \Pi_x \, \mathrm{d}\varphi(x) \, \mathrm{d}u(x)}. \tag{15.5}$$

The integration over the field of heavy particles $\varphi(x)$ can be done explicitly by means of

$$\frac{\int (\exp S)[\varphi(x_1)\varphi(x_2)\varphi(x_3)\varphi(x_4)] \Pi_x \, \mathrm{d}\varphi(x)}{\int \exp S \Pi_x \, \mathrm{d}\varphi(x)}$$
$$= G(x_1, x_2|u)G(x_3, x_4|u) + G(x_1, x_3|u) \times \tag{15.6}$$
$$\times \ G(x_2, x_4|u) + G(x_1, x_4|u)G(x_2, x_3|u);$$

$$\frac{\int \exp S \Pi_x \, \mathrm{d}\varphi(x)}{\int \exp S[g = 0] \Pi_x \, \mathrm{d}\varphi(x)} = \det(\hat{M}/\hat{M}_0), \tag{15.7}$$

where $\hat{M}$ is the operator

$$\hat{M} = -\nabla^2 + m^2 + gu(x); \tag{15.8}$$

and $\hat{M}_0$ is the operator $\hat{M}$ for $g = 0$; $G(x, y|u)$ is the Green's function of the operator $\hat{M}$.

Putting the factor $\det(\hat{M}/\hat{M}_0)$ equal to one in the first approximation (which means neglecting the vacuum polarization effects) reduces the calculation of the Green's function (15.4) to the averaging of (15.6) over the light particle field with the weight

$$\exp\left\{ -\tfrac{1}{2} \int ((\nabla u)^2 + \mu^2 u^2) \, \mathrm{d}^4 x \right\}. \tag{15.9}$$

For the approximate averaging we exploit the asymptotical form of Green's function $G(x, y|u)$ as $|x - y| \to \infty$. It can be obtained from the following considerations. For $g = 0$ the Green's function of the operator $\hat{M}$ is

$$\frac{1}{(2\pi)^4} \int \frac{\exp[i(k, x - y)]}{k^2 + m^2} \, \mathrm{d}^4 k. \tag{15.10}$$

In the limit $x - y \rightarrow \infty$ this expression acquires the asymptotic form

$$m^{1/2}(2\pi|x - y|)^{-3/2} \exp(-m|x - y|). \tag{15.11}$$

If $g \neq 0$, then for the calculation of the Green's function of operator (15.8) the expression $gu(x)$ can be considered a correction to the squared mass m^2. We speak about the Green's function of the particle with the variable mass

$$m(x) = (m^2 + gu(x))^{1/2} \approx m + \frac{gu(x)}{2m}. \tag{15.12}$$

Supposing that the mass correction $gu(x)/(2m)$ is small compared to the mass itself we can obtain the asymptotics of the Green's function of the particle with the variable mass by replacing

$$m|x - y| = m \int ds \rightarrow \int m(x)\, ds \approx m \int ds + \frac{g}{2m} \int u(x)\, ds \tag{15.13}$$

in the exponent. In this way we obtain the approximate formula

$$G(x, y|u) \approx G_0(x - y) \exp\left(-\frac{g}{2m} \int_x^y u(z)\, ds(z)\right). \tag{15.14}$$

The integral $\int_x^y$ is supposed to be performed over the straight line connecting x and y

After inserting (15.14) into the right-hand side of (15.6), the second term acquires the form

$$G_0(x_1 - x_3)G_0(x_2 - x_4) \times$$
$$\times \exp\left\{-\frac{g}{2m}\left(\int_{x1}^{x_3} u(z)\, ds(z) + \int_{x2}^{x_4} u(z)\, ds(z)\right)\right\}. \tag{15.15}$$

The first and third terms are obtained from the second one by permuting arguments. Averaging (15.15) over the field $u(x)$ with the weight (15.9) reduces to a Gaussian integration that yields

$$G_0(x_1 - x_3)G_0(x_2 - x_4) \exp\frac{g^2}{8m^2(2\pi)^4} \int \frac{d^4k}{k^2 + \mu^2} \times$$
$$\times \left|\int_{x1}^{x_3} \exp(ikz)\, ds(z) + \int_{x2}^{x_4} \exp(ikz)\, ds(z)\right|^2. \tag{15.16}$$

The next tasks are Fourier transforms and continuation to the mass shell. We shall investigate the amplitude corresponding to a transferred momentum that is small in relation to the momentum in the centre of the mass system or, in other words, we shall explore the small angle scattering. In this case, the main contribution comes from only one of the three terms on the right-hand side of (15.16), namely the term describing direct scattering without an exchange. Let such main term be (15.26).

The exponent of (15.16) can be decomposed into four terms. One of them describes the virtual interaction of light particles with the heavy particle, moving along the trajectory $x_1 \rightarrow x_3$. The second term corresponds to the analogous interaction for the second particle moving along the trajectory $x_2 \rightarrow x_4$. These terms contribute to the mass renormalization and to the renormalization of the residuum in the pole of the heavy particle Green's function.

In the first approximation we shall not consider the renormalization terms.

The scattering effect depends on cross-terms describing an exchange of a light particle between two heavy ones. Their sum can be written in the form

$$
\frac{g^2}{8m^2(2\pi)^4} \int \frac{d^4k}{k^2+\mu^2} \left(\int_{x_1}^{x_3} \exp(ikz_1)\,ds(z_1) \int_{x_2}^{x_4} \exp(-ikz_2)\,ds(z_2) + \right.
$$
$$
\left. + \int_{x_1}^{x_3} \exp(-ikz_1)\,ds(z_1) \int_{x_2}^{x_4} \exp(ikz_2)\,ds(z_2) \right). \tag{15.17}
$$

At this place it is convenient to pass to the Euclidean metrics changing

$$
\left. \begin{array}{ll} d^4k \rightarrow i\,d^4k, & k^2 \rightarrow -k^2 - i0, \\ ds(z) \rightarrow i\,ds(z). & \end{array} \right\} \tag{15.18}
$$

(15.17) transforms to

$$
\frac{ig^2}{8m^2(2\pi)^4} \int \frac{d^4k}{k^2-\mu^2+i0} \left[\int_{x_1}^{x_3} \exp(ikz_1)\,ds(z_1) \int_{x_2}^{x_4} \exp(-ikz_2)\,ds(z_2) + \right.
$$
$$
\left. + \int_{x_1}^{x_3} \exp(-ikz_1)\,ds(z_1) \int_{x_2}^{x_4} \exp(ikz_2)\,ds(z_2) \right]. \tag{15.19}
$$

Integrals over $ds(u)$ can be calculated taking into account that the heavy particles move along straight lines with the velocities v_1, v_2. Then

$$
ds_i(z) = \sqrt{1-v_i^2}\,dt_i; \qquad z_i = z_i + v_i t_i; \tag{15.20}
$$

$$\int_{x_1}^{x_3} \exp{(ikz_1)}\,ds(z_1) = \exp{(ikz_1)}\sqrt{1-\mathbf{v}_1^2}\int \exp{[i(kv_1)t]}\,dt$$

$$= 2\pi \exp{(ikz_1)}\sqrt{1-\mathbf{v}_1^2}\,\delta(kv_1), \tag{15.21}$$

and (15.19) shows to be equal to

$$\frac{ig^2\sqrt{1-\mathbf{v}_1^2}\sqrt{1-\mathbf{v}_2^2}}{4m^2(2\pi)^2}\int \frac{d^4k}{k^2-\mu^2+i0}\,\delta(kv_1)\,\delta(kv_2)\cos{(k,z_1-z_2)} \tag{15.22}$$

$$= \frac{ig^2}{4(2\pi)^2}\int \frac{d^4k}{x^2-\mu^2+i0}\,\delta(kp_1)\,\delta(kp_2)\cos{(k,z_1-z_2)}.$$

This expression can be easily evaluated in the centre of mass system where

$$p = (p_0, p_1, 0, 0); \qquad p_2 = (p_0, -p_1, 0, 0). \tag{15.23}$$

The integral over the components r_0, k_1 of the momentum k gives the factor

$$(p_0^2 + p_1^2)^{-1} = ((s/2) - m^2)^{-1}, \tag{15.24}$$

where s is the centre of mass energy. The integral over the transverse variable

$$\frac{ig^2}{(2s-4m^2)(2\pi)^2}\int \frac{d^2k_\perp}{k_\perp^2+\mu^2}\cos{(k_\perp, z_1-z_2)}$$

$$= \frac{ig^2}{2s-4m^2}N_0(\mu|z_1-z_2|) = i\chi, \tag{15.25}$$

can be expressed over the Neumann function $N_0(\mu|z_1-z_2|)$ having the asymptotics

$$N_0(x) \approx -\frac{1}{2\pi}\ln{x}. \tag{15.26}$$

for small arguments.

In this way, the averaging of the exponent

$$\exp{-\frac{g}{2m}\left(\int_{x_1}^{x_3} u(z)\,ds(z) + \int_{x_2}^{x_4} u(z)\,ds(z)\right)} \tag{15.27}$$

over $u(x)$ and the passage to the pseudoeuclidean metrics provide the so-called *eikonal phase factor*

$$\exp i\chi, \tag{15.28}$$

where χ is defined by (15.25). For Green's function (15.4) the approximate formula

$$G_0(x_1 - x_3)G_0(x_2 - x_4)\exp{(i\chi)}. \tag{15.29}$$

is obtained. Subtracting the part $G_0(x_1 - x_3)G_0(x_2 - x_4)$ which describes particle propagation without scattering we obtain

$$G_0(x_1 - x_3)G_0(x_2 - x_4)(\exp{(i\chi)} - 1). \tag{15.30}$$

To obtain the scattering amplitude it is necessary to make the Fourier transform and evaluate the coefficient at $\delta(p_1 p_2 p_3 p_4)$. It is obvious that the amplitude f is proportional to the Fourier transform of $\exp{(i\chi)} - 1$ which does not depend on the transverse variable

$$f = a \int \exp{[ip_\perp(z_1 - z_2)_\perp]}(\exp{(i\chi)} - 1)\,\mathrm{d}^2(z_1 - z_2)_\perp, \tag{15.31}$$

where the parameter $p_\perp$ stands for the transferred momentum. The proportionality coefficient a in the first approximation does not depend on the coupling constant g and can be found when performing the limit $g \to 0$ on both left- and right-hand sides of (15.31). The limit of the left-hand side is easy to find, realizing that in the limit $g \to 0$, the amplitude is given by the simplest diagram

$$
\begin{array}{c}
p_1 \longrightarrow\!\!\bullet\!\!\longrightarrow p_3 \\
\vert \\
\vert \\
p_2 \longrightarrow\!\!\bullet\!\!\longrightarrow p_4
\end{array}
\tag{15.32}
$$

corresponding to

$$\frac{g^2}{(p_1 - p_3)^2 - \mu^2} = -\frac{g^2}{p_\perp^2 + \mu^2}, \tag{15.33}$$

where $p_\perp^2$ is the square of the transferred momentum in the centre of mass system. The right-hand side of (15.31) for $g \to 0$ tends to

$$ia \int \exp\left[ip_\perp(z_1 - z_2)_\perp\right]\chi \, d^2(z_1 - z_2)_\perp$$

$$= -\frac{iag^2}{(2s - 4m^2)(2\pi)^2} \int \exp\left[ip_\perp(z_1 - z_2)_\perp\right] \times \qquad (15.34)$$

$$\times \frac{\cos k_\perp(z_1 - z_2)_\perp}{k_\perp^2 + \mu^2} \, d^2(z_1 - z_2)_\perp \, d^2 k_\perp$$

$$= \frac{iag^2}{(2s - 4m^2)(p_\perp^2 + \mu^2)}.$$

Comparing (15.34) and (15.33) we obtain

$$a = i(2s - 4m^2). \qquad (15.35)$$

The final formula of the eikonal approximation for the scattering amplitude is

$$f = i(2s - 4m^2) \int d^2 z_\perp \left(\exp\left[i\chi(z_\perp)\right] - 1\right) \exp\left(ip_\perp z_\perp\right). \qquad (15.36)$$

The formula reckons with a multiple exchange of light particle quanta between the heavy particles moving in the first approximation along straight lines. The derivation presented here is similar to that of the infrared asymptotics of electrodynamical Green's functions in Chapter 4 and to the derivation of double logarithmic asymptotics in the previous section.

Other derivations of eikonal approximation in the functional integral formalism can be found in the works of many authors [67–71]. It is not difficult to obtain higher corrections to the first approximation of straight lines. In References [70, 71] the contribution of vacuum polarization corrections is explored and a method for the evaluation of contributions from exchanges of complicated complexes between two basic particles is developed.

We are not going to deal with those problems here. We shall stay in the framework of approximation corresponding to the light particle exchange leading to eikonal formula (15.36).

CHAPTER 6

SUPERFLUIDITY

16. PERTURBATION THEORY FOR SUPERFLUID BOSE SYSTEMS

The exploitation of functional integrals for derivation of the perturbation theory in statistical physics was presented in Section 6. Such a perturbation theory is applicable, in general, only for sufficiently high temperatures, higher than a phase transition temperature. In a system of interacting Bose particles (*Bose system*) the transition to the superfluid system is possible and it is related to the appearances of the *condensate*, i.e. the microscopically-occupied quantum ground state.

The problem of Bose gas with a weak interaction was solved for the first time by Bogoliubov in 1947 [73]. The higher approximations were investigated in the work of Bogoliubov and Zubarev [74] by means of collective variables and by Tserkovnikov [75] and Tolmachev [76] who used the temperature Green's functions.

Method of quantum field theory were applied to Bose gas of small density and $T = 0$ for the first time by Beliaev [77] and by Hugenholtz and Pines[78].

The problem of Bose gas with $T \neq 0$ was investigated by means of pseudopotential in the works of Lee and Yang [80]. We recall moreover the works [81–88] devoted to various questions of superfluid Bose systems.

As has been explained in works [90–93], the characteristic attribute of the superfluidity is not the condensate itself but existence of long-range correlations. That means that the averages

$$\langle \psi(\mathbf{x}, \tau)\bar{\psi}(\mathbf{y}, \tau_1) \rangle \tag{16.1}$$

do not decrease for $r = |x - y| \to \infty$ exponentially but polynomially. Especially, this behaviour of averages occurs in the two-dimensional model of Bose gas at low temperatures and in the one-dimensional model at $T = 0$. In the three-dimensional superfluid system the average (16.1) tends to the constant ρ_0 (as $r \to 0$) which is the condensate density. In this chapter the functional integral method is applied to various problems of Bose systems. We shall start by investigating the three-dimensional Bose system in which the creation of the condensate is possible.

123

The simplest variant of the perturbation theory that reckons with the condensate presence is obtained in the functional integral formalism when a shift in the action functional (6.1) is performed

$$\psi(\mathbf{x}, \tau) \to \psi(\mathbf{x}, \tau) + \alpha; \qquad \bar{\psi}(\mathbf{x}, \tau) \to \bar{\psi}(\mathbf{x}, \tau) + \bar{\alpha}, \qquad (16.2)$$

($|\alpha|^2 = \rho_0$) is the condensate density together with construction of such a perturbation theory where the quadratic form of the new variables $\psi, \bar{\psi}$ is considered unperturbed and the forms of other degrees are the perturbations. In the operator formalism, this variant of diagram technique was investigated in the works of Faddeev and Popov [94].

The scheme of the perturbation theory that enables us to obtain the thermodynamical functions of the systems, as well as Green's functions, converges poorly for small momenta and frequencies; separate terms for the perturbation theory series are singular for $\mathbf{k}, \omega \to 0$. Among them only Green's functions at small $\mathbf{k}, \omega$ or more precisely, their analytic continuations $i\omega \to E$ for small $\mathbf{k}, E$ contain the essential information about the properties of the system determining, especially its energy spectrum at low frequencies. Therefore, after explaining the simplest scheme of the perturbation theory and its applications to small density Bose gas in Section 17, we shall investigate its modification adopted to evaluate the low-frequency asymptotics of Green's functions in Section 18. In Section 19, a method for the construction of a hydrodynamic Hamiltonian is developed in the functional integral formalism and this method is used in the theory of one-dimensional superfluidity in Section 20, as well as for the description of quantum vortices in Bose systems in Section 21.

The most convenient way is to construct the perturbation theory in the $\mathbf{k}, \omega$ representation passing from the functions $\psi(\mathbf{x}, t)$ to their Fourier coefficients $a(\mathbf{k}, \omega)$ according to formulae (6.3). We shall denote the family of the variables $\mathbf{k}, \omega$ by p and write down the shift (16.2) for a, a^+ in the form

$$a(p) \to b(p) + \alpha(\beta V)^{1/2} \delta_{p0}; \qquad a^+(p) \to b^+(p) + \bar{\alpha}(\beta V)^{1/2} \delta_{p0}, \qquad (16.3)$$

proceeding to new Fourier coefficients $b(p), b^+(p)$. Here $\delta_{p0} = 1$ for $p = 0$ and 0 for $p \neq 0$; α, β are arbitrary constant complex numbers. After shift (16.3), action S expressed in terms of b, b^+ is

$$S = \beta V C_0 - \sqrt{\beta V} \left(\bar{\gamma} b(0) + \gamma b^+(0) \right) + \sum_p \left(i\omega - \frac{k^2}{2m} + \lambda \right) \times$$

$$\times b^+(p)b(p) - \sum_p |\alpha|^2 (u(0) + u(\mathbf{k})) b^+(p)b(p) -$$

$$-\tfrac{1}{2}\sum_p u(\mathbf{k})\alpha^2 b^+(p)b^+(-p) + \bar{\alpha}^2 b(p)b(-p)) - \frac{1}{2(\beta V)^{1/2}} \times$$

$$\times \sum_{p_1+p_2=p_3} (u(\mathbf{k}_1)+u(\mathbf{k}_2))(\alpha b^+(p_1)b^+(p_2)b(p_3) +$$

$$+ \bar{\alpha}b^+(p_3)b(p_2)b(p_3)) - \frac{1}{4\beta V}\sum_{p_1+p_2=p_3+p_4}(u(\mathbf{k}_1-\mathbf{k}_3)+$$

$$+ u(\mathbf{k}_1-\mathbf{k}_4))b^+(p_1)b^+(p_2)b(p_3)b(p_4). \tag{16.4}$$

Here

$$C_0 = \lambda|\alpha|^2 - \tfrac{1}{2}u(0)|\alpha|^2; \qquad \gamma = \alpha(-\lambda)+u(0)|\alpha|^2. \tag{16.5}$$

For the unperturbed action we take the quadratic form corresponding to the ideal Bose gas

$$\sum_p \left(i\omega - \frac{k^2}{2m}+\lambda\right)b^+(p)b(p). \tag{16.6}$$

The other forms of the first, second and fourth degree will be regarded as perturbations. The perturbation theory is described in terms of diagrams built from the following elements which correspond to the written expressions:

$$\longrightarrow \qquad \left(i\omega - \frac{k^2}{2m}+\lambda\right)^{-1};$$

$$p=0 \qquad \gamma;$$

$$p=0 \qquad \bar{\gamma};$$

$$\overset{p \quad p}{\longrightarrow\bullet\longrightarrow} \qquad |\alpha|^2(u(0)+u(\mathbf{k}));$$

$$\overset{p \qquad -p}{\longleftarrow\bullet\longrightarrow} \qquad \alpha^2 u(\mathbf{k});$$

$$\overset{p \qquad -p}{\longrightarrow\bullet\longleftarrow} \qquad \bar{\alpha}^2 u(\mathbf{k});$$

$$\tag{16.7}$$

$$\alpha(u(\mathbf{k}_1)+u(\mathbf{k}_2));$$

$$\bar{\alpha}(u(\mathbf{k}_1)+u(\mathbf{k}_2));$$

$$u(\mathbf{k}_1-\mathbf{k}_3)+u(\mathbf{k}_1-\mathbf{k}_4).$$

One type of line and eight types of vertices corresponding to separate perturbations are presented here. Diagrams with one incoming line and one outgoing line having no parts that are not connected with at least one of the external lines contribute to the one-particle Green's function

$$G(p) = - \langle b(p)b^+(p) \rangle. \tag{16.8}$$

The expressions corresponding to such diagrams are obtained when the product of all expressions corresponding to separate elements is composed, the conservation law of the four momentum $p = (\mathbf{k}, \omega)$ in every vertex used, the sum over all independent momenta performed and the result multiplied by

$$r^{-1}(-1/\beta V)^{v+1-l}, \tag{16.9}$$

where v is the number of vertices, l is the number of lines and r is the order of the diagram symmetry group. Note that $v + l - 1 = c$ is the number of independent diagram loops.

Among the diagrams composed of the presented elements there are such that contain parts without external lines connected to the rest of the diagram by one line. This line has $p = 0$. We shall call those parts of the diagram insertions of zero momentum. We choose the constant α such that the sum of all zero momentum insertions vanishes. Graphically

$$\tag{16.10}$$

This condition is equivalent to the relations

$$\langle b(0) \rangle = \langle b^+(0) \rangle = 0, \tag{16.11}$$

so that it is satisfied by

$$\langle a(0) \rangle = \alpha(\beta V)^{1/2}; \qquad \langle a^+(0) \rangle = \bar{\alpha}(\beta V)^{1/2}, \tag{16.12}$$

and the relation between the averages $\langle a, a^+ \rangle$ and $\langle b, b^+ \rangle$ is very simple

$$\langle a(p)a^+(p) \rangle = \langle b(p)b^+(p) \rangle + \beta V |\alpha|^2 \delta_{p0}. \tag{16.13}$$

From (16.13) follows that $|\alpha|^2$ has the meaning of the condensate density.

If condition (16.10) is satisfied, then all diagrams with zero momentum insertions can be left out when a Green's function is calculated. As was shown in Section 6, Green's function can be expressed via the irreducible self-energy part. We have two anomalous functions

$$G_1(p) = -\langle a(p)a(-p)\rangle, \qquad \bar{G}_1(p) = -\langle a^+(p)a^+(-p)\rangle, \qquad (16.14)$$

and two anomalous self-energy parts $B(p), \bar{B}(p)$ besides the normal Green's function (16.8) for the superfluid system. The Green's functions G, G_1 can be expressed via the self-energy parts A, B which are solutions of

$$(16.15)$$

or in analytical form

$$G(p) = G_0(p) + G_0(p)A(p)G(p) + G_0(p)B(p)G_1(-p);$$

$$G_1(p) = G_0(-p)\bar{B}(p)G(p) + G_0(-p)A(-p)G_1(p). \qquad (16.16)$$

The solution of system (16.16) is of the form $(\bar{B}(p) = B(-p))$

$$G(p) = \frac{i\omega + (k^2/2m) - \lambda + A(-p)}{(i\omega + (k^2/2m) - \lambda + A(-p))(i\omega - (k^2/2m) + \lambda - A(p)) + |B(p)|^2};$$

$$(16.17)$$

$$G_1(p) = -\frac{\bar{B}(p)}{(i\omega + (k^2/2m) - \lambda + A(-p))(i\omega - (k^2/2m) + \lambda - A(p)) + |B(p)|^2}.$$

We shall explore in detail conditions (16.10) for the choice of the parameter α. We denote Γ the sum of contributions of diagrams with one incoming or outgoing line. Condition (16.10) can be written as the algebraic relation

$$G_0(0)(\gamma + \Gamma) = 0. \qquad (16.18)$$

The first factor is different from zero and for the second we shall prove the relation

$$\gamma + \Gamma = \alpha(A(0) - B(0) - \lambda), \qquad (16.19)$$

that holds if α is chosen real so that we come to the equation

$$\alpha(A(0) - B(0) - \lambda) = 0. \qquad (16.20)$$

This equation has a trivial solution $\alpha = 0$ but can also have a nontrivial one if the second factor vanishes. In the following, we shall show by means of the perturbation theory that the solution exists for sufficiently low temperatures. It is necessary to deal especially with the nontrivial solution as the trivial one leads to contradiction. When the temperature increases, the nontrivial solution disappears and it is necessary to accept the expression for the Green's function calculated according to the ordinary perturbation theory.

Note that if the nontrivial solution of (16.20) exists, then Green's functions (16.17) are similar at $p = 0$ as their denominators vanish at that point. This is related to the fact that the one-particle excitation spectrum has a photon character.

Let us prove (16.19). We notice that the differentiation of an arbitrary element of a diagram with respect to $\alpha(\bar{\alpha})$, except on the first-order vertices, attaches to the element the incoming (outgoing) line. For example:

$$\text{(16.21)}$$

The differentiation of an arbitrary diagram with respect to $\alpha(\bar{\alpha})$ results therefore in attaching the incoming (outgoing) line to the diagram in all possible ways.

We shall investigate diagrams for the pressure $p = -\Omega/V$. A contribution to p gives the constant C_0, the pressure p_0 of the ideal Bose gas and a sum of all connected vacuum diagrams. By a direct comparison of diagrams one can check that the following relation holds:

$$\gamma + \Gamma = -\frac{\partial p}{\partial \alpha}; \qquad A(0) - \lambda = -\frac{\partial^2 p}{\partial \alpha \partial \bar{\alpha}}; \qquad B(0) = -\frac{\partial^2 p}{\partial \alpha^2}.$$
$$\text{(16.22)}$$

It is obvious that p depends on α and $\bar{\alpha}$ via the product $\alpha \bar{\alpha} = z$. Therefore the relations can be rewritten in the form

$$\gamma + \Gamma = -\bar{\alpha}\frac{\partial p}{\partial z}; \qquad A(0) - \lambda = -\frac{\partial p}{\partial z} - z\frac{\partial^2 p}{\partial z^2};$$

$$B(0) = -\bar{\alpha}^2\frac{\partial^2 p}{\partial z^2}. \tag{16.23}$$

Relation (16.19) follows from (16.23) if $\alpha = \bar{\alpha} = \sqrt{\rho_0}$. Eventually, let us note that the accepted condition of the choice of ρ_0 implies the extremality of the fundamental thermodynamical function – the pressure p as a function of the parameter ρ_0

$$\frac{\partial p}{\partial \rho_0} = 0. \tag{16.24}$$

17. BOSE GAS OF SMALL DENSITY

We shall find Green's function and the thermodynamic functions of Bose gas by means of the just-explained perturbation theory. The small density condition means that the range a of the potential between the Bose particles is small compared to the mean particle distance, i.e., the gas parameter θ is small

$$\theta \equiv \rho^{1/3}a \ll 1, \tag{17.1}$$

where $\rho = N/V$ is the density.

The state of the system is characterized by temperature T, besides the density ρ, and in dependence on the relation between the density and the temperature several regions in which different asymptotic dependences on the density hold can be determined. We shall explore the region[*]

$$T \sim \rho^{2/3}. \tag{17.2}$$

As will be shown below, especially a phase transition exists in this region. For temperatures greater in order than $\rho^{2/3}$, the system is in a normal state. The low temperature region will be investigated in Section 19. We shall not deal with the intermediate regions.

Let us start investigating diagrams that contribute to the pressure p

$$p = C_0 + p_0 + D, \tag{17.3}$$

[*] In this paragraph we use the system of units $2m = 1$, where m is the mass of the Bose particle.

where C_0 is the constant determined by (16.5); p_0 is the pressure of the ideal Bose gas and D is the sum of all connected vacuum diagrams.

The contribution of each diagram is diminished by the following factors:

(1) The presence of the second- or third-order vertices that contain the factors $\rho_0, \sqrt{\rho_0}$, since ρ_0 is a small parameter which does not exceed the full density ρ.

(2) The presence of loops that can be passed through moving along the lines and in the arrow directions of the diagram, since summation over frequencies provides the factor $(\exp[\beta\varepsilon(\mathbf{k})] - 1)^{-1}$ (where $\varepsilon(\mathbf{k}) = k^2 - \lambda$) which cut off the domain of summation over momenta at the boundary $k \sim \sqrt{T}$, smaller than the boundary $k \sim a^{-1}$ at which the sums are cut off due to the potential.

That is why the main contribution to the sum of vacuum diagrams comes from the diagrams with the minimal number of the mentioned factors.

$$(17.4)$$

These diagrams contain either two vertices of the second order or two independent loops, or one independent loop with either one second-order vertex or two third-order vertices. These diagrams form three series. Their summation reduces to the summation of the four-kernel series

$$(17.5)$$

which is contained within the diagrams. The summation is equivalent to the solution of the equation

$$(17.6)$$

The main contribution to the sum over internal momenta in (17.6) comes from $k \sim a^{-1} \gg \sqrt{T} \gg \sqrt{\lambda}$. We can, therefore, neglect λ in Green's function and replace the sum over frequencies in this equation by an integral so that Equation (17.6) is transformed to the equation for the t matrix (the scattering amplitude)

$$t(\mathbf{k}_1, \mathbf{k}_2, z) \tag{17.7}$$

$$= u(\mathbf{k}_1 - \mathbf{k}_2) - (2\pi)^{-3} \int d^3 k_3 u(\mathbf{k}_1 - \mathbf{k}_3)(2k_3^2 - z)^{-1} t(\mathbf{k}_3, \mathbf{k}_2, z),$$

and the four-kernel (17.5) is represented by

$$t\left(\frac{\mathbf{k}_1 - \mathbf{k}_2}{2}, \frac{\mathbf{k}_3 - \mathbf{k}_4}{2}; i\omega_1 + i\omega_2 - \frac{(\mathbf{k}_1 + \mathbf{k}_2)^2}{2}\right) +$$

$$+ t\left(\frac{\mathbf{k}_1 - \mathbf{k}_2}{2}, \frac{\mathbf{k}_4 - \mathbf{k}_3}{2}; i\omega_1 + i\omega_2 - \frac{(\mathbf{k}_1 + \mathbf{k}_2)^2}{2}\right). \tag{17.8}$$

We shall investigate the values of the four kernel for momenta and frequencies that are smaller in order than $k \sim a^{-1}, \omega \sim a^{-2}$. For such $\mathbf{k}$ and ω the t matrix can be considered constant.

$$t(0, 0, 0) \equiv t_0. \tag{17.9}$$

In the result we find that the sum of the term $(\tfrac{1}{2})\rho_0^2 u(0)$ and of series a (17.4) is equal to $(\tfrac{1}{2})\rho_0^2 t_0$, series b is equal to $t_0 \rho_1^2$, and for series c we obtain the expression $2t_0 \rho_0 \rho_1$ where ρ_1 is the contribution of

$$\tag{17.10}$$

representing the density ρ_1 of particles 'above the condensate' (with momenta $\neq 0$). In the first approximation we get ($\varepsilon \to +0$)

$$\rho_1 = -\frac{T}{V} \sum_p \exp(i\omega\varepsilon)(i\omega - k^2)^{-1} = \frac{1}{V} \sum_k (\exp(\beta k^2) - 1)^{-1}$$

$$= (2\pi)^{-3} \int \frac{d^3 k}{\exp(\beta k^2) - 1} = (4\pi)^{-3/2} \zeta(3/2) T^{3/2}, \tag{17.11}$$

where $\zeta(s)$ is the ζ function.

Finally, the pressure p_0 of the ideal gas is given by

$$-\frac{T}{V}\sum_{\mathbf{k}}\ln\left[1-\exp\left(-\beta(k^2-\lambda)\right)\right]=\zeta(\tfrac{5}{2})(4\pi)^{-3/2}T^{5/2}+\lambda\zeta(\tfrac{3}{2})\times$$
$$\times(4\pi)^{-3/2}T^{3/2}. \tag{17.12}$$

Inserting the obtained expression into (17.3), we get

$$p=(4\pi)^{-3/2}\zeta(\tfrac{5}{2})T^{5/2}+\lambda(4\pi)^{-3/2}\zeta(\tfrac{3}{2})T^{3/2}+$$
$$+\lambda\rho_0-\frac{t_0}{2}\rho_0^2-t_0\rho_1^2-2t_0\rho_0\rho_1, \tag{17.13}$$

where the expression ρ_1 depends only on temperature T. Now we shall determine ρ_0 from condition (16.24)

$$\rho_0=\frac{\lambda}{t_0}-2\rho_1=\frac{\lambda}{t_0}-2(4\pi)^{-3/2}\zeta(\tfrac{3}{2})T^{3/2}\equiv\frac{\Lambda}{t_0}. \tag{17.14}$$

From this formulae it follows that the positive solution for the condensate density ρ_0 exists only for $\Lambda>0$. In the first approximation the equation

$$\Lambda=0 \tag{17.15}$$

gives the equation of the phase transition curve in the $\lambda-T$ plane. From (17.11) and (17.14) follows the formula for the full density

$$\rho=\rho_0+\rho_1=\frac{\lambda}{t_0}-(4\pi)^{-3/2}\zeta(\tfrac{3}{2})T^{3/2}. \tag{17.16}$$

The expression of the pressure p in terms of λ and T is

$$p=(4\pi)^{-3/2}\zeta(\tfrac{5}{2})T^{5/2}+\frac{\lambda^2}{2t_0}-\lambda(4\pi)^{-3/2}\zeta(\tfrac{3}{2})\times$$
$$\times T^{3/2}+(4\pi)^{-3}\zeta^2(\tfrac{3}{2})t_0T^3. \tag{17.17}$$

Here, the first term is the leading one, the others are of the same order, and their relation to the leading term is of the order

$$t_0T^{1/2}\sim t_0\rho^{1/3}=\theta\ll1.$$

Expression (17.16) of ρ is obtained by a differentiation of (17.17) with respect to λ, according to the well-known thermodynamical relation $\rho=\partial p/\partial\lambda$. The formulae for the pressure and density in the normal state are

$$p = (4\pi)^{-3/2}\zeta(\tfrac{5}{2})T^{5/2} + (4\pi)^{-3/2}\zeta(\tfrac{3}{2})\lambda\bar{T}^{3/2} - (4\pi)^{-3}\zeta^2(\tfrac{3}{2})t_0 T^3;$$

$$(17.18)$$

$$\rho = \rho_1 = (4\pi)^{-3/2}\zeta(\tfrac{3}{2})T^{3/2}.$$

The considerations determining the main diagram series for p are also applicable to the self-energy parts of Green's functions A, B. Here the main series are

$$(17.19)$$

They contain the second-order vertex and all diagrams with two third-order vertices without loops (the series a), further all diagrams with one loop without vertices of the second or third order (b), and finally one second order vertex and diagrams with one vertex without loops (c).

It is easy to see that the summation of those series is reduced to replacing the potential by a t matrix that can be considered equal to the constant t_0 in the essential momentum region $k \lesssim T^{1/2}$. In the result we obtain

$$A \approx 2t_0(\rho_0 + \rho_1); \qquad B \approx t_0\rho_0 = \Lambda. \qquad (17.20)$$

Inserting these expressions into (16.17) gives the normal and anomalous Green's functions for $\Lambda > 0$

$$G(p) = -\frac{i\omega + k^2 + \Lambda}{\omega^2 + k^4 + 2\Lambda k^2}; \qquad G_1(p) = \frac{\Lambda}{\omega^2 + k^4 + 2\Lambda k^2}. \qquad (17.21)$$

For $\Lambda < 0$ (above the transition) only the normal Green's function and the corresponding self-energy part exist

$$A = 2t_0\rho_1; \qquad G = (i\omega - k^2 + \Lambda)^{-1}. \qquad (17.22)$$

Putting $i\omega = E$ in (17.21) and (17.22) enables one to find the energy spectrum determined by the Green's function poles

$$E = (k^4 + 2\Lambda k^2)^{1/2}, \quad \Lambda > 0;$$
$$E = k^2 - \Lambda = k^2 + |\Lambda|, \quad \Lambda < 0.$$

(17.23)

The first of these formulae is Bogoliubov's spectrum which transcends to the free-particle spectrum at $k \gg \sqrt{\Lambda} = k_c$ and is linear at $k \ll k_c$. It is natural to call the momentum $k = k_c$ correlational because at $k \sim k_c$ the transition from the free particles to the linear photon spectrum takes place.

Note that for $\Lambda = 0$ the normal and superfluid Green's functions and energy spectra coincide.

Knowledge of Green's functions (17.21) and (17.22) allows one to determine with high accuracy the density ρ_1 of the particles above the condensate which for $\Lambda < 0$ is equal to the full density. For $\Lambda < 0 (\varepsilon \to +0)$ we have

$$\rho = \rho_1 = -\frac{T}{V} \sum_p \exp(i\omega\varepsilon)(i\omega - k^2 - |\Lambda|)^{-1}$$
$$= \frac{1}{V} \sum_k (\exp \beta(k^2 + |\Lambda|) - 1)^{-1}$$
$$= (4\pi)^{-3/2} \zeta(\tfrac{3}{2}) T^{3/2} - (4\pi)^{-1} T |\Lambda|^{1/2};$$

(17.24)

and for $\Lambda > 0$

$$\rho_1 = \frac{T}{V} \sum_p \exp(i\omega\varepsilon)(i\omega + k^2 + \Lambda)(\omega^2 + k^4 + 2\Lambda k^2)^{-1}$$
$$= \frac{1}{2V} \sum_k \left(\frac{k^2 + \Lambda}{\varepsilon(\mathbf{k})} \coth \frac{\beta\varepsilon(\mathbf{k})}{2} - 1 \right)$$
$$= (4\pi)^{-3/2} \zeta(\tfrac{3}{2}) T^{3/2} - (8\pi)^{-1} T (2\Lambda)^{1/2},$$

(17.25)

where $\varepsilon(\mathbf{k})$ is Bogoliubov's energy spectrum (17.23). More complicated is the calculation of the condensate density ρ_0 in the second approximation. It can be done if the second approximation of self-energy parts $A(p)$, $B(p)$ is found. For that purpose we shall investigate the largest diagrams that were not considered in the first approximation. The diagrams that were considered shall be evaluated with higher accuracy. It is convenient to replace the Green's function occurring in the diagrams by the Green's functions of the first approximation (17.21). Then we can omit all diagrams containing the self-energy parts. That way, the self-energy parts $A(p)$, $B(p)$

in the second approximation are

$$(17.26)$$

It can be shown that the contributions to A and B that come from the diagrams with either two third-order vertices or more than one closed loop of normal lines or more than two anomalous lines, contribute to higher orders (in all cases for $k \gtrsim k_c$ and it is not necessary to incorporate them. In the diagram equalities (17.26) the thin lines correspond to the Green's function of the first approximation (17.20), the circle with four arrows corresponds to series (17.5), but the same series multiplied by $\alpha(\tilde{\alpha})$ with one of the external momenta equal to zero is depicted by the circle with three arrows. The thick line of the diagram b_1 corresponds to the full anomalous Green's function so that the relation represents an equation.

The diagrams $a_i, b_i (i = 0, 1, \dots, 6)$ are in fact equal to diagrams that were considered by Beliaev in the case $T = 0$. In the given temperature region, the expressions corresponding to the diagrams $a_i (i \geq 3), b_i (i \geq 2)$ are quite different. In the region $k \sim k_c$ we can replace the expressions corresponding to circles with three arrows by the constant factors $2\sqrt{\rho_0}\, t_0$ and obtain

$$a_3(p) = a_4(p) = b_3(p) = b_4(-p)$$
$$= -4\rho_0 t_0^2 TV^{-1} \sum_{p_1} G_1(p_1)G(p - p_1);$$

$$a_5(p) = b_5(p) = -4\rho_0 t_0^2 TV^{-1} \sum_{p_1} G(p_1)G(p_1 - p); \qquad (17.27)$$

$$a_6(p) = b_6(p) = 2b_2(p) = -4\rho_0 t_0^2 TV^{-1} \sum_{p_1} G_1(p_1)G_1(p_1 - p).$$

The expression $a_0(p) + a_2(p)$ is equal to series (17.5) multiplied by ρ_0 when one of the outcoming momenta is equal to zero. For $k \ll a^{-1}$

$$a_0(p) + a_2(p)$$

$$\approx 2\rho_0 t_0 - 2\rho_0 t_0^2 V^{-1} \sum_{k_1} \left[T \sum_{\omega_1} G(p_1)G(p - p_1) - 2|k_1 - k_2|^{-1} \right],$$

(17.28)

where $k_2 = k - k_1$.

For the evaluation of a_1 it is sufficient to connect one entry and exit of the four-kernel by a normal line and sum over the four-momentum p_1 of the line. The main contribution of the summation comes from the region $k_1 \sim \sqrt{T}$ in which we can neglect the dependence of the four-kernel on p. Using the expression ρ_1 of the density of particles above the condensate we obtain

$$a_1(p) \approx 2t_0\rho_1 = 2t_0((4\pi)^{-3/2}\zeta(\tfrac{3}{2})T^{3/2} - (8\pi)^{-1}T(2\Lambda)^{1/2}). \quad (17.29)$$

for $k \ll a^{-1}, \omega \ll a^{-2}$.

We shall find $b_0 + b_1$. It holds

$$B(p) = b_0 + b_1 + F = \psi(k) + F, \quad (17.30)$$

where F are additions, most important of which are described by the diagrams (17.26)

$$\psi(k) = b_0 + b_1 = \rho_0 u(k) - \frac{T}{V}\sum_{p_1} u(k - k_1)G_1(p_1)$$

$$= \rho_0 u(k) - \frac{T}{V}\sum_{p_1} u(k - k_1)\frac{\psi(k_1) + F}{M},$$

(17.31)

where M is the denominator of Green's function G_1 in (17.21). Contributions of F to the sum over k_1 are essential for $k_1 \sim k_c$. From the smallness of the volume of this region of k space it follows that the contribution of F is of a higher order. Putting $M = \omega^2 + k^4 + 2\Lambda k^2$, summing over frequencies and adding the same expression to both parts of (17.31) gives the equation

$$\psi(k) + \frac{1}{2V}\sum_{k_1} u(k - k_1)k_1^{-2}\psi(k_1)$$

$$= \rho_0 u(k) + \frac{1}{2V}\sum_{k_1} u(k - k_1)\psi(k_1)\left(\frac{1}{k_1^2} - \frac{1}{\varepsilon(k_1)}\coth\frac{\beta\varepsilon(k_1)}{2}\right),$$

(17.32)

where $\varepsilon(\mathbf{k})$ is the Bogoliubov energy spectrum (17.23). The operator acting at the left-hand side of this equation is invertible. Actually, the left-hand side of (17.32) can be obtained from the left-hand side of Equation (17.7) for the t matrix. Therefore it is necessary to shift the integral term to the left-hand side and put $z = 0$. At the right-hand side of (17.32) both $u(\mathbf{k})$ and $u(\mathbf{k} - \mathbf{k}_1)$ occur under the summation sign over $\mathbf{k}_1$. Thus, we solve (17.32) when $u(\mathbf{k})$ is replaced by $t(\mathbf{k}, 0, 0)$ and $u(\mathbf{k} - \mathbf{k}_1)$ by $t(\mathbf{k}, \mathbf{k}_1, 0)$

$$\psi(\mathbf{k}) = \rho_0 t(\mathbf{k}, 0, 0) + \frac{1}{2V} \sum_{\mathbf{k}_1} t(\mathbf{k}, \mathbf{k}_1, 0) \psi(\mathbf{k}_1) \times$$

$$\times \left(\frac{1}{k_1^2} - \frac{1}{\varepsilon(\mathbf{k}_1)} \coth \frac{\beta \varepsilon(\mathbf{k}_1)}{2} \right).$$

The main contribution to the sum over $\mathbf{k}_1$ comes from the region $k_1 \lesssim k_c$ in which $t(\mathbf{k}, \mathbf{k}_1, 0) \approx t(\mathbf{k}, 0, 0)$. In the result we obtain

$$\psi \approx t_0 \left[\rho_0 + \frac{\rho_0 t_0}{2V} \sum_{\mathbf{k}_1} \left(\frac{1}{k_1^2} - \frac{1}{\varepsilon(\mathbf{k}_1)} \coth \frac{\beta \varepsilon(\mathbf{u}_1)}{2} \right) \right]$$

$$= t_0 (\rho_0 - (8\pi)^{-1} T (2\Lambda^{1/2}). \tag{17.33}$$

for $k \sim k_c$.

The described procedure proves to be useful and will be employed later. Now, knowing the contribution of all diagrams $a_i, b_i (i = 0, 1, \ldots, 6)$, we can determine ρ_0 from the equation

$$\lambda = A(0) - B(0), \tag{17.34}$$

which follows from (16.20) for $\alpha \neq 0$.

We obtain

$$\rho_0 = \frac{\Lambda}{t_0} + \frac{3}{8\pi} T (2\Lambda)^{1/2}. \tag{17.35}$$

Adding ρ_0 to the density ρ_1 of the particles above the condensate which has been established by (17.24), we obtain

$$\rho = \rho_0 + \rho_1 = \frac{\lambda}{t_0} - (4\pi)^{-3/2} \zeta(\tfrac{3}{2}) T^{3/2} + (4\pi)^{-1} T (2\Lambda)^{1/2}. \tag{17.36}$$

Using the formula $\rho = \partial p / \partial \lambda$ we can find the correction to the expression of p (17.17) associated with the correction $(4\pi)^{-1} \Gamma (2\Lambda)^{1/2}$ to the density

$$\Delta p = (12\pi)^{-1} T (2\Lambda)^{3/2}. \tag{17.37}$$

For the normal state, the corresponding correction to (17.18) is

$$(6\pi)^{-1}T(\Lambda)^{3/2}. \tag{17.38}$$

We shall explore the superfluid phenomenon now. We take the Bose system in the coordinate system moving with the velocity **v**. In such a case the system is described by the action functional

$$S + \int_0^\beta d\tau \int dx \mathbf{v} \frac{i}{2m} (\nabla \bar{\psi} \psi - \bar{\psi} \nabla \psi), \tag{17.39}$$

where S is functional (6.1).

In the normal state the mean values of particle numbers and momentum are related by

$$\mathbf{K} = m\mathbf{v}N = m\mathbf{v}V\rho, \tag{17.40}$$

which means that the whole system moves with the velocity **v**. When the condensate appears, the system may be regarded as consisting of two components: the normal one ρ_u moving with the velocity **v** and the superfluid one $\rho_s = \rho - \rho_u$ moving with the velocity of condensate $\mathbf{v}_0$ which may be different from **v**. For example, for the condensate at rest we obtain a smaller value

$$\mathbf{K} = m\mathbf{v}V\rho_n. \tag{17.41}$$

instead of (17.40).

In the first approximation, the Green's function in the moving coordinate system is obtained from (17.21) by the shift $i\omega \to i\omega + (\mathbf{v}\mathbf{k})$.

$$G(p) = \frac{i\omega + k^2 + (\mathbf{k}\mathbf{v}) + \Lambda}{(i\omega + (\mathbf{k}\mathbf{v}) + k^2 + \Lambda)(i\omega + (\mathbf{k}\mathbf{v}) - k^2 - \Lambda) + \Lambda^2}, \tag{17.42}$$

The energy spectrum

$$E(\mathbf{k}) = -(\mathbf{v}\mathbf{k}) + (k^4 + 2\Lambda k^2)^{1/2} \tag{17.43}$$

is stable ($E > 0$) for $v < (2\Lambda)^{1/2}$. Knowing Green's function (17.42), we can find the momentum mean value

$$\mathbf{K} = -T \sum_p \exp(i\omega\varepsilon)\mathbf{k}G(p) \tag{17.44}$$
$$= m\mathbf{v}V[(4\pi)^{-3/2}\zeta(\tfrac{3}{2})T^{3/2} - (6\pi)^{-1}T(2\Lambda)^{1/2}].$$

Comparing it with (17.41), we get

$$\rho_n = (4\pi)^{-3/2}\zeta(\tfrac{3}{2})T^{3/2} - (6\pi)^{-1}T(2\Lambda)^{1/2};$$
$$\rho_s = \rho - \rho_n = \frac{\Lambda}{t_0} + \frac{5}{12\pi}T(2\Lambda)^{1/2}. \tag{17.45}$$

The formulae can be derived in another way, too, namely from the asymptotic formula

$$G(\mathbf{k}, 0) \approx -\frac{\rho_0 m}{\rho_s k^2}. \tag{17.46}$$

This formula was first obtained by Bogoliubov [81]. Its proof in the functional integral formalism is presented in the following section. It follows from (17.46) that

$$\frac{\rho_0}{\rho_s} = -\lim_{\mathbf{k}\to 0} 2k^2 G(\mathbf{k}, 0)$$

$$= \lim_{\mathbf{k}\to 0} \frac{2(A(\mathbf{k}, 0) + k^2 - \lambda)}{A(\mathbf{k}, 0) + k^2 - \lambda + B(\mathbf{k}, 0)} \frac{k^2}{k^2 + A(\mathbf{k}, 0) - B(\mathbf{k}, 0) - \lambda}. \tag{17.47}$$

The first factor at the right-hand side of the formula tends to 1. Let

$$A(\mathbf{k}, 0) - \lambda = \Lambda + A_1(\mathbf{k}, 0); \qquad B(\mathbf{k}, 0) = \Lambda + B_1(\mathbf{k}, 0),$$

where $A_1(\mathbf{k}, 0), B_1(\mathbf{k}, 0)$ are the corrections of the second order determined by diagrams (17.26). These corrections are

$$A_1(\mathbf{k}, 0) = \frac{t_0 T(2\Lambda)^{1/2}}{16\pi}\left[8 - 9x^{-1}\arctan\frac{x}{2} - 2x^{-1}\arctan x - \frac{\pi}{2x}\right];$$

$$B_1(\mathbf{k}, 0) = \frac{t_0 T(2\Lambda)^{1/2}}{16\pi}\left[4 - 9x^{-1}\arctan\frac{x}{2} - 2x^{-1}\arctan x - \frac{\pi}{2x}\right], \tag{17.48}$$

where $x = k/\sqrt{2\Lambda}$. Inserting the expressions into (17.47) we obtain

$$\frac{\rho_s}{\rho_0} = \lim_{\mathbf{k}\to 0} k^{-2}(k^2 + A(\mathbf{k}, 0) - B(\mathbf{k}, 0) - \lambda) = 1 + \frac{t_0 T}{12\pi}(2\Lambda)^{1/2}. \tag{17.49}$$

This formula exhibits the difference of ρ_0 from ρ_s and for ρ_n and ρ_s leads again to expressions (17.45).

At the end of this section we shall investigate the limits of applicability of the explained perturbation theory. The self-energy part in the first approximation is of the order Λ and the corrections of the second approximation are of the order $t_0 T \Lambda^{1/2}$. In the neighbourhood of the phase transition $(\Lambda \to 0)$ the ratio of the second to the first approximation ceases to be small and the perturbation theory series ceases to converge.

Besides that, even far from the phase transition, the perturbation series converges poorly for small momenta $\mathbf{k}$. This can be seen, e.g., from formula (17.48) for the corrections of the second approximation. They grow unboundedly in their absolute values for $\mathbf{k} \to 0$, and for $\mathbf{k} \sim t_0 T$ they are comparable to the self-energy parts in the first approximation. It is clear that when physical quantities are evaluated, the divergencies must cancel. This happens, e.g., at the calculation of the limit (17.49). Nevertheless, the appearance of singularities in separate diagrams of the perturbation theory is unwanted.

A reformulation of the perturbation theory that excludes the singularities is desirable. Such a reformulation in the functional integral formalism is investigated in the following section.

18. APPLICATION OF FUNCTIONAL INTEGRALS
TO THE DERIVATION OF LOW-ENERGY ASYMPTOTIC BEHAVIOUR
OF THE GREEN'S FUNCTIONS

The perturbation theory developed in Section 16 is based on the transformation (16.2) which defines the condensate. It has been shown that convergence of the perturbation series is slow for small momenta and frequencies. In this section we shall introduce a modified perturbation theory suitable for calculating the low-frequency asymptotic behaviour of Green's functions [97], [149]. The basic idea is shown in Section 12 and consists of successive integration first over rapidly oscillating, then over slowly oscillating fields, using different perturbation schemes for these different steps. We shall use the term 'the slowly oscillating part $\psi_0(\mathbf{x}, \tau)$ of the function $\psi(\mathbf{x}, \tau)$' for the sum of terms in the expansion.

$$\psi(\mathbf{x}, \tau) = (\beta V)^{-1/2} \sum_{\mathbf{k}, \omega} \exp\left[-i(\mathbf{k}\mathbf{x} - \omega\tau)\right] a(\mathbf{k}, \omega) \qquad (18.1)$$

with momenta $\mathbf{k}$ less than some k_0. The difference $\psi(\mathbf{x}, \tau) - \psi(\mathbf{x}_0, \tau)$ will then be called 'the rapidly oscillating part $\psi_1(\mathbf{x}, \tau)$'. We shall not distinguish explicitly between rapid and slow frequencies ω. Thus, we have

$$\psi(\mathbf{x}, \tau) = \psi_0(\mathbf{x}, \tau) + \psi_1(\mathbf{x}, \tau). \qquad (18.2)$$

The quantity k_0 which separates the 'small' momenta from the 'large' ones depends on the particular Bose system and only its order of magnitude can be estimated. For the Bose gas, the order of magnitude of k_0 is estimated at the end of this section.

If one looks at the low-frequency asymptotic behaviour of a one-particle Green's function – the mean value of the product $\psi(\mathbf{x}, \tau)\bar{\psi}(\mathbf{y}, \tau_1)$ – then only the mean value of the product of slowly oscillating fields $\psi_0(\mathbf{x}, \tau)\bar{\psi}_0(\mathbf{y}, \tau_1)$, i.e., the integral of this product weighted by $\exp S$ over all fields, contributes to the Green's function. As the first step one has to carry out the integral

$$\int \exp S \, d\bar{\psi}_1 \, d\psi_1 = \exp \tilde{S}[\bar{\psi}_0, \psi_0], \tag{18.3}$$

which can be interpreted as the statistic sum of the system of 'rapid' particles described by the field ψ_1, which are located in a slowly-oscillating field ψ_0.

To carry out the integral (18.3) one can develop a perturbation theory that has no divergencies at small momenta, as the sums over momenta are cut at the lower limit k_0.

At the second stage, that is integration over 'slow' fields $\psi_0(\mathbf{x}, \tau), \bar{\psi}_0(\mathbf{x}, \tau)$, it is convenient to introduce new variables, densities $\rho(\mathbf{x}, \tau)$ and phases $\varphi(\mathbf{x}, \tau)$ according to formulae

$$\begin{aligned}
\psi_0(\mathbf{x}, \tau) &= \sqrt{\rho(\mathbf{x}, \tau)} \exp\left[i\varphi(\mathbf{x}, \tau)\right]; \\
\bar{\psi}_0(\mathbf{x}, \tau) &= \sqrt{\rho(\mathbf{x}, \tau)} \exp\left[-i\varphi(\mathbf{x}, \tau)\right],
\end{aligned} \tag{18.4}$$

while

$$d\bar{\psi}_0(\mathbf{x}, \tau) \, d\psi_0(\mathbf{x}, \tau) = d\rho(\mathbf{x}, \tau) \, d\varphi(\mathbf{x}, \tau). \tag{18.5}$$

The presence of a condensate at low temperatures demonstrates itself in the fact that the main contribution to the integral over the density $\rho(\mathbf{x}, \tau)$ is given by values close to some positive ρ_0, which has the meaning of the condensate density. The quantity ρ_0 can be determined from the extremality condition applied to the functional $\tilde{S}[\bar{\psi}_0, \psi_0]$, which for $\psi_0 = \bar{\psi}_0 = \sqrt{\rho_0} = \text{const}$ becomes a function of ρ_0. Thus, the relation

$$\frac{\partial \tilde{S}}{\partial \rho_0} = 0. \tag{18.6}$$

must hold. The same equation was obtained in Section 16 (Equation (16.24)). For the calculation of $\tilde{S}[\rho_0]$ one can use the perturbation theory which differs from that developed in Section 16 in one respect – namely, that the summation over internal momenta is cut at some lower limit k_0. This is the reason why $\tilde{S}$, and, consequently, also ρ_0, depend on k_0. We shall call the quantity $\rho_0(k_0)$ *the bare condensate*. In the case of the three-dimensional Bose system, $\rho_0(k_0)$ is only slightly different from the exact ρ_0 defined as the limit of the mean values $\langle \psi(\mathbf{x}, \tau)\bar{\psi}(\mathbf{y}, \tau_1) \rangle$ as $|\mathbf{x} - \mathbf{y}| \to \infty$. In Section 19 it will be shown that in a two-dimensional system at $T > 0$, the bare condensate $\rho_0(k_0) > 0$ and the superfluidity may exist even when the exact $\rho_0 = 0$. Assuming that the (bare) condensate exists it is suitable to introduce the variable $\pi(\mathbf{x}, \tau)$ by the formula

$$\pi(\mathbf{x}, \tau) = \rho(\mathbf{x}, \tau) - \rho_0(k_0) \tag{18.7}$$

and to investigate $\tilde{S}$ as the functional of variables $\varphi(\mathbf{x}, \tau)$ and $\pi(\mathbf{x}, \tau)$.

Let us now look in greater detail at the calculation of the functional $\tilde{S}$. After the substitution (18.2), the original functional S takes the form

$$\int_0^\beta d\tau \int d\mathbf{x} \left(\bar{\psi}_0 \, \partial_\tau \psi_0 - \frac{\nabla \bar{\psi}_0 \nabla \psi_0}{2m} + \lambda \bar{\psi}_0 \psi_0 \right) +$$

$$+ \int_0^\beta d\tau \int d\mathbf{x} \left(\bar{\psi}_1 \, \partial_\tau \psi_1 - \frac{\nabla \bar{\psi}_1 \nabla \psi_1}{2m} + \lambda \bar{\psi}_1 \psi_1 \right) -$$

$$- \int_0^\beta d\tau \int d\mathbf{x} \, d\mathbf{y} u(\mathbf{x} - \mathbf{y}) \bar{\psi}_0(\mathbf{x}, \tau) \bar{\psi}_0(\mathbf{y}, \tau) \psi_0(\mathbf{y}, \tau) \psi_0(\mathbf{x}, \tau) +$$

$$+ 2\bar{\psi}_1(\mathbf{x}, \tau) \psi_1(\mathbf{x}, \tau) \bar{\psi}_0(\mathbf{y}, \tau) \psi_0(\mathbf{y}, \tau) +$$

$$+ 2\psi_1(\mathbf{x}, \tau) \psi_1(\mathbf{y}, \tau) \bar{\psi}_0(\mathbf{y}, \tau) \psi_0(\mathbf{x}, \tau) +$$

$$+ \bar{\psi}_1(\mathbf{x}, \tau) \bar{\psi}_1(\mathbf{y}, \tau) \psi_0(\mathbf{x}, \tau) \psi_0(\mathbf{y}, \tau) +$$

$$+ \bar{\psi}_0(\mathbf{x}, \tau) \bar{\psi}_0(\mathbf{y}, \tau) \psi_1(\mathbf{x}, \tau) \psi_1(\mathbf{y}, \tau) +$$

$$+ 2\bar{\psi}_0(\mathbf{x}, \tau) \bar{\psi}_1(\mathbf{y}, \tau) \psi_1(\mathbf{y}, \tau) \psi_1(\mathbf{x}, \tau) +$$

$$+ 2\bar{\psi}_1(\mathbf{x}, \tau) \bar{\psi}_1(\mathbf{y}, \tau) \psi_1(\mathbf{y}, \tau) \psi_0(\mathbf{x}, \tau) +$$

$$+ \bar{\psi}_1(\mathbf{x}, \tau) \bar{\psi}_1(\mathbf{y}, \tau) \psi_1(\mathbf{y}, \tau) \psi_1(\mathbf{x}, \tau). \tag{18.8}$$

Here we have assumed that the crossing terms in the quadratic form $\sim \bar{\psi}_0 \psi_1, \bar{\psi}_1 \psi_0$ vanish when integrated over the volume V. In order to express the explicit dependence of the functional $\tilde{S}$ on the phase $\varphi(\mathbf{x}, \tau)$, we perform the transformation of variables in the integral over $\psi_1, \bar{\psi}_1$

$$\psi_1(\mathbf{x}, \tau) \to \psi_1(\mathbf{x}, \tau) \exp \left[i\varphi(\mathbf{x}, \tau) \right];$$
$$\bar{\psi}_1(\mathbf{x}, \tau) \to \bar{\psi}_1(\mathbf{x}, \tau) \exp \left[-i\varphi(\mathbf{x}, \tau) \right], \tag{18.9}$$

where $\varphi(\mathbf{x}, \tau)$ is the phase of the slowly oscillating field. After substituting (18.9) into (18.8) the part of the action that contains the potential actually depends only on $\rho(\mathbf{x}, \tau) = \rho_0(\mathbf{k}_0) + \pi(\mathbf{x}, \tau)$, the modulus squared of the function $\psi_0(\mathbf{x}, \tau)$. The quadratic form of the variables $\psi_1, \bar{\psi}_1$ in (18.8) is given by

$$
\int_0^\beta d\tau \int d\mathbf{x} \bigg(\bar{\psi}_1 \, \partial_\tau \psi_1 - \frac{1}{2m} \nabla \bar{\psi}_1 \nabla \psi_1 +
$$

$$
+ \frac{i}{2m} (\bar{\psi}_1 \nabla \psi_1 - \nabla \bar{\psi}_1 \psi_1) \nabla \varphi + \bar{\psi}_1 \psi_1 \bigg(\lambda - \frac{(\nabla \varphi)^2}{2m} + i \, \partial_\tau \varphi \bigg) \bigg).
\tag{18.10}
$$

The perturbation theory can be constructed regarding the quadratic form

$$
\int_0^\beta d\tau \int d\mathbf{x} \bigg(\bar{\psi}_1 \, \partial_\tau \psi_1 - \frac{1}{2m} \nabla \bar{\psi}_1 + \lambda \bar{\psi}_1 \psi_1 \bigg),
\tag{18.11}
$$

which describes the ideal gas of Bose particles with momenta $k > k_0$ as the nonperturbed action, while other parts of the action S depending on $\psi_1, \bar{\psi}_1$ as a perturbation. Thus, we get a perturbation theory which differs from that introduced in Section 16 in the following respects:

(1) The second- and third-order vertices, which appear in (16.7) and describe the interaction with the slowly oscillating field, now depend on the functional variable $\rho(\mathbf{x}, \tau)$.
(2) The second-order vertices describing the interaction with the φ-field emerge. They are determined by φ-dependent terms in the quadratic form in formula (18.10).
(3) All summations (integrations) over internal momenta are cut at some lower limit k_0.

Let us expand the expression of $\tilde{S}$ into the functional series of variables $\varphi(\mathbf{x}, \tau), \pi(\mathbf{x}, \tau)$:

$$
\tilde{S}[\varphi, \pi] = \tilde{S}[0, 0] + \tfrac{1}{2} \int d\mathbf{x} \, d\mathbf{y} \, d\tau \, d\tau' [a_{11}(\mathbf{x}, \tau | \mathbf{y}, \tau_1) \varphi(\mathbf{x}, \tau) \varphi(\mathbf{y}, \tau_1) +
$$

$$
+ 2a_{12}(\mathbf{x}, \tau | \mathbf{y}, \tau_1) \varphi(\mathbf{x}, \tau) \pi(\mathbf{y}, \tau_1) +
$$

$$
+ a_{22}(\mathbf{x}, \tau | \mathbf{y}, \tau_1) \pi(\mathbf{x}, \tau) \pi(\mathbf{x}, \tau_1)] + \cdots
\tag{18.12}
$$

There are no linear terms in this expansion. Indeed, as we have seen, $\tilde{S}$ depends on the gradients of the phase. The coefficient functions of $\nabla \varphi$ and $\partial_\tau \varphi$ are translationally invariant and, thus, vanish upon integration with $\nabla\varphi, \partial_\tau\varphi$. The coefficient function at $\pi(\mathbf{x}, \tau)$ is proportional to $\partial\tilde{S}/\partial\rho_0$ and is equal to zero thanks to the choice of ρ_0 according to (18.6).

If the momentum k_0 is infinitesimally small, one can restrict oneself to the terms of second order in φ and π in the expansion (18.12). In that case, the low frequency asymptotic behaviour of the Green's function is determined by coefficient functions a_{11}, a_{12} and a_{22}. If the condensate is present, it is suitable to subtract from the Green's function its asymptotic part as $|\mathbf{x} - \mathbf{y}| \to \infty$ and to investigate the expression

$$\tilde{G}(\mathbf{x}, \tau|\mathbf{y}, \tau_1) = -\langle \psi(\mathbf{x}, \tau)\bar{\psi}(\mathbf{y}, \tau_1)\rangle + \langle \psi(\mathbf{x}, \tau)\rangle\langle \bar{\psi}(\mathbf{y}, \tau_1)\rangle$$

$$= \rho_0\left[1 - \frac{\langle\sqrt{\rho(\mathbf{x}, \tau)\rho(\mathbf{y}, \tau_1)}\exp[i(\varphi(\mathbf{x}, \tau) - \varphi(\mathbf{y}, \tau_1))]\rangle}{\langle\sqrt{\rho(\mathbf{x}, \tau)}\exp[i\varphi(\mathbf{x}, \tau)]\rangle\langle\sqrt{\rho(\mathbf{y}, \tau_1)}\exp[-i\varphi(\mathbf{y}, \tau_1)]\rangle}\right].$$

$$(18.13)$$

Let us restrict ourselves to the terms of maximally second order in the limit $k_0 \to 0$ in expansion (18.12). For sufficiently large $|\mathbf{x} - \mathbf{y}|$ one can neglect the difference between the variables $\rho(\mathbf{x}, \tau), \rho(\mathbf{y}, \tau_1)$ and the bare condensate density $\rho_0(k_0)$. Indeed, the mean values $\langle\varphi\pi\rangle, \langle\pi\pi\rangle$ with different arguments $(\mathbf{x}, \tau), (\mathbf{y}, \tau_1)$, when calculated perturbatively, vanish for $|\mathbf{x} - \mathbf{y}| \to \infty$ faster than $\langle\varphi\varphi\rangle$ (this follows from formulae (19.16)), and the mean values $\langle\varphi\pi\rangle, \langle\pi\pi\rangle$ with identical arguments (either $(\mathbf{x}, \tau)$ or $(\mathbf{y}, \tau_1)$) appear both in the numerator and in the denominator and therefore mutually cancel. Finally, we obtain the formula

$$\tilde{G}(\mathbf{x}, \tau|\mathbf{y}, \tau_1) \approx \rho_0\left(1 - \frac{\langle\exp[i(\varphi(\mathbf{x}, \tau) - \varphi(\mathbf{y}, \tau_1))]\rangle}{\langle\exp[i\varphi(\mathbf{x}, \tau)]\rangle\langle\exp[-i\varphi(\mathbf{y}, \tau_1)]\rangle}\right) \quad (18.14)$$

containing the functional integrals of the Gaussian type. The values of these integrals are determined by coefficient functions a_{11}, a_{12} and a_{22} from (18.22) which can be regarded as matrix elements of an operator A. They can be expanded in terms of matrix elements of the inverse operator A^{-1}:

$$\langle\exp[i(\varphi(\mathbf{x}, \tau) - \varphi(\mathbf{y}, \tau_1))]\rangle$$
$$= \exp\{-\tfrac{1}{2}(\hat{A}^{-1})_{11}(\mathbf{x}, \tau|\mathbf{x}, \tau) - \tfrac{1}{2}(\hat{A}^{-1})_{11}(\mathbf{y}, \tau_1|\mathbf{y}, \tau_1) +$$
$$+ \tfrac{1}{2}\hat{A}_{11}^{-1},(\mathbf{x}, \tau|\mathbf{y}, \tau_1) + \tfrac{1}{2}\hat{A}_{11}^{-1}(\mathbf{y}, \tau_1|\mathbf{x}, \tau)\};$$
$$\langle\exp[i\varphi(\mathbf{x}, \tau)]\rangle \qquad\qquad (18.15)$$
$$= \exp\{-\tfrac{1}{2}(\hat{A}^{-1})_{11}(\mathbf{x}, \tau|\mathbf{x}, \tau_1)\};$$
$$\langle\exp[-i\varphi(\mathbf{y}, \tau_1)]\rangle$$
$$= \exp\{-\tfrac{1}{2}(\hat{A}^{-1})_{11}(\mathbf{y}, \tau_1|\mathbf{y}, \tau_1)\}.$$

The Green's function G in the limit $|\mathbf{x} - \mathbf{y}| \to \infty$ then takes the form

$$\rho_0\{1 - \exp\left(\tfrac{1}{2}(\hat{A}^{-1})_{11}(\mathbf{x}, \tau | \mathbf{y}, \tau_1) + \tfrac{1}{2}(\hat{A}^{-1})_{11}(\mathbf{y}, \tau_1 | \mathbf{x}, \tau)\right)\}$$

$$\approx -\frac{\rho_0}{2}\{(\hat{A}^{-1})_{11}(\mathbf{x}, \tau | \mathbf{y}, \tau_1) + (\hat{A}^{-1})_{11}(\mathbf{y}, \tau_1 | \mathbf{x}, \tau)\} \tag{18.16}$$

(Here, the inequality $|(A^{-1})_{11}(\mathbf{x}, \tau | \mathbf{y}, \tau_1)| \ll 1$, valid for large $|\mathbf{x} - \mathbf{y}|$, has been used.)

The formula analogous to (18.16) for the Fourier coefficients of the function G can be written as

$$\tilde{G}(\mathbf{k}, \omega) \approx -\frac{\rho_0 a_{22}(\mathbf{k}, \omega)}{a_{11}(\mathbf{k}, \omega)a_{22}(\mathbf{k}, \omega) - a_{12}(\mathbf{k}, \omega)a_{12}(-\mathbf{k}, -\omega)}. \tag{18.17}$$

The Fourier coefficients of functions a_{11}, a_{12} and a_{22}, which appear in this formula, can be obtained by double differentiation of the functional $\tilde{S}[\varphi, \pi]$ with respect to φ, π.

The following exact diagrammatic equations determine each of the Fourier coefficients a_{11}, a_{12} and a_{22} as a sum of an infinite number of diagrams of the temperature diagrammatic technique, which differs from the diagrammatic technique investigated in Sections 16 and 17, by the fact that the integrals over internal momenta are cut at some lower limit k_0.

$$ia_{12}(\mathbf{k}, \omega) = -ia_{12}(-\mathbf{k}, -\omega) \tag{18.18}$$

The internal lines of the diagrams correspond to full normal and anomalous Green's functions. They can be distinguished by arrows at the ends of the lines which point in the same direction in the case of normal Green's functions and in opposite directions in the case of anomalous ones. The diagrams (18.18) should be understood with summation over all possible combinations of arrow directions accomplished.

After the analysis of the arrow directions in diagrams (18.18) different vertex parts of the following form arise:

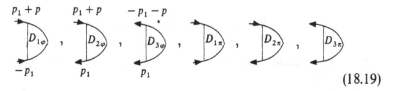

$$\text{(18.19)}$$

They can be obtained by differentiating normal and anomalous self-energy parts, calculated in the presence of an external field φ_0, with respect to arguments φ, π. The results of differentiation with respect to φ are denoted $D_{j\varphi}(j = 1, 2, 3)$, and with respect to π—by $D_{j\pi}$. The contributions of irreducible (in the sense that the interaction vertex with external field cannot be disconnected by cutting one or two internal lines) diagrams to $D_{j\varphi}$ and $D_{j\pi}$, will be denoted $d_{j\varphi}$ and $d_{j\pi}$, respectively. The contribution of irreducible (in the same sense) diagrams to the expression a_{22}, which can be obtained by double differentiating of $\tilde{S}[\varphi, \pi]$ with respect to π, will be denoted t. Let us mention that the expressions for irreducible diagrams $d_{j\varphi}$ are known exactly – they are given by the formulae

$$d_{1\varphi} = d_{3\varphi} = 0; \qquad d_{2\varphi} = i\omega - (\mathbf{k}\mathbf{k}_1)/m, \qquad \text{(18.20)}$$

where $(\mathbf{k}, \omega)$, $(\mathbf{k}_1, \omega_1)$ are the four momenta of the external field φ and of the internal line of the diagram incoming to the vertex $d_{2\varphi}$, respectively.

Expression (18.20) is determined by the quadratic form of $\psi_1, \bar{\psi}_1$ (18.10). The diagram $\bigcirc$ in (18.18), which defines the density of condensed particles ρ_1, is determined by the integral of the expression $\bar{\psi}_1 \psi_1 (\nabla \varphi)^2$. To ρ_1 one must add the condensate density ρ_0 which has its origin in the expression $-(2m)^{-1} \nabla \bar{\psi}_0 \nabla \psi_0$ after getting rid of the component proportional to $(\nabla \varphi)^2$. The sum $\rho_0 + \rho_1$ is the full density ρ. The term $i\omega$ in formula (18.18) expressing $i a_{12}$ corresponds to the term $\bar{\psi}_0 \partial_\tau \psi_0$ in the action integral.

Let us stress that the diagrammatic equations (18.18) do not take into account the low-density assumption and, thus, in this sense they are exact. What is actually used is only the smallness of momentum k with respect to k_0.

From the formulae derived so far one can easily obtain the asymptotic behaviour of Green's functions at $\omega = 0$. Indeed, the expression $a_{12}(p)$ is proportional to k^2 for $\omega = 0$ and small k, while a_{22} goes to a constant $a_{22}(0)$. As $a_{11}(\mathbf{k}, 0) \approx k^2$, $G(\mathbf{k}, 0) \approx -\rho_0/a_{11}(\mathbf{k})$ for small $\mathbf{k}$. The second term in formula (18.18) for a_{11} is for $\omega = 0$ equal to $-(mV)^{-1}(\mathbf{k}\mathbf{K})$, where $\mathbf{K}$

is the average momentum in the coordinate system moving with the velocity **v**. According to the definition of the normal component (17.14) we obtain $\mathbf{K} = V_{\rho n}\mathbf{k}$. Finally, the formula for a_{11} takes the form

$$a_{11} \approx \frac{k^2}{m}(\rho - \rho_n) = \frac{\rho_s k^2}{m}, \qquad (18.21)$$

where $\rho_s = \rho - \rho_n$ is the superfluid component density. This yields the well-known Bogoliubov asymptotic formula for the function $\tilde{G}$ [81]

$$\tilde{G}(\mathbf{k}, 0) \approx -\frac{\rho_0 m}{\rho_s k^2}. \qquad (18.22)$$

From this formula and from formula (6.49) for the average number of particles with momentum **k** it follows that for $\mathbf{k} \to 0$

$$N_\mathbf{k} = \frac{T\rho_0 m}{\rho_s k^2} + O(1), \qquad (18.23)$$

as the terms with $\omega \neq 0$ in (6.49) give a finite contribution for $\mathbf{k} \to 0$. From (18.23) it follows that the condensate cannot emerge in two- and one-dimensional systems for the temperatures $T \neq 0$, as the function k^{-2} is not integrable at small momenta in one- and two-dimensional k space.

The investigation of the function $G(\mathbf{k}, \omega)$ for $\omega \neq 0$ and, in particular, its analytic continuation $i\omega \to E$ to the domain $E\tau, kl \lesssim 1$ (where τ is the relaxation time and l is the mean free path) is more complicated. In this case the summation of diagrams (18.18) is equivalent to solving the equation of kinetic type [97]. In the hydrodynamical domain $E\tau, kl \ll 1$ one obtains an expression for the Green's function $G(k, E)$ that contains poles corresponding to the first and second sounds. In the following section we shall derive kinetic equations for a Bose gas of small density in the low temperature limit $T \to 0$.

Now we can estimate the order of magnitude of the momentum k_0 which distinguishes between 'large' and 'small' momenta. In a general theory k_0 can be regarded as arbitrarily small. However, if one wants to carry out particular calculations, it is convenient to choose such k_0 that the perturbation theory introduced in Section 16 converges well for $k \sim k_0$. For example, in the domain $T \sim \rho^{2/3}$ investigated in the previous section, the second-order corrections (17.48) of the perturbation theory behave like $\Lambda t_0 T k^{-1}$ for $k \ll \Lambda^{1/2}$. Requiring that for $k \sim k_0$ the order of magnitude of the corrections remains less than the magnitude of the self-energy parts in the first approximation, which are of order Λ, we obtain the condition

$k_0 \gg t_0 T$. We shall estimate k_0 from above, requiring that the perturbation theory from Section 6 is still valid for $k \sim k_c = \Lambda^{1/2}$. Finally we obtain

$$t_0 T \ll k_0 \ll \Lambda^{1/2} \tag{18.24}$$

valid in the temperature domain $T \sim \rho^{2/3}$. This condition is expressed in the units adopted in section 17 ($2m = 1$). The inequality $k_0 T \ll \Lambda^{1/2}$ following from (18.24) is valid everywhere except a small domain near the phase transition.

19. AN HYDRODYNAMICAL LAGRANGIAN OF NONIDEAL BOSE GAS

The microscopic theory of superfluidity was developed for the first time by Bogoliubov [73]. The former semi-phenomenological theory of Landau [72] leaned on an *ad hoc* postulated Hamiltonian that described the hydrodynamics of a quantum liquid.

The essence of the method of this *hydrodynamical Hamiltonian* can be obtained by means of microscopic theory. We shall construct a hydro-dynamical Hamiltonian for a nonideal Bose gas using our method of separate integration over rapidly and slowly oscillating fields. At first we do not obtain the hydrodynamical Hamiltonian, but the functional of *hydrodynamical action*, that is a functional of the fields $\psi_0(\mathbf{x}, \tau), \bar{\psi}_0(\mathbf{x}, \tau)$. It is suitable to rewrite the functional in new variables, density and phase, introducing the polar coordinates $\rho(\mathbf{x}, \tau), \varphi(\mathbf{x}, \tau)$ according to formulae (18.4). The transformation to polar coordinates is, in fact, also used in the method of collective variables (see the paper by Bogoliubov and Zubarev [74]). The perturbation theory constructed in terms of Green's functions of the variables ρ and φ does not contain low-energy and low-momentum divergencies. It is suitable for the calculation of low-frequency energy spectrum and for deriving the kinetic equations. In the following section, the method of construction of a hydrodynamical Hamiltonian will be extended to two- and one-dimensional Bose systems.

Now we shall investigate a low-temperature Bose system in the limit $T \to 0$.

The hydrodynamical Hamiltonian can be obtained by integration of the functional exp S over rapidly oscillating fields $\psi_1(\mathbf{x}, \tau)$. Such an integral (18.3) has the meaning of the statistical sum of a system of particles located in a slowly varying field ψ_0. Let us denote the result of such integration exp $\tilde{S}$. We can expand $\tilde{S}$ into a functional series of variables $\varphi(\mathbf{x}, \tau)$ and $\pi(\mathbf{x}, \tau) = \rho(\mathbf{x}, \tau) - \rho_0(k_0)$. At low temperatures the coefficient functions of this expression can be put equal to the corresponding functions at temperature equal to zero. The leading components, which depend on the variables φ, π of $\tilde{S}$, are given as a quadratic form of the variables φ, π. The coefficient functions of the quadratic form can be expressed in terms of thermodynamical functions. This can be accomplished not only for a Bose gas, but for an arbitrary Bose system. So as to convince ourselves of this, we can make the substitution (18.9) in the integral over the rapid variables, after which the quadratic form of $\psi_1, \bar{\psi}_1$ takes the form of (18.10). The same situation occurs when a system of particles described by the fields $\psi_1, \bar{\psi}_1$ and interacting with an inhomogeneous condensate $\rho(\mathbf{x}, \tau)$, is inserted into the field of an inhomogeneous chemical potential

$$\lambda(\mathbf{x}, \tau) = \lambda + i\,\partial_\tau \varphi(\mathbf{x}, \tau) - \frac{(\nabla\varphi(\mathbf{x}, \tau))^2}{2m} \tag{19.1}$$

and the field of velocities

$$\mathbf{v}(\mathbf{x}, \tau) = \frac{1}{m}\nabla\varphi(\mathbf{x}, \tau). \tag{19.2}$$

In the first (quasihomogeneous) approximation, the inhomogeneities induced by the slowly oscillating functions $\varphi(\mathbf{x}, \tau), \rho(\mathbf{x}, \tau)$ can be incorporated by the formula

$$\tilde{S} = \int d\tau\, d\mathbf{x}\, p(\lambda(\mathbf{x}, \tau), \rho(\mathbf{x}, \tau), \mathbf{v}(\mathbf{x}, \tau)), \tag{19.3}$$

where $p(\lambda, \rho_0, \mathbf{v})$ is the pressure of a homogeneous system with chemical potential λ and rest condensate density $\rho_0(k_0)$ in a coordinate system moving with the velocity $\mathbf{v}$.

Let us expand p according to different powers of inhomogeneity. The coefficient at $\mathbf{v}^2$ is proportional to the normal component density ρ_n and vanishes in the limit $T \to 0$. The coefficient at $\mathbf{v}$ is equal to zero due to the translation invariance. Thus, restricting ourselves to the terms of not higher than second order in $\mathbf{v}$, we come to the conclusion that $\tilde{S}$ is $\mathbf{v}$-independent. Using the fact that $p_{\rho_0} = 0$, we can write the expansion of $\tilde{S}$ in the form

$$\tilde{S} = p(\lambda\rho_0(k_0), 0)\beta V +$$
$$+ \int d\tau\, d\mathbf{x} \left\{ p_\lambda \left(i\,\partial_\tau \varphi - \frac{(\nabla\varphi)^2}{2m} \right) + \frac{p_{\lambda\lambda}}{2} \left(i\,\partial_\tau \varphi - \frac{(\nabla\varphi)^2}{2m} \right)^2 + \right. \tag{19.4}$$
$$\left. + p_{\lambda\rho_0} \left(i\,\partial_\tau \varphi - \frac{(\nabla\varphi)^2}{2m} \right) \pi + \frac{p_{\rho_0\rho_0}}{2} \pi^2 \right\} + \cdots$$

Concentrating on the quadratic form of φ and π in (19.4), we obtain

$$\int d\tau \, dx \left\{ -\frac{p_\lambda}{2m}(\nabla\varphi)^2 - \frac{p_{\lambda\lambda}}{2}(\partial_\tau\varphi)^2 + ip_{\lambda\rho_0}\pi\,\partial_\tau\varphi + \frac{p_{\rho_0\rho_0}}{2}\pi^2 \right\}. \quad (19.5)$$

The coefficients of the quadratic form in (19.5) can be expressed in terms of the derivatives of $p(\lambda, \rho_0)$, i.e., in terms of thermodynamical functions.

Let us now concentrate on some of the terms in the expansion of $\tilde{S}$ that have not been taken into account by formula (19.5). In the case of a Bose gas, the most important of these terms have their origin in the expression

$$-\int d\tau \, d^3x \frac{1}{2m}\nabla\bar\psi_0\nabla\psi_0 \qquad\qquad (19.6)$$

in the functional S. This expression describes the kinetic energy of a slowly oscillating field. Writing (19.6) in terms of the variables φ, π:

$$-\int d\tau \, d^3x \left(\frac{1}{2m}\rho_0(\nabla\varphi)^2 + \frac{\pi(\nabla\varphi)^2}{2m} + \frac{(\nabla\pi)^2}{8m(\rho_0 + \pi)} \right), \qquad (19.7)$$

we come to the conclusion that the first term in (19.7) contributes to the quadratic form (19.5). In the third component, π from the expression $\rho_0 + \pi$ can be neglected, which means that the third term itself represents a correction to the quadratic form leading to the deviation of the energy spectrum from linearity. Adding the contribution of the second and third terms of (19.7) to the functional (19.5) we get

$$\int d\tau \, d^3x \left\{ ip_{\lambda\rho_0}\pi\,\partial_\tau\varphi - \frac{p_\lambda}{2m}(\nabla\varphi)^2 - \frac{p_{\lambda\lambda}}{2}(\partial_\tau\varphi)^2 + \right.$$
$$\left. + \frac{p_{\rho_0\rho_0}}{2}\pi^2 - \frac{(\nabla\pi)^2}{8m\rho_0} - \frac{\pi(\nabla\varphi)^2}{2m} \right\}. \qquad (19.8)$$

The expression for $p(\lambda, \rho_0)$ at $T = 0$ can be in the first approximation calculated without difficulties employing the scheme that was used in Section 17 for the temperature domain $T \sim \rho^{2/3}$. At $T = 0$, the particle density above the condensate is less in order of magnitude than the condensate density. Diagrams (17.4) contain one or two loops of normal lines that can be passed through in the direction of arrows (the sequences b, c in (17.4) are less in the order of magnitude than the sequence a). For the same reason one can neglect the contribution of the pressure of an ideal Bose gas p_0. The first approximation for the pressure p at $T = 0$ is finally obtained from formula (17.13) by inserting $T = 0$ and neglecting ρ_1 :

$$p = \lambda_{\rho_0} - (t_0/2)\rho_0^2. \tag{19.9}$$

From (19.9) it follows that

$$p_{\lambda_{\rho_0}} = 1; \qquad p_\lambda = \rho = \rho_0; \qquad p_{\lambda\lambda} = 0; \qquad p_{\rho_0\rho_0} = -t_0, \tag{19.10}$$

the functional (19.8) takes the form

$$\int d\tau\, d^3x \left(i\pi\, \partial_\tau \varphi - \frac{(\rho_0 + \pi)}{2m}(\nabla\varphi)^2 - \frac{t_0}{2}\pi^2 - \frac{(\nabla\varphi)^2}{8m\rho_0} \right) \tag{19.11}$$

and can be interpreted as the action corresponding to the Hamiltonian

$$\int d^3x \left(\frac{m}{2}\rho v^2 + \frac{t_0}{2}(\rho - \rho_0)^2 + \frac{(\nabla\rho)^2}{8m\rho_0} \right), \tag{19.12}$$

where $v = (1/m)\nabla\varphi$. This Hamiltonian represents a particular realization of the Landau hydrodynamical Hamiltonian. It is easy to construct a hydrodynamical Hamiltonian of a Bose gas in higher density approximations, too.

The hydrodynamical action (19.8) enables one to construct a perturbation theory without divergencies at small energies and momenta. We take the quadratic form in φ and π as a nonperturbed action while the trilinear interaction proportional to $\pi(\nabla\varphi)^2$ is regarded as a perturbation. Let us point out that the perturbation has its origin in the term $(\nabla\bar\psi_0, \nabla\psi_0)$, which is completely included in the nonperturbed action of the perturbation theory described in Section 16.

Passing now from the functions φ and π to their Fourier coefficients according to formulae similar to (6.3), we can write the action in the form

$$\frac{1}{2}\sum_p \left\{ -\left(\frac{p_\lambda}{m}k^2 + p_{\lambda\lambda}\omega^2 \right)\varphi(p)\varphi(-p) - \right.$$

$$- 2p_{\lambda\rho_0}\omega\varphi(p)\pi(-p) + \left(p_{\rho_0\rho_0} - \frac{k^2}{4m\rho_0} \right)\pi(p)\pi(-p) + \tag{19.13}$$

$$\left. + \frac{1}{\sqrt{\beta V}} \sum_{p_1 + p_2 - p_3 = 0} \frac{(\mathbf{k}_1 \mathbf{k}_2)}{2m}\varphi(p_1)\varphi(p_2)\pi(p_3) \right\},$$

where p is the four vector $(\mathbf{k}, \omega)$. Notice the analogy of this formula with formula (2.6) of Reference [74].

In the p-representation the elements of the diagrammatic technique have the form

$$\underset{p\qquad\qquad -p}{\rule{3cm}{0.4pt}} = \langle \varphi(p)\varphi(-p)\rangle_0 \,;$$

$$\underset{p\qquad\qquad -p}{\rule{3cm}{1pt}} = \langle \pi(p)\pi(-p)\rangle_0 \,;$$

$$\underset{p\qquad\qquad -p}{\rule{3cm}{}} = \langle \varphi(p)\pi(-p)\rangle_0 \,;$$

$$\underset{p\qquad\qquad -p}{\rule{3cm}{}} = \langle \pi(p)\varphi(-p)\rangle_0 \,;$$

$$(19.14)$$

$$p_3 \mathrel{<}\begin{array}{c} p_1 \\[4pt] p_2 \end{array} = \frac{(\mathbf{k}_1\mathbf{k}_2)}{m}$$

In these diagrams, the field φ is denoted by a single line and the field π by a double line.

The nonperturbed functions $\langle\varphi(p)\varphi(-p)\rangle_0$, $\langle\varphi(p)\pi(-p)\rangle_0$, $\langle\pi(p)\varphi(-p)\rangle_0$, $\langle\pi(p),\pi(-p)\rangle_0$ form a matrix $G_0(p)$ inverse to the matrix

$$G_0^{-1} = \begin{pmatrix} \dfrac{p_\lambda}{m}k^2 + p_{\lambda\lambda}\omega^2, & -p_{\lambda\rho_0}\omega \\[12pt] p_{\lambda\rho_0}\omega, & -p_{\rho_0\rho_0} + \dfrac{k^2}{4m\rho_0} \end{pmatrix}. \qquad (19.15)$$

Let us write the nonperturbed functions in the approximation where the derivatives p_λ, $p_{\lambda\lambda}$, $p_{\lambda\rho_0}$ and $p_{\rho_0\rho_0}$ are replaced by their first approximations according to (19.10):

$$\langle\varphi(p)\varphi(-p)\rangle_0 = \frac{t_0 + (k^2)/(4m\rho_0)}{\omega^2 + \varepsilon^2(\mathbf{k})} \qquad (19.16)$$

$$\langle\pi(p)\pi(-p)\rangle_0 = \frac{(\rho_0/m)k^2}{\omega^2 + \varepsilon^2(\mathbf{k})};$$

$$\langle\varphi(p)\pi(-p)\rangle_0 = -\langle\pi(p)\varphi(-p)\rangle = \frac{\omega}{\omega^2 + \varepsilon^2(\mathbf{k})}, \qquad (19.17)$$

where

$$\varepsilon^2(\mathbf{k}) = \left(\frac{k^2}{2m}\right)^2 + \frac{t_0 \rho_0}{m} k^2$$

is the square of the Bogoliubov spectrum.

The perturbation theory constructed above has no divergencies at small momenta. Indeed, the singularity proportional to k^{-2} that appears in the function $\langle \varphi(p)\varphi(-p) \rangle$ at $\omega = 0$ is compensated by the momenta of each of the two third-order vertices connected by the φ-line.

The low-frequency energy spectrum is after the analytic continuation $i\omega \to E$ given by the equation

$$\det G^{-1}(p) = \det (G_0^{-1}(p) - \Sigma(p)) = 0, \tag{19.18}$$

where $\Sigma(p)$ is the self-energy matrix. If the matrix Σ and the term $k^2/4m\rho_0$ in G_0^{-1} are neglected, Equation (19.18) takes the form

$$\left(\frac{p_\lambda}{m}k^2 - p_{\lambda\lambda}E^2\right)(-p_{\rho_0\rho_0}) - p_{\lambda\rho_0}^2 E^2 = 0. \tag{19.19}$$

Its solution

$$E^2 = k^2 \frac{p_\lambda}{m}\left(p_{\lambda\lambda} - \frac{p_{\lambda\rho_0}^2}{p_{\rho_0\rho_0}}\right)^{-1} \equiv \frac{k^2}{m}\frac{dp}{d\rho} \tag{19.20}$$

describes the acoustic excitation and the sound velocity is given by the formula $c^2 = (1/m)(dp/d\rho)$ (for $\rho_{\rho_0} = 0$, the expression $p_{\lambda\lambda} - (p_{\lambda\rho_0}^2/p_{\rho_0\rho_0})$ can be put equal to the total derivative $d^2p/d\lambda^2$ of the function $p(\lambda, \rho_0(\lambda))$ with respect to the variable λ).

The term $k^2/4m\rho_0$ in the G_0^{-1} matrix (19.15) gives the deviation of the spectrum from linearity. When the three-dimensional interaction is investigated, the imaginary part, and for temperatures different from zero also the sound branch, appear in the spectrum.

Let us now study the other diagrams of the matrix

$$\tag{19.21}$$

The contribution to the imaginary part of the spectrum is given by the set of all diagrams (19.21) except a. Finally we obtain

$$\Delta(\mathbf{k}) = \frac{9(1 - \exp(-\beta \varepsilon_1))}{128\pi^2 \rho mc^2} \int \varepsilon_1 \varepsilon_2 \varepsilon_3 n_2 n_3 [2\exp(\beta \varepsilon_3) \delta(1+2-3) + \\ + \exp(\beta \varepsilon_1) \delta(1-2-3)] d^3 k_2 \, d^3 k_3, \tag{19.22}$$

where

$$\delta(1 \pm 2 - 3) \equiv \delta(\mathbf{k}_1 \pm \mathbf{k}_2 - \mathbf{k}_3) \delta(\varepsilon_1 \pm \varepsilon_2 - \varepsilon_3);$$

$$n_i = \exp(\beta \varepsilon_i - 1)^{-1}; \qquad \varepsilon_i^2 = \varepsilon^2(\mathbf{k}_i) = \left(\frac{k_i^2}{2m}\right)^2 + \frac{t_0 \rho_0}{m} k_i^2; \quad \mathbf{k}_1 = \mathbf{k}, \tag{19.23}$$

and c is the sound velocity (19.20).

If $T = 0$, then (19.22) yields

$$\Delta(\mathbf{k}) = \frac{9}{128\pi^2 \rho mc^2} \int \varepsilon_1 \varepsilon_2 \varepsilon_2 \delta(1-2-3) d^3 k_2 \, d^3 k_3 = \frac{3k^5}{640\pi m\rho}. \tag{19.24}$$

for small momenta, i.e., we have rederived the well-known result by Beliaev [77]. However, in the perturbation theory described above, the number of second-order diagrams is less than in the Beliaev formalism and the diagrams themselves do not diverge at small momenta.

At $T \neq 0$ and small $\mathbf{k}$ the expression (19.22)

$$\Delta(\mathbf{k}) = \frac{3\pi^3 T^4 k}{40 \rho mc^4} \tag{19.25}$$

is linear in k. The functional dependence in (19.25) ceases to be linear in the kinetic domain $kl \lesssim 1, E\tau \lesssim 1$, where l denotes the mean free path and τ is the relaxation time. The correct result can be obtained only if one performs the summation of diagrams leading to the solution of equations of a kinetic type.

Let us investigate the sum of diagrams given by the diagrammatic equations

$$\tag{19.26}$$

The first of the equations expresses the self-energy Σ in terms of Σ_0, i.e., the contribution of diagrams in which one cannot separate the incoming and outgoing parts by cutting any two internal lines, and the full vertex D. The second equation (19.26) is an approximation to the irreducible four point function K. The internal lines of diagrams (19.26) correspond to the full Green's functions $\langle \varphi\varphi \rangle$, $\langle \pi\pi \rangle$, $\langle \pi\rho \rangle$. To give the schematical equations (19.26) any concrete meaning, it is necessary to sum over all possible internal lines of the diagrams. The vertex parts can be classified according to the lines connected to them. We have altogether only six different vertex parts:

$$D_{\varphi\varphi,\varphi}; \quad D_{\varphi\pi,\varphi}; \quad D_{\pi\pi,\varphi}; \quad D_{\varphi\varphi,\pi}; \quad D_{\varphi\pi,\pi}; \quad D_{\pi\pi,\pi}. \tag{19.27}$$

Here, the first two indices show which lines are connected to the internal points of the diagram, while the third determines the type of line connected to the external point.

As mentioned above, Equations (19.26) can be transformed into kinetic ones. We shall now concentrate on the basic points of the derivation of kinetic equations.

The vertex parts (19.27) depend on the two frequencies ω and ω_1. After the analytic continuation $i\omega_1 \to z$, they are defined in the complex z-plane with two cuts $\text{Im}\, z = 0$ and $\text{Im}\, z = \omega$. Equations (19.26) can be transformed into equations for boundary values of analytic functions $D(\omega, z)$ alongside the cuts. Let us denote D^e as the boundary values of D on the outer cuts $\text{Im}\, z = -0, \omega + 0$ by D^e, and the boundary values on the inner cuts $\text{Im}\, z = +0, \omega - 0$ by D^i. In the first approximation the functions D^e turn out to be equal to the corresponding irreducible parts

$$D^e \approx d. \tag{19.28}$$

On the other hand, after the analytic continuation $i\omega \to E$ into the domain $E \lesssim \tau^{-1}$ we come to the conclusion that the system of equations for D^i – the boundary values of functions D on the inner cuts – is nontrivial. The leading contribution to the integral operators acting on the functions D^i is for $E \lesssim \tau^{-1}$ determined by some neighbourhoods of nearby located singularities of the Green's functions. These singularities have the character of poles, and the Green's functions themselves are in the neighbourhoods of the singularities proportional to the function

$$(x - \varepsilon(\mathbf{k}_1) \pm i\Delta(\mathbf{k}_1))^{-1}, \tag{19.29}$$

where x is the energy variable and $\varepsilon(\mathbf{k}_1)$ is the energy spectrum with the imaginary part $\Delta(\mathbf{k}_1)$. In the integrals over the energy variable x we need not integrate any functions except the singular ones (19.29). It is sufficient to put the values of the nonsingular functions in the point $x = \varepsilon(\mathbf{k}_1)$ in front of the integration symbol. The remaining product of the singular functions can be integrated according to the formula

$$\int (x - \varepsilon(\mathbf{k}_1) + i\Delta(\mathbf{k}_1))^{-1}(x + E - \varepsilon(\mathbf{k}_1 + \mathbf{k}) - $$
$$- i\Delta(\mathbf{k}_1 + \mathbf{k}))^{-1} dx \approx 2\pi i Z^{-1} \tag{19.30}$$

Here

$$Z = E - \varepsilon(\mathbf{k}_1 + \mathbf{k}) + \varepsilon(\mathbf{k}_1) + 2i(\Delta(\mathbf{k}_1) + \Delta(\mathbf{k}_1 + \mathbf{k})) \approx$$
$$\approx E - 2\varepsilon'_1(\mathbf{k}_1\mathbf{k}) + 2i\Delta(\mathbf{k}_1), \tag{19.31}$$

where

$$\varepsilon'_1 = \frac{\partial \varepsilon(\mathbf{k}_1)}{\partial k_1^2}.$$

One can show that after the integration over x described above the vertex functions D^i enter the equations in the combinations

$$h(\mathbf{k}_1, \mathbf{k}, E) = \frac{1}{2Z}\left\{ D^i_{\varphi\pi,\varphi} - D^i_{\pi\varphi,\varphi} + i\left(t_0 + \frac{k_1^2}{2m\rho_0} \right)\varepsilon_1^{-1} D^i_{\varphi\varphi,\varphi} + \right.$$
$$\left. + i\frac{\rho_0}{m}k_1^2\varepsilon_1^{-1} D^i_{\pi\pi,\varphi} \right\}\Bigg|_{\substack{i\omega_1 \to \varepsilon_1 - i0 \\ i\omega \to E + i0}} ; \tag{19.32}$$

$$g(\mathbf{k}_1, \mathbf{k}, E) = -\frac{\rho_0}{2Z}\left\{ i(D^i_{\pi\varphi,\pi} - D^i_{\varphi\pi,\pi}) + \right.$$
$$\left. + \left(t_0 + \frac{k_1^2}{2m\rho_0} \right)\varepsilon_1^{-1} D^i_{\varphi\varphi,\pi} + \frac{\rho_0 k_1^2}{m}\varepsilon_1^{-1} D^i_{\pi\pi,\pi}) \right\}\Bigg|_{\substack{i\omega_1 \to \varepsilon_1 - i0 \\ i\omega \to E + i0}} ,$$

where Z is defined above by relation (19.31). In particular, the combinations h and g satisfy the equations

$$(E - 2\varepsilon'_1(\mathbf{k}\cdot\mathbf{k}_1))h + iI(h) = \frac{1}{m}(\mathbf{k}\cdot\mathbf{k}_1); \tag{19.33}$$

$$(E - 2\varepsilon'_1(\mathbf{k}\cdot\mathbf{k}_1))g + iI(g) = \varepsilon(\mathbf{k}_1).$$

In (19.33) the expression

$$I(h) = \frac{9(1 - \exp(-\beta\varepsilon_1))}{64\pi^2 \rho m c^2} \int \varepsilon_1 \varepsilon_2 \varepsilon_3 n_2 n_3 \times$$
$$\times [2 \exp(\beta\varepsilon_3)\delta(1 + 2 - 3)(h_1 + h_2 - h_3) + \quad (19.34)$$
$$+ \exp(\beta\varepsilon_1)\delta(1 - 2 - 3)(h_1 - h_2 - h_3)] d^3k_2 d^3k_3$$

is denoted by $I(h)$. Here $h_i = h((k_i k), E)$. The remaining cancellations are explained above.

As far as the form is concerned, Equations (19.33) represent the linearized kinetic equations, in which the expressions $I(h), I(g)$ have the meaning of the collision integrals. Notice that the inhomogeneous terms $(1/m)(k \cdot k_1)$, $\varepsilon(k_1)$ in Equations (19.33) are the integrals of motion which cancel the collision integrals.

In the hydrodynamical domain ($E\tau, kl \ll 1$), the Chapman–Enscope–Hilbert method [99] can be applied to Equations (19.33). The basic idea of this method is to look for the solution of kinetic equations (19.33) in the form

$$h = a\varepsilon(k_1) + (b \cdot k_1) + \delta h;$$
$$g = a_1\varepsilon(k_1) + (b_1 \cdot k_1) + \delta_g, \quad (19.35)$$

i.e., in the form of the sum of linear combinations of the integrals of motion $\varepsilon(k_1), k_1$, and the additional terms $\delta h, \delta g$, which are small with respect to the integrals of motion. The leading terms do not contribute to the collision integrals, while out of the integrals we may neglect the contributions of $\delta h, \delta g$.

We can now apply the condition of orthogonality of the collision integral I and the integrals of motion $\varepsilon(k_1), k_1$ weighted by

$$\frac{\exp[\beta\varepsilon(k_1)]}{(\exp[\beta\varepsilon(k_1)] - 1)^2}. \quad (19.36)$$

This condition leads to a system of linear equations for the coefficients at the integrals of motion in (19.35) a, a_1, b, b_1. Solving this system, we obtain the following formulae for the first approximation of the Chapman–Enscope–Hilbert method (this approximation is usually called 'acoustic')

$$h = \frac{\frac{2}{3}\langle\!\langle \varepsilon\varepsilon' k^2 \rangle\!\rangle / \langle\!\langle \varepsilon^2 \rangle\!\rangle k^2 \varepsilon(k_1) + E(k \cdot k_1)}{m(E^2 - u^2k^2)};$$

$$g = \frac{E\varepsilon(k_1) - 2\langle\!\langle \varepsilon\varepsilon' k^2 \rangle\!\rangle / \langle\!\langle k^2 \rangle\!\rangle (k \cdot k_1)}{E^2 - u^2k^2}, \quad (19.37)$$

where

$$\langle\!\langle f \rangle\!\rangle = (2\pi)^{-3} \int f(\mathbf{k}_1) \exp\left[\beta\varepsilon(\mathbf{k}_1)\right] (\exp\left[\beta\varepsilon(\mathbf{k}_1)\right] - 1)^2 \, d\mathbf{k}_1 ;$$

$$u^2 = \frac{4}{3} \frac{\langle\!\langle \varepsilon\varepsilon' k^2 \rangle\!\rangle^2}{\langle\!\langle k^2 \rangle\!\rangle \langle\!\langle \varepsilon^2 \rangle\!\rangle}. \tag{19.38}$$

The expressions of self-energy parts $\Sigma_{\varphi\varphi}, \Sigma_{\pi\varphi}, \Sigma_{\pi\pi}$ in terms of the functions h and g have the form

$$\Sigma_{\varphi\varphi} = \frac{11\beta^2 \langle\!\langle k^2 \rangle\!\rangle k^2}{24m^2} - \beta E \left\langle\!\!\left\langle \frac{(\mathbf{k}\cdot\mathbf{k}_1)}{m} h \right\rangle\!\!\right\rangle ;$$

$$\Sigma_{\varphi\pi} = -\Sigma_{\pi\varphi} = \frac{i\beta E}{2\rho_0} \langle\!\langle \varepsilon_1 h \rangle\!\rangle = \frac{i\beta E}{2\rho_0} \left\langle\!\!\left\langle \frac{(\mathbf{k}\cdot\mathbf{k}_1)}{m} g \right\rangle\!\!\right\rangle ; \tag{19.39}$$

$$\Sigma_{\pi\pi} = \frac{1}{4\rho_0^2} \left(\frac{5}{4}\beta \langle\!\langle \varepsilon^2 \rangle\!\rangle - \beta E \langle\!\langle \varepsilon_1 g \rangle\!\rangle \right).$$

Let us put here the values (19.37) instead of the symbols h and g. The resulting expressions can be used for calculation of the spectrum according to formula (19.18). Equation (19.18) then takes the form

$$\left(E^2 - c_0^2 k^2 + \frac{37}{16}\frac{\rho_n}{\rho} c_0^2 k^2 \right)(E^2 - u^2 k^2) - \frac{11}{4}\frac{\rho_n}{\rho} E^2 c_0^2 k^2 = 0, \tag{19.40}$$

where

$$c_0^2 = \frac{1}{m}\left(\frac{dp}{d\rho} \right)_{T=o} ; \qquad \rho_n = \frac{\beta \langle\!\langle k^2 \rangle\!\rangle}{3m} = \frac{2\pi^2 T^4}{45mc^5}. \tag{19.41}$$

Equation (19.40) is identical with the well-known Landau equation for sound velocities (Equation (20.13) of Reference [100]). It can be applied to the model of Bose gas and its solutions have the form

$$E_1^2(\mathbf{k}) = c_1^2 k^2 = c_0^2\left(1 + \frac{29}{16}\frac{\rho_n}{\rho} \right)k^2 ;$$

$$E_2^2(\mathbf{k}) = c_2^2 k^2 = u^2 k^2 = \frac{4}{3}\frac{\langle\!\langle \varepsilon\varepsilon' k^2 \rangle\!\rangle^2}{\langle\!\langle k^2 \rangle\!\rangle \langle\!\langle \varepsilon^2 \rangle\!\rangle} k^2. \tag{19.42}$$

Formulae (19.42) are valid at low temperatures (in the phonon domain of temperatures). They express the temperature corrections to the sound velocities, to which all second-order diagrams (19.21) contribute.

In the second (viscosity) approximation of the Chapman–Enscope–Hilbert method the additional terms of formula (19.35) $\delta h, \delta g$ can be formally obtained by inverting the operator of the collision integral. The coefficients $a, a_1, \mathbf{b}, \mathbf{b}_1$ can again be obtained from the orthogonality conditions just as in the first approximation but with higher precision. Finally, the branches of spectrum linear in k (19.42) acquire imaginary parts proportional to k^2. The coefficients of k^2 terms can be expressed in terms of kinetic coefficients of first viscosity, second viscosity, and thermoconductivity. At low temperatures the most important of them is the first viscosity coefficient η.

In the viscosity approximation the formulae for the sound branches have the form

$$E_1(\mathbf{k}) = c_1 k - i\frac{27}{8}\frac{\eta}{m\rho}k^2;$$

$$E_2(\mathbf{k}) = c_2 k - i\frac{2}{3}\frac{\eta}{m\rho_n}k^2$$

(19.43)

Here, only the contribution of the first viscosity coefficient defined by the equation

$$\eta = \frac{2\beta}{15}\langle\!\langle \varepsilon' k^4 f \rangle\!\rangle,$$

(19.44)

has been taken into account. The function $f(k^2)$ in (19.44) is the solution of equation

$$I\left(k_i k_j - \frac{k^2}{3}\delta_{ij}\right)f = 2\varepsilon_1'\left(k_{1i}k_{1j} - \frac{k_1^2}{3}\delta_{ij}\right).$$

(19.45)

In the limit $T \to 0$, the first viscosity coefficient of a Bose gas can be expressed explicitly in terms of quadratures

$$\eta = \frac{(4\pi)^7\rho}{3^9 5^3 J}\left(\frac{T}{mc^2}\right)^{-5},$$

(19.46)

where J is the integral

$$J = \int_0^\infty dx \int_0^\infty dy\, xy(x+y)^4[(\exp(x)-1)(\exp(y)-1) \times$$
$$\times (1 - \exp(-x-y))]^{-1}.$$

(19.47)

The temperature dependence of the first viscosity coefficient proportional to T^{-5} turns out to be the same as in the Landau–Khalatnikov theory [100], where the main contribution to viscosity is given by four-phonon processes.

The contribution of the kinetic coefficients of the second viscosity and thermoconductivity to the spectrum damping, contrary to the contribution of the first viscosity, turns out to be a quantity proportional to T^8. This is the reason why it need not be taken into account in the limit $T \to 0$.

20 SUPERFLUIDITY OF TWO-DIMENSIONAL AND ONE-DIMENSIONAL BOSE SYSTEMS

The superfluidity of a three-dimensional Bose system is connected with the macroscopic filling of the lowest quantum level – the condensate. In the two- and one-dimensional systems at $T \neq 0$ the condensate does not appear. This statement follows from the Bogoliubov k^{-2}-theorem [81] and has been proved in Section 18.

The fact that the condensate is not present, however, does not rule out the possibility of the existence of superfluidity.

Conjectures suggesting such a possibility can be found in several papers. They are based either on variational calculation of the ground-state wave function (the method of collective variables [89], [80]), or on the method of coherent states [91].

Berezinski [92] developed a method which uses the transformation from the continuous system to the discrete set of planar rotators.

Let us apply to two- and one-dimensional systems our method of successive functional integration over rapidly and slowly oscillating fields. The results confirm that there is a superfluidity in two-dimensional Bose systems, and indicate the possibility that superfluidity exists, even in one-dimensional systems.

The basic physical idea is that particles with small momenta behave like a 'bare condensate' that is responsible for the creation of a superfluid density ρ_s with an order of magnitude equal to that of the full density of the system ρ. The method of functional integration enables us to turn this idea into reality and to construct an hydrodynamical action characteristic for superfluid systems.

Let us consider a two-dimensional Bose system, in which the existence of condensate at zero temperature is possible. The condensate density is determined by the equation $\partial \tilde{S}/\partial \rho_0 = 0$, where $\tilde{S}$ is the hydrodynamical action of the system, which can be obtained by integration over rapidly oscillating fields.

We shall now investigate the expansion of the functional $\tilde{S}$ into the functional series of the variables $\varphi(\mathbf{x}, \tau)$, $\pi(\mathbf{x}, \tau) = \rho(\mathbf{x}, \tau) - \rho_0(k_0)$. In the limit $T \to 0$, k_0 fixed, one can suppose that the coefficient functions are equal to their values at $T = 0$.

The derivation of the hydrodynamical action carried out in the previous paragraph can also be extended to a two-dimensional system. The leading contribution to $\tilde{S}$, which depends on φ and π, is given by the quadratic form

$$\int d\tau\, d^2x \left(-\frac{p_\lambda}{2m}(\nabla\varphi)^2 - \frac{p_{\lambda\lambda}}{2}(\partial_\tau\varphi)^2 \right.$$
$$\left. + ip_{\lambda\rho_0}\pi\, \partial_\tau\varphi + \tfrac{1}{2}p_{\rho_0\rho_0}\pi^2 \right), \tag{20.1}$$

where $p_\lambda, p_{\lambda\lambda}, p_{\lambda\rho_0}$ and $p_{\rho_0\rho_0}$ are the partial derivatives of the function $p(\lambda, \rho_0)$ and $p_{\rho_0} = 0$. Expression (20.1) describes the noninteracting phonon excitations, the energy of which depends linearly on the momentum. The next-to-leading terms in the expansion describe the deviation of the spectrum from linearity and the interaction of excitations.

In the case of a low density Bose gas, the leading terms of the expansion $\tilde{S}$ which do not appear in the quadratic form, can be incorporated in a functional corresponding to the kinetic energy of the slowly oscillating field:

$$-\int d\tau\, d^2x \frac{|\nabla\psi_0|^2}{2m}$$
$$= -\int \frac{d\tau\, d^3x}{2m}\left(\rho_0(\nabla\varphi)^2 + \pi(\nabla\varphi)^2 + \frac{(\nabla\pi)^2}{4(\rho_0 + \pi)} \right). \tag{20.2}$$

The first term in the integrand on the right-hand side of (20.2) contributes to the quadratic form (20.1). The second and third terms describe the addition to the quadratic form; in the third term, the contribution of π is negligible with respect to ρ_0.

The quadratic form (20.1) together with the additional terms mentioned above determines the functional

$$\int d\tau\, d^2x \left(-\frac{p_\lambda}{2m}(\nabla\varphi)^2 - \frac{p_{\lambda\lambda}}{2}(\partial_\tau\varphi)^2 + ip_{\lambda\rho_0}\pi\partial_\tau\varphi + \right.$$
$$\left. +\frac{1}{2}p_{\rho_0\rho_0}\pi^2 - \frac{\pi(\nabla\varphi)^2}{2m} - \frac{(\nabla\pi)^2}{8m\rho_0} \right),$$

(20.3)

which has the meaning of 'hydrodynamical action'.

Assuming that the integral of the expression $-\pi(\nabla\varphi)^2/2m$ is a perturbation with respect to the quadratic forms of the variables φ and π, one can develop a perturbation theory which, as well as the theory of a three-dimensional system constructed in Section 19, has no divergencies at small momenta. The application of this perturbation theory to the calculation of the low-energy spectrum yields for $T \neq 0$ a spectrum with two sound branches, which is typical of a superfluid Bose system.

Let us now find the asymptotic behaviour of the one-particle Green's function

$$-\langle\psi_0(\mathbf{x},\tau)\bar\psi_0(\mathbf{y},\tau_1)\rangle$$
$$= -\langle(\rho(\mathbf{x},\tau)\rho(\mathbf{y},\tau_1))^{1/2}\exp i(\varphi(\mathbf{x},\tau) - \varphi(\mathbf{y},\tau_1))\rangle,$$

(20.4)

where $\langle\cdots\rangle$ denotes the mean value over slowly oscillating fields weighted by $\exp S$ in the limit $r = |\mathbf{x} - \mathbf{y}| \to \infty$.

Suppose that the momentum k_0 which distinguishes between 'large' and 'small' momenta is sufficiently small, so that we could neglect in the first approximation calculation of the mean value (20.4) the fact that the variables $\rho(\mathbf{x},\tau), \rho(\mathbf{y},\tau_1)$ are different from the zero-temperature condensate density ρ_0, and put $\bar S$ equal to the quadratic form (20.1). In this case the integral becomes Gaussian and for the Green's function (20.4) we obtain the expression

$$-\rho_0\exp\left\{-\tfrac{1}{2}\langle(\varphi(\mathbf{x},\tau) - \varphi(\mathbf{y},\tau_1))^2\rangle\right\}.$$

(20.5)

where

$$\tfrac{1}{2}\langle(\varphi(\mathbf{x},\tau) - \varphi(\mathbf{y},\tau_1))^2\rangle$$
$$= \frac{1}{2\beta V}\sum_{\omega,k<k_0}\frac{m}{\rho(k^2 + \omega^2/c^2)} \times$$
$$\times |\exp[i(\mathbf{k}\cdot\mathbf{x} - \omega\tau)] - \exp[i(\mathbf{k}\cdot\mathbf{y} - \omega\tau_1)]|^2.$$

(20.6)

The formula

$$\langle\varphi(\mathbf{k},\omega)\varphi(-\mathbf{k},-\omega)\rangle = \frac{m}{\rho(k^2 + (\omega^2/c^2))}$$

(20.7)

has been used for the calculation of the mean value $\langle \varphi\varphi \rangle$ in the $(\mathbf{k}, \omega)$-representation (see the previous section). Here, c^2 is the sound velocity squared:

$$c^2 = \frac{p_\lambda}{m}\left(p_{\lambda\lambda} - \frac{p_{\lambda\rho 0}^2}{p_{\rho 0\rho 0}}\right)^{-1} = \frac{1}{m}\frac{dp}{d\rho}. \tag{20.8}$$

The large distances $(r \to \infty)$ asymptotic behaviour of expression (20.6) is

$$\frac{mc^2}{2\rho(2\pi)^2}\int_{k<k_0} d^2k \frac{1 - \cos(\mathbf{k}, \mathbf{x} - \mathbf{y})}{ck}\coth\frac{\beta ck}{2}$$

$$= \frac{m}{2\pi\beta\rho}\ln\frac{r}{\beta c} + \text{const.} \tag{20.9}$$

From this expression it follows that for $r \to \infty$ the Green's function vanishes like $r^{-\alpha}$ with the exponent

$$\alpha = \frac{m}{2\pi\beta\rho}. \tag{20.10}$$

Now we shall show that the power-like asymptotic behaviour of the Green's function remains valid even if the temperature increases. The exponent then has the form

$$\alpha = \frac{m}{2\pi\beta\rho_s}, \tag{20.11}$$

which differs from (20.10) by replacing the full density by the superfluid component density ρ_s. In the limit $T \to 0$ $\rho = \rho_s$ and (20.11) becomes equivalent to (20.10).

For $T \neq 0$, the solution of the equation $\partial\tilde{S}/\partial\rho_0 = 0$ has the meaning of the bare condensate density $\rho_0(k_0)$. As we shall see, $\rho_0(k_0) \sim k_0^\alpha$. Let us investigate the expansion of $\tilde{S}$ into the functional series of variables $\varphi(\mathbf{x}, \tau), \pi(\mathbf{x}, \tau) = \rho(\mathbf{x}, \tau) - \rho_0(k_0)$. The coefficient of $(\nabla\varphi)^2/2m$ in the quadratic form (20.1) is for $T \neq 0$ equal to

$$-(\rho_0(k_0) + \rho_1(k_0) - \rho_n(k_0)), \tag{20.12}$$

where $\rho_0(k_0)$ and $\rho_1(k_0)$ are the densities of particles with $k < k_0$ and $k > k_0$, respectively, and ρ_n is the particle number density of the normal component with $k > k_0$. The derivation of Equation (20.12) is analogous to that carried out in Section 18 for a three-dimensional system. In the limit $k_0 \to 0$ expression (20.12) can be put equal to $\rho_n - \rho = -\rho_s$, where ρ_s is the superfluid component density.

Let us return to the Green's function. The ideology applied in the passage from (20.4) to (20.5) can be used here, too. It is only necessary to replace ρ_0 by $\rho_0(k_0)$ in formula (20.5) and ρ by ρ_s in (20.6) and (20.7). The leading contribution to the sum over frequencies (20.6) is given by the term with $\omega = 0$. Formula (20.9) should be replaced by the formula

$$\frac{m}{\beta \rho_s (2\pi)^2} \int_{k < k_0} d^2 k \frac{1 - \cos(\mathbf{k}, \mathbf{x} - \mathbf{y})}{kr}$$

$$= \frac{m}{2\pi \beta \rho_s} \ln k_0 r + c_1, \tag{20.13}$$

where c_1 is a k_0-independent constant. The asymptotic behaviour of the Green's function turns out to be

$$G \approx -\rho_0(k_0)(k_0 r)^{-\alpha} a \exp(-c_1). \tag{20.14}$$

In (20.14), the Green's function is k_0-independent if one assumes that $\rho_0(k_0) \sim k_0^\alpha$. Such an assumption is self-consistent. It leads to the formula

$$G \approx -a_1 r^{-\alpha} \tag{20.15}$$

with a_1 independent of k_0. From (20.15), the asymptotic formula for the particle number $N(\mathbf{k})$ in the limit of small k follows. It has the form

$$N(\mathbf{k}) \approx a_2 k^{\alpha - 2}. \tag{20.16}$$

This formula leads to the expression for the density $\rho_0(k_0)$

$$\rho_0(k_0) = (2\pi)^{-2} \int_{k < k_0} N(\mathbf{k}) d^2 k = a_3 k_0^\alpha \tag{20.17}$$

which has been supposed to hold.

The above studies show that the one-particle Green's function (20.4) of a two-dimensional Bose system vanishes for sufficiently low temperature as $r^{-\alpha}$ when $|\mathbf{x} - \mathbf{y}| \to \infty$.

In a three-dimensional case, the Green's function converges to the condensate density ρ_0. In two-dimensional systems, one can speak about the long-range correlations – the one-particle Green's function vanishes as an inverse power of r, and not exponentially as in the case of a sufficiently high temperature. The temperature of transition from the exponential to the power-like behaviour of the Green's function is evidently the temperature of phase transition to the superfluid state.

So far the excitations of the quantum vortex type have not been taken into account. A description of vortices in the formalism of functional integral is given in the following section. In particular, it is shown that the role of vortices is not important if the exponent α (20.11) remains small. If $\alpha \ll 1$, then the Green's functions preserve their power-like behaviour even when the vortices are taken into account.

Can the superfluidity occur in one-dimensional Bose systems? Let us consider a one-dimensional gas of Bose particles interacting through a repulsive δ-function potential

$$u(x - y) = g\,\delta(x - y). \tag{20.18}$$

For such a system, the exact expression for the ground state energy is known (Lieb and Liniger [101], [102]). The thermodynamical functions found by the authors of Reference [103] have no singularities, and, consequently, in such system the phase transition does not occur for nonzero temperatures. The possibility of superfluidity in this model at $T = 0$ has been studied by Sonin [89].

We shall now show that in this system long-range correlations vanishing like a power of r are possible at $T = 0$. Particles with small momenta create the bare condensate which gives rise to a superfluid density of the same order as full density ρ. One can convince oneself of this by applying the method of derivation of the hydrodynamical action to the one-dimensional case. In this case, the equation $\partial \tilde{S}/\partial \rho_0 = 0$ determines the bare condensate density $\rho_0(k_0)$ which has the power-like asymptotic behaviour

$$\rho_0(k_0) \sim k_0^{\gamma}; \quad \gamma = \frac{mc}{2\pi\rho}. \tag{20.19}$$

The exponent is expressed in terms of the density and the sound velocity. The asymptotic behaviour of Green's function for $r \to \infty$ and low temperatures (in particular, at $T = 0$) is given by the formula

$$G \approx -\rho_0(k_0)\exp\left\{ -\frac{m}{2\rho\beta V}\sum_{\omega,\mathbf{k}<k_o}\left(k^2 + \frac{\omega^2}{c^2}\right)^{-1} \times \right.$$
$$\left. \times |\exp[i(\mathbf{k}x - \omega\tau)] - \exp[i(\mathbf{k}y - \omega\tau_1)]|^2 \right\}, \tag{20.20}$$

analogous to (20.5) and (20.6). At $T = 0$ formula (20.20) takes the form

$$G \approx -\rho_0(k_0)\exp\left\{ -\frac{mc}{4\pi\rho}\ln k_0^2(k^2 + c^2\delta^2) + \text{const} \right\}, \tag{20.21}$$

where $\delta = \tau - \tau_1$, or, equivalently,

$$G \approx -a(r^2 + c^2\delta^2)^{-\gamma/2}. \tag{20.22}$$

The self-consistence of the assumption $\rho_0(k_0) \sim k_0^\gamma$ is evident. Indeed, if $\rho_0(k_0) \sim k_0^\gamma$, then the coefficient at (20.22) does not depend on k_0 and from (20.22) we obtain that $N(k) \sim k^{\gamma-1}, \rho_0(k_0) \sim k_0^\gamma$. If $T \neq 0$, then expression (20.20) yields the exponentially vanishing Green's function

$$G \sim \exp\left(-\frac{m}{2\beta\rho}r\right). \tag{20.23}$$

in the limit $r \to \infty$.

We can see that in a one-dimensional Bose system, the longe-range correlations which vanish like a power of energy (and not exponentially) may exist only if $T = 0$. It is therefore natural to conclude that a one-dimensional system can posses the property of superfluidity only at zero temperature. This conclusion is in agreement with the results of the exact thermodynamical calculations of Reference [103] which do not indicate the violation of analyticity of thermodynamical functions for nonzero values of temperature.

Now we shall find the Green's function, the thermodynamical functions and the equation of the phase transition curve for a two-dimensional nonideal Bose gas of low density.

In the three-dimensional case, the answer can be expressed by virtue of t_0 – the zero energy and zero momentum t matrix which describes the scattering of two Bose particles in vacuum.

In the case of two dimensions the t matrix defined by the formula

$$t(\mathbf{k}_1, \mathbf{k}_2, z) + (2\pi)^{-2} \int d^2 k_3 u(\mathbf{k}_1 - \mathbf{k}_3)\left(\frac{k_3^2}{m} - z\right)^{-1} t(\mathbf{k}_3, \mathbf{k}_2, z)$$
$$= u(\mathbf{k}_1 - \mathbf{k}_2), \tag{20.24}$$

has for small $\mathbf{k}_1, \mathbf{k}_2$ and z the asymptotic form

$$t \approx \frac{4\pi}{m \ln(\varepsilon_0/-z)} \tag{20.25}$$

and vanishes in the limit $\mathbf{k}_1, \mathbf{k}_2, z \to 0$. The asymptotic formula (20.25) is valid provided $|\mathbf{k}_1|, |\mathbf{k}_2| \ll r_0^{-1}, |z| \ll m^{-1}r_0^{-2}$, where r_0 is the effective radius of the potential. The quantity ε_0 in (20.25) is of order $m^{-1}r_0^{-2}$, and, consequently, $|\ln(\varepsilon_0/-z)| \gg 1$.

So to prove the validity of (20.25), we shift the second term in the left-hand side of (20.24) to the right-hand side and then add the following expression to both sides:

$$t(\mathbf{k}_1, \mathbf{k}_2, z) + (2\pi)^{-2} \int d^2 k_3 u(\mathbf{k}_1 - \mathbf{k}_3) \left(\frac{k_3^2}{m} - z_0 \right)^{-1} t(\mathbf{k}_3, \mathbf{k}_2, z)$$

$$= u(\mathbf{k}_1 - \mathbf{k}_2) + (2\pi)^{-2} \int d^2 k_3 u(\mathbf{k}_1 - \mathbf{k}_3) \times \qquad (20.26)$$

$$\times \left[\left(\frac{k_3^2}{m} - z_0 \right)^{-1} - \left(\frac{k_3^2}{m} - z \right)^{-1} \right] t(\mathbf{k}_3, \mathbf{k}_2, z).$$

Here, z_0 is an arbitrary parameter. The operator acting on the t matrix in the left-hand side of (20.26) can be inverted. This is equivalent to replacing the potential on the right-hand side by the t matrix

$$t(\mathbf{k}_1, \mathbf{k}_2, z) = t(\mathbf{k}_1, \mathbf{k}_2, z_0) + (2\pi)^{-2} \int d^2 k_3 t(\mathbf{k}_1, \mathbf{k}_3, z_0) \times$$

$$\times \left[\left(\frac{k_3^2}{m} - z \right)^{-1} - \left(\frac{k_3^2}{m} - z \right)^{-1} \right] t(\mathbf{k}_3, \mathbf{k}_2, z). \qquad (20.27)$$

The passage from (20.24) to (20.27) represents the transformation of the t matrix equation

$$t_z + u R_z t_z = u \qquad (20.28)$$

to the Hilbert identity

$$t_z - t_{z_0} = t_{z_0}(R_{z_0} - R_z)t_z, \qquad (20.29)$$

where $R_z = (H_0 - z)^{-1}$ is the resolvent of the unperturbed Schrödinger operator (without a potential), t_z and u are the operators of the t matrix and of the potential energy.

In the asymptotic domain $|\mathbf{k}_1|, |\mathbf{k}_2| \ll r_0^{-1}, |z|, |z_0| \ll m^{-1} r_0^{-2}$ one can assume that t is momentum independent. This brings formula (20.27) to the form

$$t(z) - t(z_0) \approx (2\pi)^{-2} t(z_0) t(z) \int d^2 k_3 \times$$

$$\times \left[\left(\frac{k_3^2}{m} - z_0 \right)^{-1} - \left(\frac{k_3^2}{m} - z \right)^{-1} \right] \qquad (20.30)$$

$$= \frac{m}{4\pi} t(z_0) t(z) \ln \frac{z}{z_0}$$

from which (20.25) follows.

For the separation of the potential and the transition to the t matrix, we adopt the following convention. We start with the integration of the functional exp S over the variables $a^+(\mathbf{k}, \omega), a(\mathbf{k}, \omega)$ with momenta $k > k_0'$, where k_0' satisfies the inequalities

$$\max(|\lambda|, T) \ll \frac{(k_0')^2}{2m} \ll \varepsilon_0. \tag{20.31}$$

The result of this integration is the functional of variables $a^+(\mathbf{k}, \omega), a(\mathbf{k}, \omega)$ with $k < k_0'$, which will be denoted exp S' in the sequel. For a system of low density, the first approximation formula for S' has the form

$$S' = \sum_{\omega, \mathbf{k} < k_0'} \left(i\omega - \frac{k^2}{2m} + \lambda \right) a^+(p)a(p) -$$

$$- \frac{1}{2\beta V} \sum_{\substack{p_1 + p_2 = p_3 + p_4 \\ \mathbf{k}_1 < k_0'}} t' a^+(p_1) a^+(p_2) a(p_3) a(p_4) \tag{20.32}$$

and differs from formula (6.7) by the fact that the summations over momenta are cut at an upper limit k_0' and that the potential is replaced by the t matrix, in definition of which (20.24) the integral over k_3 is cut at the lower limit k_0'. Such a t matrix (denoted t') has the form

$$t' = t'(\omega_1 + \omega_2) = \frac{4\pi}{m \ln\left(\varepsilon_0 / [(k_0'^2/m) - i\omega_1 - i\omega_2]\right)} \tag{20.33}$$

Let us integrate the functional exp S' over the variables $a^+(p), a(p)$ with momenta k satisfying the condition $k_0 < k < k_0'$. Here, the variables $a^+(p), a(p)$ with $k < k_0$ have the meaning of 'inhomogeneous terms' with respect to integration variables a^+ and a with $k \in [k_0, k_0']$. The order of magnitude of the quantity k_0 will be estimated at the end of this section. In any case, for all values of temperature lower than the phase transition temperature we obtain $k_0 \ll (m\lambda)^{1/2}$.

When calculating the thermodynamical functions, we assume that the condensate is homogeneous in the first approximation; we can incorporate it by transformation of variables $a^+(p), a(p)$ with $k < k_0$ in S':

$$a^+(p), a(p) \to (\rho_0(k_0)\beta V)^{1/2} \delta_{p_0}. \tag{20.34}$$

The quantity $\rho_0(k_0)$ in (20.34) has the meaning of the density of particles with $k < k_0$. This density, as well as the density of particles with $k > k_0 \rho_1(k_0)$, depends on the value of the momentum k_0 which distinguishes between the 'slow' and the 'rapid' particles. The full density $\rho = \rho_0(k_0) + \rho_1(k_0)$, however, must be k_0 independent.

The first approximation to $\rho_0(k_0)$ is determined by the condition that the components that do not contain a^+ give maximal contribution to S', and after the transformation of (20.34), i.e., of the maximum of the expression

$$\beta V\left(\lambda_{\rho_0}(k_0) - \frac{t'(0)}{2}\rho_0^2(k_0) \right),$$ (20.35)

we obtain

$$\rho_0(k_0) = \frac{\lambda}{t'(0)} = \frac{m\lambda}{4\pi} \ln \frac{m\varepsilon_0}{k_0'^2}.$$ (20.36)

After the transformation (20.34) one can use the formalism of perturbation theory with normal and anomalous Green's functions analogous to that introduced in Section 16 for three-dimensional systems. The leading contribution to the self-energy parts A, B is given by the second-order vertices. We obtain

$$A \approx 2t'(0)\rho_0(k_0) = 2\lambda; \qquad B \approx t'(0)\rho_0(k_0) = \lambda.$$ (20.37)

The normal and anomalous Green's functions are in the first approximation given by the formulae

$$G = -\frac{i\omega + k^2/2m + \lambda}{\omega^2 + \varepsilon^2(\mathbf{k})}; \qquad G_1 = \frac{\lambda}{\omega^2 + \varepsilon^2(\mathbf{k})},$$ (20.38)

where

$$\varepsilon^2(\mathbf{k}) = \left(\frac{k^2}{2m} \right)^2 + \frac{\lambda}{m}k^2.$$ (20.39)

The density $\rho_1(k_0)$ can be calculated according to the formula

$$\rho_1(k_0) = -\frac{1}{\beta V} \sum_{\substack{\omega, k > k_0 \\ \varepsilon \to +0}} e^{i\omega\varepsilon}G(\mathbf{k}, \omega)$$

$$= \frac{1}{2(2\pi)^2} \int_{k > k_0} d^2k \left(\frac{(k^2/2m) + \lambda}{\varepsilon(\mathbf{k})} \operatorname{cth}\frac{\beta\varepsilon(\mathbf{k})}{2} - 1 \right). \quad (20.40)$$

We get an integral that diverges logarithmically as $k_0 \to 0$. This divergence must cancel the divergencies of the quantity $\rho_0(k_0)$ calculated in the second order of perturbation theory. $\rho_0(k_0)$ can be determined from the equation

$$\lambda = A(0) - B(0) \qquad (20.41)$$

equivalent to the equation $\partial \tilde{S}/\partial \rho_0 = 0$. This equation has been used for the determination of the second-order approximation to ρ_0 in the case of a three-dimensional Bose gas in Section 17. Let us consider the self-energy diagrams which have the form

$$(20.42)$$

To give the diagrams (20.42) any concrete meaning, one has to attach arrows to the incoming parts of the diagrams and to sum over all possible combinations of arrow directions at the ends of internal lines. Here, the lines with arrows pointing in the same directions correspond to normal Green's functions, while those with arrows pointing in the opposite directions point to anomalous ones.

Equation (20.41) can be given the form

$$\lambda = t'(0)(\rho_0(k_0) + \rho_1(k_0)) + \frac{t'(0)}{\beta V} \times$$

$$\times \sum_{\substack{\omega \\ k_0 < k < k_0'}} G_1(\mathbf{k}, \omega) - \frac{2\rho_0(k_0)(t'(0))^2}{\beta V} \times \qquad (20.43)$$

$$\times \sum_{\substack{\omega \\ k_0 < k < k_0'}} (G(\mathbf{k}, \omega)G(-\mathbf{k}, -\omega) - G_1(\mathbf{k}, \omega)G_1(-\mathbf{k}, -\omega)).$$

It leads to the expression for $\rho_0(k_0)$

$$\rho_0(k_0) = \frac{m\lambda}{4\pi}\left(\ln\frac{\varepsilon_0}{\lambda} - 2\right) - \qquad (20.44)$$

$$- \frac{1}{(2\pi)^2}\int_{k > k_0} d^2k \frac{k^2/2m + \lambda}{\varepsilon(k)}\frac{1}{\exp(\beta\varepsilon(k)) - 1}.$$

Contrary to the first approximation (20.36), expression (20.44) does not depend on k_0. Summing (20.40) and (20.44), we obtain the full density

$$\rho = \frac{m\lambda}{4\pi}\left(\ln\frac{\varepsilon_0}{\lambda} - 1\right) - \qquad (20.45)$$

$$- \frac{1}{(2\pi)^2}\int d^2k \frac{k^2}{2m\varepsilon(k)}\frac{1}{\exp(\beta\varepsilon(k)) - 1}.$$

The integral in (20.45) converges even if the cut-off momentum k_0 approaches zero. Thus, for the density we obtain an expression independent of the auxiliary quantities k_0, k_0', for which only an estimate of the order of magnitude can be given.

Let us now write down the expression analogous to (20.45) for the pressure

$$p = \frac{m\lambda^2}{8\pi}\left(\ln\frac{\varepsilon_0}{\lambda} - \frac{1}{2}\right) - \frac{1}{\beta(2\pi)^2}\int d^2k \ln\left(1 - \exp\left(-\beta\varepsilon(\mathbf{k})\right)\right), \quad (20.46)$$

and the formulae for the normal and superfluid component densities

$$\rho_n = \frac{\beta}{2m(2\pi)^2}\int d^2k\, k^2 \frac{\exp\left(\beta\varepsilon(\mathbf{k})\right)}{\left(\exp\left(\beta\varepsilon(\mathbf{k})\right) - 1\right)^2};$$

$$\rho_s = \frac{m\lambda}{4\pi}\left(\ln\frac{\varepsilon_0}{\lambda} - 1\right) - \frac{1}{(2\pi)^2}\int\frac{d^2k}{2m} \times \quad (20.47)$$

$$\times\left(\frac{k^2}{\varepsilon(\mathbf{k})}\frac{1}{\exp\left(\beta\varepsilon(\mathbf{k})\right) - 1} - \frac{\beta k^2 \exp\left(\beta\varepsilon(\mathbf{k})\right)}{\left(\exp\left(\beta\varepsilon(\mathbf{k})\right) - 1\right)^2}\right).$$

The density ρ_n can be easily calculated by evaluating the average value of momentum in the system of coordinates moving with the velocity $\mathbf{v}$. Such a calculation is completely analogous to that accomplished in Section 17 for three dimensions by means of the Green's functions obtained from (20.38) after the transformation

$$i\omega \to i\omega + (\mathbf{v}\mathbf{k}). \quad (20.47a)$$

Formulae (20.45)–(20.47) determine the thermodynamics below the phase transition. The equation of the phase transition curve can be found from the condition

$$\rho = \rho_n, \quad (20.48)$$

indicating that the superfluid density ρ_s vanishes. The solution of Equation (20.48) may exist only if $T \gg \lambda$, because if $T \lesssim \lambda$, then $\rho_n \lesssim m\lambda$, $p \sim m\lambda \ln(\varepsilon_0/\lambda)$, and $\rho \gg \rho_n$. For $T \gg \lambda$, Equation (20.48) can be written in the form

$$\frac{m\lambda}{4\pi}\ln\frac{\varepsilon_0}{\lambda} = \frac{mT}{\pi}\ln\frac{T}{\lambda}. \quad (20.49)$$

From this formula we obtain the following expression for the transition temperature T_c

$$T_c \approx \frac{\lambda}{4} \frac{\ln(\varepsilon_0/\lambda)}{\ln\ln(\varepsilon_0/\lambda)}. \tag{20.50}$$

If $T > T_c$, then the system is in the normal state and one can use the ordinary perturbation theory for calculation of the Green's functions and the thermodynamical function. The one-particle Green's function is given by the formula

$$G = \left(i\omega - \frac{k^2}{2m} + \Lambda \right)^{-1}, \tag{20.51}$$

where

$$\Lambda = \lambda - \frac{2}{\pi m \ln(\varepsilon_0/|\Lambda|)} \int \frac{d^2k}{\exp\left[\beta(k^2/2m + |\Lambda|)\right] - 1}, \tag{20.52}$$

and the additional term in (20.52) is determined by the diagram B from (20.42).

Let us now estimate the order of magnitude of the momentum k_0 for the two-dimensional Bose gas. Proceeding as in Section 18 where k_0 was estimated in the three-dimensional model, we require that for $k > k_0$ the second-order approximation to the self-energy

$$\tag{20.53}$$

is not larger than the leading term which is of order λ.

Putting for simplicity the external frequency ω of this diagram equal to zero, we consider in our estimation of the order of k_0 only the term $\omega_1 = 0$ in the sum over internal frequencies.

The expression corresponding to diagram (20.53) can be then estimated by

$$\rho_0(k_0)(t'(0))^2 T \int G(\mathbf{k}_1, 0)G(\mathbf{k} - \mathbf{k}_1, 0) \, d^2k_1. \tag{20.54}$$

Here $G(\mathbf{k}, 0) \sim mk^{-2}$ due to (20.38). Considering that the integration in (20.54) is carried out over the domain $k_1 > k_0, |k - k_1| > k_0$, we can estimate the integral by $m^2 k_0^{-2}$. As $\rho_0(k_0) \lesssim m\lambda \ln(\varepsilon_0/\lambda), t'(0) \sim (m\ln(\varepsilon_0/\lambda))^{-1}$, we arrive at the conclusion that diagram (20.53) is less in the order of magnitude than

$$\frac{mT\lambda}{k_0^2 \ln \dfrac{\varepsilon_0}{\lambda}}.$$ (20.55)

The ratio of diagram (20.53) to the first approximation function is less in order of magnitude than $mT(k_0^2 \ln (\varepsilon_0/\lambda))^{-1}$. From these relations, the lower estimate of k_0 follows:

$$\frac{k_0^2}{m} \gg \frac{T}{\ln (\varepsilon_0/\lambda)}.$$ (20.56)

To estimate k_0 from above, one has to require that the 'ordinary' perturbation theory with normal and anomalous Green's functions is still applicable for $k^2/m \sim \lambda$. Finally, we obtain for k_0 the restricting conditions

$$\frac{T}{\ln (\varepsilon_0/\lambda)} \ll \frac{k_0^2}{m} \ll \lambda.$$ (20.57)

The inequality $T \ll \lambda \ln (\varepsilon_0/\lambda)$ following from these conditions holds for all temperatures lower than the phase transition temperature T_c, as, according to (20.50), $T_c \ll \lambda \ln (\varepsilon_0/\lambda)$.

The approach to the calculation of Green's functions and thermodynamical functions of a two-dimensional Bose gas developed above can be applied provided the quantity $|1 - T/T_c|$ is sufficiently large. For temperatures $T < T_c$, the condition $\alpha \ll 1$ indicating that the probability of creation of quantum vortices is small must hold in order to ensure the applicability of the perturbation theory.

21. Quantum Vortices in a Bose Gas

In addition to the acoustic excitations described above, specific excitations called 'quantum vortices' may exist in a superfluid Bose system.

Attempts to incorporate the vortex excitations have already been made in the original version of Landau's theory [72]. The existence of a periodical lattice of quantum vortices in a super-conductor located in a magnetic field was conjectured by Abrikosov in 1952 [104]. The idea of quantum vortices occurring in rotating helium below the λ-point was introduced independently by Onsager [139] and Feynman [140].

In this section, the description of quantum vortices in the formalism of continual integral will be presented. We will concentrate our attention mainly upon the two-dimensional Bose systems and show that the set of phonons and vortices occurring in such systems at low temperatures is equivalent to two-dimensional relativistic electrodynamics, in which the photons correspond to phonons and the vortices play the role of charged particles. At sufficiently low temperatures the vortices may exist only in the form of coupled pairs with opposite signs. It will be shown that the power behaviour of one-particle Green's functions at large distances remains untouched, even if the vortex pairs are taken into account.

Afterwards we shall study the influence of quantum vortices on the phase transition between the superfluid and the normal state. In a two-dimensional system, such a phase transition is connected with a dissociation of the coupled vortex pairs. Similar analysis of the three-dimensional system leads to the conclusion that in that case long vortex lines are created during phase transition.

Our approach is again based on the idea of successive integration over rapidly and slowly oscillating fields with different perturbation schemes employed at these two stages.

After integration over the rapidly oscillating fields we obtain the functional

$$\int \exp S \, d\bar\psi_1 \, d\psi_1 = \exp \tilde S[\bar\psi_0, \psi_0] \tag{21.1}$$

which can be interpreted as the statistic sum of the system of 'rapid' particles in a slowly oscillating field ψ_0. The functional $\tilde S$ has the meaning of 'hydrodynamical action'. In the integral over $\psi_0, \bar\psi_0$, it proves convenient to transform the variables into ρ and φ by the prescription $\psi_0 = \sqrt{\rho} \exp(i\varphi), \bar\psi_0 = \sqrt{\rho} \exp(-i\varphi)$.

The expression $\tilde S$ in the model of two-dimensional Bose gas is given (in the limit $T \to 0$) by formula (20.3). Notice that the coefficient of $-((\nabla\varphi)^2/2m)$ in the hydrodynamical action has the meaning of superfluid component density ρ_s, which in the limit $T \to 0$ becomes equivalent to the full density $\rho = p_\lambda$. This allows us to replace p_λ by ρ_s in the sequel.

During the derivation of the hydrodynamical action of a two-dimensional Bose gas (20.3) we considered only the functions $\psi_0, \bar\psi_0$ different from zero. Let us now evaluate the contribution of functions $\psi_0, \bar\psi_0$ that are (for any fixed τ) equal to zero at some discrete set of points of the x plane. Circumventing each such point the phase acquires an increment $2\pi n$ (where n is an integer). We shall consider only the points with $n = \pm 1$, which will be called the centres of quantum vortices rotating in positive and negative directions. The points with $|n| > 1$ can be regarded as bounded systems of $|n|$ vortices rotating with the same orientation. Such systems are not stable and decay to single vortices with $|n| = 1$.

From the above discussion, it is clear that incorporating of vortices is equivalent to the integration over the functions $\varphi(\mathbf{x}, \tau)$ that acquire an increment $\pm 2\pi$ after circumventing the 'singularities', i.e., zeros of the functions $\psi_0, \bar{\psi}_0$. One has to integrate over the variables including the density $\rho(\mathbf{x}, \tau)$ and the phase $\varphi(\mathbf{x}, \tau)$ determined by the ambiguity conditions, as well as over the paths of the vortex centres in the $(\mathbf{x}, \tau)$ space.

In the integrand of (20.3) we can neglect the last two terms (proportional to $(\nabla \pi)^2$ and $(\nabla \varphi)^2$) that lead to the deviation of the phonon spectrum from linearity and to the phonon–phonon interaction. Integrating then $\exp \tilde{S}$ over the variable π, we give the action the form

$$- \int d\tau \, d^2x \frac{\rho_s}{2m} \left((\nabla \varphi)^2 + \frac{1}{c^2} (\partial_\tau \varphi)^2 \right). \tag{21.2}$$

where c^2 is the sound velocity squared. The substitution $p_\lambda = \rho \to \rho_s$ is explained above. It ensures the validity of the formulae even if ρ_s is considerably different from ρ.

Expression (21.2) represents the action of a relativistic system in Euclidean variables, where the light velocity has been replaced by the sound velocity c. Continuing the analogy with relativity, we shall show that if vortices are taken into account, the action (21.2) coincides in principle with the action of two-dimensional (i.e., $(2 + 1)$-dimensional) relativistic electrodynamics, where photons are replaced by phonons and charged particles by quantum vortices.

Expressed in the variables $x_1, x_2, x_3 = c\tau$, expression (21.2) has the form

$$- \frac{\rho_s}{2mc} \int (\nabla \varphi)^2 \, d^3x, \tag{21.3}$$

where $\nabla \varphi$ is the three-dimensional gradient of the phase φ. The integration over φ can be accomplished using the phase shift.

$$\varphi(\mathbf{x}) \to \varphi(\mathbf{x}) + \varphi_0(\mathbf{x}) \tag{21.4}$$

where the function $\varphi_0(\mathbf{x})$ is the solution of the three-dimensional Laplace equation which takes into account the phase ambiguity. To find the function $\varphi_0(\mathbf{x})$ we notice that its three-dimensional gradient $\nabla \varphi_0(\mathbf{x}) = \mathbf{h}(\mathbf{x})$ is the solution of the magnetostatic problem in three-dimensional space defined by the equation

$$\text{rot}\,\mathbf{h} = 2\pi\mathbf{j}; \qquad \text{div}\,\mathbf{h} = 0. \tag{21.5}$$

Here, $\mathbf{j}$ is the sum of unimodular linear currents flowing along the trajectories of the vortex centres. The function $\varphi_0(\mathbf{x})$ is a nonunique scalar potential of the magnetic field $\mathbf{h}$ induced by a system of linear currents. The gradient squared $(\nabla\varphi)^2$ in formula (20.3) transforms by the translation (21.4) to the sum $(\nabla\varphi)^2 + (\nabla\varphi_0)^2$. The integral of the first term describes the noninteracting field and is irrelevant to our discussion. The integral of $(\nabla\varphi_0)^2 = h^2$ is proportional to the energy of the magnetic field of the system of linear currents.

The magnetostatic problem (21.5) is usually solved by means of the vector potential $\mathbf{a}(\mathbf{x})(\mathbf{h} = \text{rot}\,\mathbf{a}, \text{div}\,\mathbf{a} = 0)$. For a system of linear currents the vector potential is the sum of contributions from the current

$$\mathbf{a}(\mathbf{x}) = \frac{1}{2}\sum_i \int \frac{d\mathbf{l}_i(\mathbf{y})}{|\mathbf{x} - \mathbf{y}|}. \tag{21.6}$$

The action obtained from (21.3) after the transformation $\varphi \rightarrow \varphi_0$ can be expressed as a double sum of contributions from the currents

$$-\frac{\pi\rho_s}{2mc}\sum_{i,k}\int\int \frac{d\mathbf{l}_i(\mathbf{x})\,d\mathbf{l}_k(\mathbf{y})}{|\mathbf{x} - \mathbf{y}|}. \tag{21.7}$$

The terms with $i = k$ in (21.7) diverge as $\mathbf{x}$ is close to $\mathbf{y}$. The divergence stems from the character of our approximation in which the vortices are regarded as point-like objects and the corresponding currents as linear. So as to remove the divergencies, it is necessary to take into account the finite volume of vortices. To accomplish this, we surround the centres of vortices by circles of radius r_0, which is greater than the vortex core, but less than the average vortex distance. The influence of the finite vortex volumes leads to the substitution

$$\frac{\pi\rho_s}{2mc}\sum_i \iint\limits_{|\mathbf{x}-\mathbf{y}|<r_0} \frac{d\mathbf{l}_i(\mathbf{x})d\mathbf{l}_i(\mathbf{y})}{|\mathbf{x} - \mathbf{y}|} \rightarrow \frac{E_B(r_0)}{c}\sum_i \int ds_i, \tag{21.8}$$

Here $d_0 = |d\mathbf{l}| = \sqrt{ds^2}$ and $E_B(r_0)$ is a part of the vortex energy located in the circle of radius r_0. The expression for $E_B(r_0)$ depends on r_0 logarithmically:

$$E_B(r_0) = \frac{\pi\rho_s}{m}\ln\frac{r_0}{a}. \tag{21.9}$$

It is natural to call the quantity a in (21.9) the vortex core radius. It is of the same order as $(\lambda m)^{-1/2}$, where λ is the chemical potential and m is the Bose particle mass. In order to determine a, one can use, for example, the solution of Ginzburg–Landau equations which describes the vortex structure given by Pitayevski [106].

Let us transform the integrals over $|\mathbf{x} - \mathbf{y}| > v_0$ in (21.7). We introduce a new vector potential $\mathbf{A}(\mathbf{x})$, the expansion of which

$$\mathbf{A}(\mathbf{x}) = \int_{k < \tilde{k}_0} \exp{(i\mathbf{k}\mathbf{x})}\mathbf{a}(\mathbf{k})\, d^3k \tag{21.10}$$

is bounded by momenta less than $\tilde{k}_0 \sim r_0^{-1}$. The action (20.7) can be given the form

$$S' = -m_B(r_0)c \sum_i \int ds_i - iq \int (\mathbf{A} \cdot \mathbf{j})\, d^3x - \frac{1}{2c} \int (\mathrm{rot}\, \mathbf{A})^2\, d^3x, \tag{21.11}$$

where

$$m_B(r_0) \equiv E_B(r_0)\overline{c^2} \tag{21.12}$$

is the vortex 'mass' and the coefficient

$$q = 2\pi\sqrt{\frac{\rho_s}{mc^2}} \tag{21.13}$$

has the meaning of a coupling constant.

The action (21.11) written in Euclidean variables describes the system of charged particles interacting with the electromagnetic field $\mathbf{A}(\mathbf{x})$ (momenta of this field are bounded by the upper limit $k_0 \sim r_0^{-1}$). The functional $\exp S'$ must be integrated over the field $\mathbf{A}(\mathbf{x})$ and over the paths of the charged particles. This is the correct procedure for quantization of a system described by the action (21.11). The integration of $\exp S'$ over the field $\mathbf{A}(\mathbf{x})$ can be performed exactly using the translation $\mathbf{A} \to \mathbf{A} + \mathbf{A}_0$ which makes the linear form of $\mathbf{A}$ in (21.11) vanish. We again obtain the action (21.7) and prove the correctness of expansion (21.11).

If we want to describe the motion of vortices with velocities much less than c, it is suitable to make a nonrelativistic approximation to the action (21.11). In this approximation we get

$$\frac{E_B(r_0)}{c} \int ds \approx E_B(r_0)\left(\int_0^\beta \left(1 + \frac{1}{2c^2}\left(\frac{d\mathbf{x}}{d\tau}\right)^2\right) d\tau\right)$$

$$= \beta E_B(r_0) + \int_0^\beta \frac{m_B(r_0)}{2}\mathbf{v}^2(\tau)\, d\tau. \tag{21.14}$$

The contribution of the scalar potential A_0 to the action (21.11) can be transformed to the term describing the direct interaction of charged particles through logarithmic potential. Finally the action (21.11) in nonrelativistic approximation acquires the form

$$-\sum_i (\beta E_B(r_0)) + \int_0^\beta d\tau \left(\frac{m_B(r_0)}{2} v_i^2 + \frac{ig_i}{c}(v_i \cdot A_i) \right) -$$

$$-\frac{1}{2} \int d\tau \, d^2x \left((\partial_1 A_2 - \partial_2 A_1)^2 + \frac{1}{c^2}(\partial_\tau A)^2 \right) + \qquad (21.15)$$

$$+\frac{1}{4\pi} \int d\tau \, d^2x \, d^2y j_0(x, \tau) j_0(y, \tau) \ln |x - y|.$$

Here, A_i is the vector potential in the centre of the ith vortex moving with the velocity v_i and having the charge g_i, where

$$\frac{g^2}{4\pi} = \frac{\pi \rho_s}{m}, \qquad (21.16)$$

and the function

$$j_0(x, \tau) = \sum_i g_i \delta(x - x_i(\tau)) \qquad (21.17)$$

describes the charge density.

Let us now study some of the consequences of the equivalence between the system of phonons and vortices and the two-dimensional electrodynamics.

If the temperatures are sufficiently low, the vortices in the system may exist only in the form of opposite sign pairs coupled by long-range logarithmic potential.

We shall show that the including of these coupled vortex pairs does not affect the power character of the asymptotic behaviour of one-particle Green's functions at large distances. Let us regard the quadratic form (21.1) (with ρ_λ replaced by ρ_s) as the hydrodynamical action $\tilde{S}$. We restrict ourselves to integration over the functions $\rho_1 \varphi$ independent of τ. One can show that the contribution of the τ-dependent functions yields in the first approximation a correction to the coefficient of $r^{-\alpha}$. We shall not make the explicit calculation of this coefficient here. We shall only calculate the exponent in the case when the bound pairs of vortices are considered. The problem leads to functional integration of the expression

$$\exp\left\{-\frac{\rho_s\beta}{m}\int d^2z(\nabla\varphi(z))^2 + i\varphi(x) - i\varphi(y)\right\}. \tag{21.18}$$

If the vortices occur, then the functions $\varphi(z)$ acquire increments $\pm2\pi$ corresponding to each circumventing of the vortex and are therefore ambiguous. We can make them unique by means of cuts connecting the vortices of each pair. Besides, it is suitable to suppose that the vortices, as well as the 'sources' x and y, are surrounded by the circles of radius a (where a is the radius of the vortex core), and integrate over the domain outside these circles.

Let us shift the integration variable
where

$$\varphi(z) \rightarrow \varphi(z) + \varphi_o(z),$$
$$\varphi_0(z) = \frac{im}{2\pi\beta\rho_s}\ln\frac{|z-y|}{|z-x|} \equiv i\alpha\chi_0(z). \tag{21.19}$$

If the vortices are present, then the expression

$$-\frac{i}{2\pi}\int \text{div}(\varphi\nabla\chi_0)d^2z \tag{21.20}$$

emerging after the transformation of the integral in (21.18) not only results in the integrals over the circles surrounding the 'sources' x and y, but also in the sum of integrals alongside the cuts. This gives rise to the additional coefficient

$$\exp\left\{i\sum_i r_i|\nabla\chi_{0i}|\sin\theta_i\right\}, \tag{21.21}$$

where r_i is the separation between the vortices of the ith pair, $|\nabla\chi_{0i}|$ is the modulus of the vector $\nabla\chi_0$ in the point of location of the pair, and θ_i is the angle between the vectors r_i and $\nabla\chi_{0i}$. Taking the average of the coefficient (21.21) with respect to the orientation of vectors r_i, we obtain

$$\exp\left\{-\frac{1}{2}\sum_i r_i^2(\nabla\chi_{0i})^2\right\}. \tag{21.22}$$

For large $r = |x - y|$ we have

$$\frac{1}{2}\sum_i r_i^2(\nabla\chi_{0i})^2 = \frac{S_B}{S}\int\frac{(\nabla\chi_0)^2}{4\pi}d^2z = \frac{S_B}{S}\ln\frac{r}{a}. \tag{21.23}$$

Here, S_B/S is the mean relative area occupied by a vortex pair, (the area occupied by the ith pair with vortices separated by r_i is assumed to be $2\pi r_i^2$).

Formula (21.23) shows that the power character of the asymptotic behaviour remains valid, even if the vortices are considered, in which case the exponent α acquires the addition

$$\Delta\alpha = S_B/S, \tag{21.24}$$

small with respect to α, provided α itself is sufficiently small ($\Delta\alpha/\alpha \to 0$ as $T \to 0$), which has the meaning of a relative area occupied by a vortex pair.

Let us now study the role of quantum vortices in the phase transition between the superfluid and the normal states. When the temprature increases, the number of coupled vortex pairs increases and the mean distance between them decreases. Finally, at some temperature T_c, the coupled pairs start to dissociate. Above the dissociation temperature, single vortices as well as the coupled pairs appear in the system and we can observe a state similar to plasma. It is natural to assume that the phase transition from the superfluid to the normal state is caused by the dissociation of the coupled pairs.

For $T > T_c$, the long-range correlations disappear in the plasma-like state due to the characteristic Debye screening. In particular, the correlation function $\langle \psi(\mathbf{x},\tau)\bar{\psi}(\mathbf{y},\tau_1)\rangle$ decreases exponentially for $T > T_c$. Notice moreover that the vanishing of the second sound for $T > T_c$ can be interpreted as a transformation of the first sound branch into the plasma oscillations branch.

At the same time the quantity ρ_s defined as the coefficient of $-((\nabla\varphi)^2/2m)$ in the hydrodynamical action does not vanish even for $T > T_c$. This coefficient is analogous to the quantity ρ_s introduced by Berezinski [92] in his model of planar rotators. For $T < T_c$, this coefficient is in fact equivalent to macroscopic superfluid density defined everywhere except a small phase transition domain, in which an intensive creation of quantum vortices occurs. Thus, we arrive at the conclusion that the description of a two-dimensional Bose system in terms of normal and superfluid components with quantum vortices is possible both below and above the phase transition point and that it becomes meaningless only if the vortex core radius is of the same order as the mean vortex separation.

The method of description of quantum vortices developed above can be extended to three-dimensional Bose systems, too. In the case, the quantum vortices again correspond to the zeros of $\psi(\mathbf{x},\tau)$ functions, much the same way as in the two-dimensional case. The complex three-dimensional function ψ acquires zero values alongside lines (sets of dimension 1), and the four-dimensional function – on two-dimensional surfaces. One has to integrate first over the functions $\psi, \bar{\psi}$ equal to zero on two-dimensional surfaces, and then over the surface configurations.

We shall not be interested in all the details of the application of our approach to three-dimensional systems, restricting ourselves only to a qualitative estimation of the role of quantum vortices in the description of phase transitions. Our conclusions are analogous to those of References [107] and [108], where the authors use the analogy with the Ising model.

At sufficiently low values of temperatures specific excitations may occur in a nonrotating Bose system which have the form of vortex rings. The closed zero lines of the functions $\psi_1 \bar{\psi}$ correspond to these rings in the formalism of the functional integral. When the temperature rises, the number of vortex rings in a volume unit increases and the mean distance between them decreases. When the mean distance between the rings becomes comparable with the mean ring length, the system exhibits the tendency to create very long rings. It is natural to assume that the phase transition between the superfluid and the normal states is connected with the creation of vortex lines of infinite length (in a real system, the lines which begin and end on the vessel walls). A vortex line of sufficient length does not emerge immediately, but rather by subsequent combinations and an increasing of vortex rings of finite length. Thus, if the number of vortex rings per volume unit is sufficiently large, the probability of the existence of infinitely long vortex lines may be anything but infinitesimally small.

Notice that the quantity ρ_s defined as the coefficient of $-((\nabla\varphi)^2/2m)$ in the hydrodynamical action is different from zero even above the transition point, much the same way as in two-dimensional systems. In other words, the description of a system in terms of normal and superfluid components filled with quantum vortices is possible above the transition, too. In connection with this analysis, an idea emerges that the difference between normal and superfluid liquid is given by the character of vortex excitations occurring in the system – in a normal system, also long vortex lines must exist besides the vortex rings.

Thus, the above qualitative estimate shows that in a three-dimensional system the phase transition is related to quantum vortices, much the same way as in two dimensions. Closed vortex rings of a three-dimensional system correspond to coupled vortex pairs of two dimensions, and long vortex lines are analogues of single vortices.

Quantum vortices in rotating liquid helium above the λ-point have been found in the experiments of Andronikashvili and Mamaladze [109]. They appear 18–20 minutes after heating the system 0.1–0.2 K above the λ point. The formalism described above offers the explanation why these vortices exist above the λ point. The decay of the ordered vortex structure occurs after increasing the temperature above the λ point due to the creation of chaotically distributed long vortex lines.

Finally, we bring a summary of the most important results obtained in this section.

(1) The perturbation theory of superfluid Bose systems can be easily constructed by means of the path integration method (see Sections 16 and 17).

(2) The modification of the method by successive integration over rapidly oscillating and slowly oscillating fields is suitable for derivation of low-frequency asymptotic behaviour of Green's functions and of kinetic equations (see Section 18).

(3) The separate integration over 'rapid' variables enables one to derive the functional of hydrodynamical action and the corresponding hydrodynamical Hamiltonian, which has to be postulated in semi-phenomenological theories of superfluidity (see Section 19).

(4) The method of hydrodynamical action can be extended to two-dimensional and one-dimensional Bose systems, in which the superfluidity may occur without the condensate. In that case, the phase transition corresponds to emerging of long-range correlations with power-like (and not exponential) asymptotic behaviour.

(5) In the formalism of continual integration, quantum vortices correspond to the zeros of functions $\psi(\mathbf{x},\tau)$, $\bar{\psi}(\mathbf{x},\tau)$ – the continual integration variables. In a superfluid state, the main contribution to the integral is given by the functions ψ and $\bar{\psi}$ without zeros, which describe the state without quantum vortices. In a normal state, the most important contribution is from the functions of the type of superposition of plane waves. In the neighbourhood of phase transition, the maximal contribution to the integral is from the functions the modulus of which is almost constant everywhere except the vortex core (see Section 21).

(6) Quantum vortices may also occur in a normal Bose system. In a two-dimensional system, the phase transition from the superfluid to the normal state corresponds to the dissociation of coupled pairs, in a three-dimensional system to the emerging of long vertex lines.

CHAPTER 7

SUPERCONDUCTIVITY

22. PERTURBATION THEORY OF SUPERCONDUCTING FERMI SYSTEMS

The effect of superconductivity in Fermi systems is of the same nature as the superfluidity in Bose systems. The main point in constructing the superconductivity theory is again the existence of long-range correlations emerging at the transition temperature. The counterpart of the Bose field ψ_B in the formalism of continual integral is the product of the Fermi fields $\psi_F \psi_F$. Further, the one-particle correlation function of Bose systems

$$\langle \psi(\mathbf{x}, \tau) \bar{\psi}(\mathbf{y}, \tau_1) \rangle \qquad (22.1)$$

corresponds to the two-particle correlation function of a Fermi system (i.e., the mean value of the product of four Fermi fields)

$$\langle \psi(\mathbf{x}_1, \tau_1) \psi(\mathbf{x}_2, \tau_2) \bar{\psi}(\mathbf{x}_3, \tau_3) \bar{\psi}(\mathbf{x}_4, \tau_4) \rangle \qquad (22.2)$$

We say that the long-range correlations exists in the system provided correlation function (22.2) goes to zero as an inverse power of $|\mathbf{x}_1 - \mathbf{x}_3|$ or if it has a finite limit as $|\mathbf{x}_1 - \mathbf{x}_3| \to \infty$, while the differences $\mathbf{x}_1 - \mathbf{x}_2, \mathbf{x}_3 - \mathbf{x}_4$ (and the quantity τ_i) are held fixed. If the limit is finite, one can suppose that there exist anomalous Green's functions

$$\langle \psi(\mathbf{x}_1, \tau_1) \psi(\mathbf{x}_2, \tau_2) \rangle; \qquad \langle \bar{\psi}(\mathbf{x}_3, \tau_3) \bar{\psi}(\mathbf{x}_4, \tau_4) \rangle. \qquad (22.3)$$

Let us investigate the perturbation theory of superconducting Fermi systems supposing that anomalous mean values (22.3) exist. Such mean values are identically zero when calculated in the framework of the ordinary perturbation theory.

One of the possibilities of obtaining a nonzero anomalous mean value is to add to the Fermi-system action the 'source functional'

$$\sum_p (\bar{\eta}(p) a(p) a(-p) + \eta(p) a^+(p) a^+(-p)) \qquad (22.4)$$

where the complex functions $\eta(p), \bar{\eta}(p), p = (\mathbf{k}, \omega)$ are infinitesimally small. Expression (22.4) does not satisfy the particle number conservation law, and if η and $\bar{\eta}$ are nonzero, then also the anomalous mean values (22.3) will be different from zero. The question is what happens in the limit $\eta, \bar{\eta} \to 0$. If the anomalous correlation functions vanish in this limit, then the system is in a normal state. If, on the other hand, the limit of the anomalous mean values is nonzero as $\eta, \bar{\eta} \to 0$, then the system is in the state of superconductivity.

The method of 'source' had been used by Bogoliubov in his formalism of 'quasi-mean values' [81] and since then it has been applied in a number of papers.

The same result can be obtained using a modification of the skeleton diagram technique, which differs from the skeleton technique introduced in Section 6 in the fact that anomalous Green's functions also occur, besides the normal ones. Recall that in the diagram technique the diagram lines correspond to the full Green's functions, but, on the other hand, only diagrams with no subdiagrams of the self-energy type are taken into account. The diagram vertices are identical with those of the ordinary diagram technique.

To give the diagram technique a concrete meaning, one has to consider the Fermi-particle spins and describe particles with different helicities s by different functions $\psi_s(\mathbf{x}, \tau)$.

Consider the situation when the anomalous mean values of fields with opposite spins

$$\langle \psi_s(\mathbf{x}, \tau)\psi_{-s}(\mathbf{y}, \tau_1) \rangle; \qquad \langle \psi_s^+(\mathbf{x}, \tau)\psi_{-s}^+(\mathbf{y}, \tau_1) \rangle \qquad (22.5)$$

are different from zero, while the mean values $\langle \psi_s\psi_s \rangle, \langle \psi_s^+ \psi_s^+ \rangle$ identically vanish.

In this case, the diagram technique contains the following elements and corresponding terms:

$$G(p, s); \qquad G_1(p, s);$$

$$\bar{G}_1(p, s);$$

$$u(\mathbf{k}_1 - \mathbf{k}_3) - u(\mathbf{k}_1 - \mathbf{k}_4); \qquad (22.6)$$

$$u(\mathbf{k}_1 - \mathbf{k}_3).$$

Here, G is the full normal Green's function and $G_1, \tilde{G}_1$ are the full anomalous Green's functions. They are *a priori* unknown and must be found by solving a system of equations. First, this system includes the Dyson–Gorkov equations which express Green's functions in terms of normal ($A(p)$) and anomalous ($B(p)$) self-energy parts. Graphically, they have the form

$$(22.7)$$

identical the Dyson–Beliaev [77] equations describing superfluid Bose systems. In an analytic form, the equations can be written as

$$G(p) = G_0(p) + G_0(p)A(p)G(p) + G_0(p)B(p)G_1(p);$$
$$G_1(p) = -G_0(-p)\tilde{B}(p)G(p) + G_0(-p)A(-p)G_1(p).$$

$$(22.8)$$

Having in mind that $G_0(p)$ is given by formula (6.34), we can write the solution of this system in the form

$$G(p) = \frac{i\omega + (k^2/2m) - \lambda + A(-p)}{(i\omega + (k^2/2m) - \lambda + A(-p))(i\omega - (k^2/2m) + \lambda - A(p)) - |B(p)|^2};$$

$$(22.9)$$

$$G_1(p) = \frac{\bar{B}(p)}{(i\omega + (k^2/2m) - \lambda + A(-p))(i\omega - (k^2/2m) + \lambda - A(p)) - |B(p)|^2}.$$

Each of the self-energy parts $A(p)$, $B(p)$ is given by the sum of the skeleton perturbation theory diagrams

$$(22.10)$$

Equations (22.7) and (22.10) form the system for determining the normal and anomalous Green's functions. To find the solution of this system, one has either to restrict oneself to the first several terms of the perturbation series, or to find the leading diagram sequence of Equations (22.10). When the expressions of Green's functions in terms of the self-energy parts (22.9) instead of the Green's functions themselves are inserted into these equations, the problem is transformed into the solution of a system of integral self-energy equations.

Let us look for the solution of the system of Equations (22.7) and (22.10) for the model of Fermi gas. Assuming that the gas parameter $\rho^{1/3}a$ is sufficiently small

$$\rho^{1/3}a \ll 1, \tag{22.11}$$

it is not difficult to separate the leading sequence of diagrams for A and B. Here, the same factors decreasing the diagram order are in operation as in the Bose gas theory (Section 17). We are speaking namely about the anomalous lines and the closed loops of normal lines that can be passed in the direction of diagram arrows. Thus, the main contribution to the self-energy parts $A(p)$ and $B(p)$ is given by diagrams in which the influence of these factors is as weak as possible. They are the diagrams without anomalous lines and with one closed loop of normal lines in the expressions of $A(p)$, and diagrams without closed loops and with one anomalous line when calculating $B(p)$ (only one diagram of the last type exists, namely the first in the diagrammatic equation (22.10)). The remaining diagrams give contributions containing higher degrees of the gas parameter than those considered above. Thus, we obtain the approximate equations

$$\tag{22.12}$$

The four-point function standing in the $A(p)$ diagram represents the diagram sequence

$$\tag{22.13}$$

Much the same way as in the Bose gas theory, the summation of this series results in substituting the potential by the t matrix which can be put equal to a constant $t_0 = t$ in the domain where $k \leq k_F = \sqrt{2m\lambda}$ ($k_1 = 0$, $k_2 = 0, z = 0$). The vertex describing the scattering of identical spin particles in (22.6) contains the antisymmetrical potential $u(\mathbf{k}_1 - \mathbf{k}_3) - u(\mathbf{k}_1 - \mathbf{k}_4)$. After the summation of (22.13) this expression transforms into the antisymmetrized t matrix vanishing after the substitution $t \rightarrow t_0$. This is the reason why only the t matrix describing the scattering of opposite spin particles gives nonzero contribution in the first approximation. For the self-energy $A(p)$ we obtain

$$A_+ = t_0 \rho_-; \qquad A_- = t_0 \rho_+. \tag{22.14}$$

Here, $A_\pm$ are the self-energies of particles with spins $\pm$ and $\rho_\pm$ are corresponding densities. If magnetic field and other forces with an opposite influence on particles with opposite spins are not present, then $A_+ = A_-$ and $\rho_+ = \rho_-$. The densities can be in the first approximation expressed in terms of the nonperturbative Green's functions $G_0(p)(\varepsilon \rightarrow +0)$

$$\rho_+ = \rho_- = \frac{T}{V} \sum_p \exp(i\omega\varepsilon) G_0(p)$$
$$= (2\pi)^{-3} \int \frac{d^3 k}{\exp[\beta((k^2/2m) - \lambda)] + 1}. \tag{22.15}$$

In the following we assume that the Fermi gas is degenerated, i.e., that $\beta\lambda \gg 1$. We shall show later that this holds for all temperatures less than the superconducting state transition temperature T_c. In a degenerated Fermi gas, the density is determined by the formula

$$\rho_\pm = (2\pi)^{-3} \int_{k < k_F} d^3 k = \frac{(2m\lambda)^{3/2}}{6\pi^2}. \tag{22.16}$$

The self-energy part $A(p)$ can be shown to be equal to the constant

$$A_\pm = \frac{t_0 (2m\lambda)^{3/2}}{6\pi^2} \tag{22.17}$$

in approximation (22.14) and can be regarded as a small correction $\Delta\lambda$ to the chemical potential λ. Indeed,

$$\frac{|\Delta\lambda|}{\lambda} \sim |t_0| m^{3/2} \lambda^{1/2} \sim a\rho^{1/3} \ll 1. \tag{22.18}$$

The second of the diagrammatic equations (22.12) is the equation of the anomalous self-energy part B, which in this approximation does not depend on ω. The anomalous Green's function is in this approximation given by the formula

$$G_1(p) = \frac{B(k)}{\omega^2 + ((k^2/2m) - \mu)^2 + B^2(k)}, \tag{22.19}$$

where the parameter

$$\mu = \lambda - A_{\pm} = \lambda - \frac{t_0 (2m\lambda)^{3/2}}{6\pi^2} \tag{22.20}$$

has the meaning of the renormalized chemical potential.

The equation for the function B can be given the form

$$B(\mathbf{k}) = -\frac{T}{V} \sum_{p_1} u(\mathbf{k} - \mathbf{k}_1) G_1(p_1) \tag{22.21}$$

$$= -\frac{1}{(2\pi)^3} \int d^3 k_1 u(\mathbf{k} - \mathbf{k}_1) B(\mathbf{k}_1) \frac{\tanh (\beta \varepsilon(\mathbf{k}_1)/2)}{2\varepsilon(\mathbf{k}_1)},$$

where

$$\varepsilon(\mathbf{k}) = \left(\left(\frac{k^2}{2m} - \mu \right)^2 + B^2(\mathbf{k}) \right)^{1/2} \tag{22.22}$$

In order to solve Equation (22.21) we adopt the convention used in Section 17 for the model of Bose gas. Adding an identical expression to both sides of (22.21), we write the equation as

$$B(\mathbf{k}) + (2\pi)^{-3} \int u(\mathbf{k} - \mathbf{k}_1) B(\mathbf{k}_1) \frac{m}{k_1^2} d^3 k_1$$

$$= (2\pi)^{-3} \int u(\mathbf{k} - \mathbf{k}_1) B(\mathbf{k}_1) \left(\frac{m}{k_1^2} - \frac{\tanh (\beta \varepsilon(\mathbf{k}_1)/2)}{2\varepsilon(\mathbf{k}_1)} \right) d^3 k_1. \tag{22.23}$$

The operator acting on $B(\mathbf{k})$ at the left-hand side of this equation can be inverted. This is equivalent to replacing the potential by the t matrix on the right-hand side. In the domain $k \lesssim k_F$ (which gives the main contribution to the integral) the t matrix can be replaced by a constant t_0. We obtain the following equation for $B(\mathbf{k})$

$$B(\mathbf{k}) = \frac{t_0}{(2\pi)^3} \int B(\mathbf{k}_1) \left(\frac{m}{k_1^2} - \frac{\tanh (\beta \varepsilon(\mathbf{k}_1)/2)}{2\varepsilon(\mathbf{k}_1)} \right) d^3 k_1. \tag{22.24}$$

The equation has a trivial solution $B(\mathbf{k}) = 0$; it may have also a nontrivial solution corresponding to the superconducting state. Due to (22.24), the function $B(\mathbf{k})$ is constant in the domain $k \lesssim k_F$. This constant can be interpreted as the gap in the energy spectrum obtained from the condition that the denominator of Green's function (22.9) should vanish as $\lambda - A \to \mu$, $B(\mathbf{k}) \to \Delta$ after the analytic continuation $i\omega \to E$. The poles of the Green's functions are at $E = \pm \varepsilon(\mathbf{k})$, where $\varepsilon(\mathbf{k})$ is given by formula

(22.22), in which $B(\mathbf{k})$ can be substituted by Δ. The quantity Δ is the minimum value of $\varepsilon(\mathbf{k})$ acquired on the Fermi level if $k^2/2m = \mu$. Assuming that $\Delta \neq 0$, we obtain the equation for Δ

$$1 = \frac{mt_0}{2\pi^2} \int_0^\infty dk \left(1 - \frac{k^2}{2m\sqrt{\xi^2 + \Delta^2}} \tanh\left(\frac{\beta}{2}\sqrt{\xi^2 + \Delta^2}\right)\right), \quad (22.25)$$

where

$$\xi(\mathbf{k}) = k^2/2m - \mu. \quad (22.26)$$

As in the following Δ is shown to be small ($\Delta \ll \lambda$), one can neglect its contribution and replace $\tanh(\beta/2)\sqrt{\xi^2 + \Delta^2}$ by unity everywhere except a small interval $[k_F - k_0, k_F + k_0]$ close to the Fermi level, where k_0 is chosen such that the inequalities

$$\Delta \ll \frac{k_0^2}{2m} \ll \lambda; \qquad \frac{k_0 k_F}{2m} \gg T. \quad (22.27)$$

hold. Finally we obtain the expressions

$$\int_{|k-k_F|>k_0} \left(1 - \frac{k^2}{2m\sqrt{\xi^2 + \Delta^2}} \tanh\left(\frac{\beta}{2}\sqrt{\xi^2 + \Delta^2}\right)\right) dk$$

$$\approx \int_{|k-k_F|>k_0} \left(1 - \frac{k^2}{|k^2 - k_F^2|}\right) dk = k_F\left(2 - \ln\frac{2k_F}{k_0}\right) - 2k_0;$$

$$\int_{|k-k_F|<k_0} \left(1 - \frac{k^2}{2m\sqrt{\xi^2 + \Delta^2}} \tanh\left(\frac{\beta}{2}\sqrt{\xi^2 + \Delta^2}\right)\right) dk$$

$$\approx 2k_0 - k_F \int_0^{(k_F k_0)/m} \frac{\tanh((\beta/2)\sqrt{\xi^2 + \Delta^2})}{\sqrt{\xi^2 + \Delta^2}} d\xi.$$

for the integrals over $|k - k_F| > k_0$ and $|k - k_F| < k_0$, respectively. Equation (22.25) can be rewritten in the form

$$-1 = \frac{mt_0 k_F}{2\pi^2}\left(-2 + \ln\frac{2k_F}{k_0} + \int_0^{(k_F k_0)/m} \frac{\tanh((\beta/2)\sqrt{\xi^2 + \Delta^2})}{\sqrt{\xi^2 + \Delta^2}} d\xi\right). \quad (22.28)$$

To find the solution of this equation, one has to ensure that t_0 is negative. This means that the interaction potential of Fermi particles must be attractive. Denote

$$\zeta = \frac{m|t_0|k_F}{2\pi^2}. \quad (22.29)$$

The parameter ζ is small ($\zeta \ll 1$) due to the small density condition (22.11).

Equation (22.28) has the solution only for sufficiently small temperatures $T = \beta^{-1}$. The transition temperature T_c can be defined as the temperature at which the solution vanishes ($\Delta = 0$). For $\Delta = 0$ we obtain

$$\int_0^{(k_0 k_F)/m} \frac{\tanh\left((\beta/2)\sqrt{\xi^2 + \Delta^2}\right)}{\sqrt{\xi^2 + \Delta^2}} d\xi$$

$$= \int_0^{(k_0 k_F)/m} \frac{\tanh x}{x} dx \approx \ln\frac{k_0 k_F}{m} - \int_0^\infty \frac{\ln x}{\cosh^2 x} dx = \ln\frac{2k_0 k_F \gamma}{\pi m T}, \tag{22.30}$$

where $\ln \gamma = C = 0.577\ldots$ is the Euler constant. The substitution of (22.30) into (22.28) yields

$$T_c = \frac{4\gamma k_F^2}{\pi m} \exp\left(-1/\zeta - 2\right). \tag{22.31}$$

We can now find the value $\Delta(0)$ of the gap Δ at $T = 0$. Inserting

$$\int_0^{(k_0 k_F)/m} \frac{d\xi}{\sqrt{\xi^2 + \Delta^2}} = \ln\frac{2k_0 k_F}{m\Delta} \tag{22.32}$$

into (22.28), we obtain

$$\Delta(0) = \frac{4k_F^2}{m} \exp\left(-1/\zeta - 2\right). \tag{22.33}$$

The final results (22.31), (22.32) do not depend on the auxiliary parameter k_0, as we have supposed. Dividing (22.31) by (22.33), we obtain the relation

$$T_c = (\gamma/\pi)\Delta(0), \tag{22.34}$$

which represents a general property of superconducting theories. It is also valid in the BCS model [141] and in the electron–phonon interaction model [38].

Finally, we notice that the Fermi gas remains degenerated until $T = T_c$, as we have supposed earlier. Indeed

$$T_c \sim \frac{k_F^2}{2m} \exp\left(-1/\zeta\right) = \lambda \exp\left(-1/\zeta\right) \ll \lambda \tag{22.35}$$

and the degeneracy condition $\beta\lambda \gg 1$ is satisfied for all $T \le T_c$.

23. Superconductivity of the Second Kind

The interaction of a superconducting electron with electromagnetic field results in some interesting effects. One of the most celebrated is the Meissner effect, the consequence of which is that a sufficiently weak magnetic field is pushed out from the superconductor. There are two possible scenarios when the magnetic field is increased:

(1) The field penetrates the superconductor, changing it into the ordinary state. Such superconductors are called 'superconductors of the first kind'.

(2) What emerges in the system is the so-called 'mixed state'. During this process the field partially penetrates the superconductor in which a periodic lattice of quantum vortices is formed. This effect is called 'superconductivity of the second kind', and the superconductors exhibiting this type of superconductivity are called the second kind superconductors.

The creation of the quantum vortex lattice had been forecast by Abrikosov [104] as early as 1952 in the framework of the pheno-menological superconductivity theory of Ginzburg and Landau [144]. Formulation of the microscopic superconductivity theory allowed one to derive the Ginzburg–Landau equation microscopically and to understand the meaning of the parameters appearing in the equations. In particular, it turned out that the effective charge of the phenomenological theory is equal to twice the charge of an electron, i.e., that it corresponds to the charge of a Cooper pair.

In this paragraph we shall apply the method of functional integration to some problems of the superconductivity of the second kind. First, we shall show that the homogeneous superconducting state cannot exist in a constant homogeneous magnetic field. Three possibilities are consistent with this fact:

(a) the magnetic field is pushed out of the superconductor (the Meissner effect),
(b) the magnetic field destroys the superconductivity,
(c) the inhomogeneous superconducting state is created.

Namely the last possibility occurs in the second kind superconductors, where the Abrikosov lattice is created. We shall show that the area of the lattice plaquette can be determined purely kinematically, from the relations of translational invariance.

Further, we shall derive the Ginzburg–Landau equations of a Fermi gas by means of subsequent continual integration over rapidly and slowly oscillating fields. These equations yield the extremality conditions for the functional that has the meaning of hydrodynamical action of a superconductor. It is well known that the Ginzburg–Landau equations are applicable near the phase transition point where the energetic gap in the one-particle excitations spectrum Δ is small with respect to the value of this gap at $T = 0$. For small temperatures and a weak electromagnetic field one can derive an equation for the function describing the deviation of the ordering parameter from the equilibrium value.

The action functional of a Fermi system in the electromagnetic field S differs from functional (6.1) by the transformation

$$\partial_\tau \to \partial_\tau + ie\varphi, \nabla \to \nabla + ie\mathbf{A}, \tag{23.1}$$

and also by an additional term describing the interaction between spin and the magnetic field. For constant magnetic field $\mathbf{H}$ we have

$$\varphi = 0; \qquad \mathbf{A} = \tfrac{1}{2}[\mathbf{H}, \mathbf{x}]; \tag{23.2}$$

$$S = \int d\tau\, dx \sum_s \left(\bar\psi_s \partial_\tau \psi_s - \frac{1}{2m}|(\nabla + ie\mathbf{A})\psi_s|^2 + \lambda\bar\psi_s\psi_s + \frac{eH}{2m}s\bar\psi_s\psi_s \right) - \tag{23.3}$$

$$- \tfrac{1}{2}\int d\tau\, dx\, dy\, u(\mathbf{x} - \mathbf{y}) \sum_{ss'} \rho_s(\mathbf{x}, \tau)\rho_{s'}(\mathbf{y}, \tau),$$

where

$$\rho_s(\mathbf{x}, \tau) = \bar\psi_s(\mathbf{x}, \tau)\psi_s(\mathbf{x}, \tau) \tag{23.4}$$

is the density of electrons with the helicity s.

Let us now examine the changes of normal and anomalous mean values

$$G(\mathbf{x}, \tau; \mathbf{y}, \tau_1) = \langle \psi_s(\mathbf{x}, \tau)\bar\psi_s(\mathbf{y}, \tau_1) \rangle; \tag{23.5}$$

$$G_1(\mathbf{x}, \tau; \mathbf{y}, \tau_1) = \langle \psi_s(\mathbf{x}, \tau)\psi_{-s}(\mathbf{y}, \tau_1) \rangle$$

under the translation $\mathbf{x} \to \mathbf{x} + \mathbf{a}$, $\mathbf{y} \to \mathbf{y} + \mathbf{a}$. The mean values (23.5) can be written in terms of continual integrals of the products of averaged anticommuting fields weighted by $\exp S$. We make the substitution $\mathbf{x} \to \mathbf{x} + \mathbf{a}$, $\mathbf{y} \to \mathbf{y} + \mathbf{a}$ in the action S. This yields

$$\mathbf{A} = \tfrac{1}{2}[\mathbf{Hx}] \to \mathbf{A}' = \tfrac{1}{2}[\mathbf{Hx}] + \tfrac{1}{2}[\mathbf{Ha}]. \tag{23.6}$$

The functions $\psi_s(\mathbf{x} + \mathbf{a}, \tau)$, $\bar\psi_s(\mathbf{x} + \mathbf{a}, \tau)$ will be considered as new continual integration variables. This leads to the conclusion that the mean values (23.5) of fields with shifted arguments are equal to the mean values of fields with original arguments but evaluated with the vector potential

A' (23.6) instead of **A** (23.2). One can return to the original vector potential **A** by phase transformation of the Fermi fields:

$$\psi_s(\mathbf{x}, \tau) \to \exp\left(\frac{ie}{2}[\mathbf{Ha}]x\right)\psi_s(\mathbf{x}, \tau);$$

$$\bar{\psi}_s(\mathbf{x}, \tau) \to \exp\left(-\frac{ie}{2}[\mathbf{Ha}]x\right)\bar{\psi}_s(\mathbf{x}, \tau).$$

(23.7)

Here we use the knowledge of action (23.3) and put the coefficient $\exp(i/2[\mathbf{Ha}](x \pm y))$ in front of the integration symbol. We obtain the formulae

$$G(\mathbf{x} + \mathbf{a}, \tau; \mathbf{y} + \mathbf{a}, \tau_1) = \exp\left(\frac{ie}{2}[\mathbf{Ha}](\mathbf{x} - \mathbf{y})\right) G(\mathbf{x}, \tau; \mathbf{y}, \tau_1);$$

$$G_1(\mathbf{x} + \mathbf{a}, \tau; \mathbf{y} + \mathbf{a}, \tau_1) = \exp\left(\frac{ie}{2}[\mathbf{Ha}], \mathbf{x} + \mathbf{y}\right) G_1(\mathbf{x}, \tau; \mathbf{y}, \tau_1).$$

(23.8)

The second of these formulae indicates that it is impossible to form a homogeneous superconducting state in a homogeneous magnetic field. Indeed, if the superconducting state is homogeneous, then formulae (23.8) hold for any **a**. Shifting one more $\mathbf{x} \to \mathbf{x} + \mathbf{b}$, we find that the function G_1 acquires the factor

$$\exp\left(ie[\mathbf{Ha}]\mathbf{b}\right)\exp\left(\frac{i}{2}e[\mathbf{H}, \mathbf{a} + \mathbf{b}], \mathbf{x} + \mathbf{y}\right)$$

(23.9)

after the translation of its arguments $\mathbf{x}, \mathbf{y}$ by $\mathbf{a} + \mathbf{b}$. If, on the other hand, we simply transform $\mathbf{a} \to \mathbf{a} + \mathbf{b}$ in (23.8), we obtain the factor

$$\exp\frac{i}{2}(e[\mathbf{H}, \mathbf{a} + \mathbf{b}]\mathbf{x} + \mathbf{y}).$$

(23.10)

instead of (23.9). Expression (23.9) can be equal to (23.10) only if the condition

$$\exp i(e\mathbf{H}[\mathbf{a}, \mathbf{b}]) = 1,$$

(23.11)

is satisfied. But this cannot be true for arbitrary **a** and **b**. Thus, the homogeneous superconducting state in the magnetic field is impossible.

Equation (23.11) can be satisfied provided the vectors **a**, **b** form a periodic lattice with the minimal plaquette area

$$\frac{2\pi n}{eH},$$

(23.12)

in the plane perpendicular to **H** (n is some positive integer). Putting $n = 1$ and assuming that the lattice is triangular, we obtain the formulae

$$a = \left(\frac{4\pi}{\sqrt{3eH}} \right)^{1/2}. \tag{23.13}$$

for the lattice spacing a.

Thus, we have shown that the existence of a periodic structure in a superconductor is possible.

Let us now derive the hydrodynamical action functional for the Fermi gas model described by the action (23.3). The extremality conditions of this functional yield equations describing the periodic structure in detail. We extend the idea of successive integration over rapid and slow variables to the Fermi systems. We shall call 'the slowly oscillating part $\psi_0(\mathbf{x}, \tau)$ of the Fermi field $\psi(\mathbf{x}, \tau)$' the sum

$$\psi(\mathbf{x}, \tau) = (\beta V)^{-1/2} \sum_{\mathbf{k}, \omega} \exp\left[i(\omega\tau - \mathbf{kx}) \right] a(p) \tag{23.14}$$

where the momenta $\mathbf{k}$ lie in the narrow layer near the Fermi level $|k - k_F| < k_0$. Consequently, the difference $\psi - \psi_0$ will be called 'the rapidly oscillating part ψ_1 of the field ψ'.

Carrying out the functional integration over rapid variables

$$\int \exp S \, d\bar{\psi}_1 \, d\psi_1 = \exp S'[\bar{\psi}_0, \psi_0], \tag{23.15}$$

we obtain a functional depending on the 'slow' variables $\psi_0, \bar{\psi}_0$. It is not difficult to calculate the functional $\exp S'$ in the case of a system with zero density. In the first approximation, we can restrict ourselves to the terms of an order not higher than the fourth. In this approximation S' has the form

$$S' = p_0\beta V + \int d\tau \, dx \times$$

$$\times \sum_s \left(\bar{\psi}_{0s} \partial_\tau \psi_{0s} - \frac{1}{2m} |(\nabla + ie\mathbf{A})\psi_{0s}|^2 + \left(\mu + ie\varphi + \frac{eH_s}{2m} \right) \bar{\psi}_{0s}\psi_{0s} \right) -$$

$$- \frac{1}{2} \int t' \bar{\psi}_{0+}\psi_{0+}\bar{\psi}_{0-}\psi_{0-} \, dx \, d\tau. \tag{23.16}$$

Contrary to the original expression (23.3), the free term $p_0\beta V$ emerges in (23.16), and the chemical potential λ as well as the interaction potential $u(\mathbf{x} - \mathbf{y})$ is replaced by corresponding renormalized quantities u and $t'\delta(\mathbf{x} - \mathbf{y})$. The pressure p_0 of nonideal Fermi gas containing no particles of momenta $|k - k_F|$ less than k_0, is given by the formula

$$p_0 = \frac{T}{(2\pi)^3} \int_{|k - k_F| > k_0} \ln\left(1 + \exp\left[-\beta\varepsilon_0(\mathbf{k})\right]\right) d^3k \times$$
$$\times (-t')\left((2\pi)^{-3} \int_{|k - k_F| > k_0} \frac{d^3k}{\exp\left[\beta\varepsilon_0(\mathbf{k})\right] + 1}\right)^2. \tag{23.17}$$

Here, $\varepsilon_0(\mathbf{k}) = (k^2/2m) - \lambda$ is the nonperturbation energy spectrum, t' is the t matrix defined below by formula (23.21). The renormalization $\lambda \to \mu$ is described by the formula

$$\mu = \lambda - \frac{t'}{(2\pi)^3} \int_{|k - k_F| > k_0} \frac{d^3k}{\exp\left[\beta\varepsilon_0(\mathbf{k})\right] + 1}. \tag{23.18}$$

Finally, the t matrix t' can be obtained as the result of summation of diagram sequence (22.13) which describes the scattering of opposite-spin particles, whereas the internal lines momenta of these diagrams acquire values outside the interval $|k - k_F| < k_0$. Such a t matrix is given by the equation

$$t'(\mathbf{k}_1, \mathbf{k}_2 ; 0) + (2\pi)^{-3} \int_{|k_3 - k_F| > k_0} u(\mathbf{k}_1 - \mathbf{k}_3) \times$$
$$\times \frac{\tanh\left(\beta\varepsilon_0(\mathbf{k}_3)/2\right)}{2\varepsilon_0(\mathbf{k}_3)} t'(\mathbf{k}_3, \mathbf{k}_2 ; 0) d^3k_3 = u(\mathbf{k}_1 - \mathbf{k}_3). \tag{23.19}$$

This equation can be rewritten in the form

$$t'(\mathbf{k}_1, \mathbf{k}_2 ; 0) + (2\pi)^{-3} \int u(\mathbf{k}_1 - \mathbf{k}_3) t'(\mathbf{k}_3, \mathbf{k}_2 ; 0) \frac{m}{k_3^2} d^3k_3$$

$$= u(\mathbf{k}_1 - \mathbf{k}_2) + (2\pi)^{-3} \int u(\mathbf{k}_1 - \mathbf{k}_3) \times \tag{23.20}$$

$$\times t'(\mathbf{k}_3, \mathbf{k}_2 ; 0)\left(\frac{m}{k_3^2} - \frac{\tanh\left(\beta\varepsilon_0(\mathbf{k}_3)/2\right)}{2\varepsilon_0(\mathbf{k}_3)}\right) d^3k_3$$

and the operator acting on t' at the left-hand side can be inverted. This trick, which has been used many times above, allows us to determine the function $t'(k_1, k_2; 0)$ which can be replaced by the constant

$$
\begin{aligned}
t' &= t_0 + \frac{mt_0^2}{(2\pi)^3} \int_{|k - k_F| > k_0} d^3k \left(\frac{m}{k^2} - \frac{\tanh(\beta \varepsilon_0(\mathbf{k})/2)}{2\varepsilon_0(\mathbf{k})} \right) \\
&\approx t_0 + \frac{mt_0^2 k_F}{2\pi^2} \left(2 - \ln \frac{2k_F}{k_0} \right),
\end{aligned}
\tag{23.21}
$$

in the domain $k \sim k_F$. This constant depends on the auxiliary parameter k_0. However, it will become evident from the following that the final results are k_0-independent. Let us make yet two remarks concerning formula (23.16).

(1) The incorporation of equal-spin particles into the description of scattering is equivalent to replacing the antisymmetric potential by the antisymmetric t matrix vanishing if t' is replaced by a constant. Thus, the interaction of equal-spin particles does not contribute to S' in the first approximation.

(2) We have neglected the A- and φ-dependences of p_0. The reasonability of this procedure will be discussed below.

The effect of superconductivity should demonstrate itself after the integration of the functional $\exp S'$ over slow variables ψ_0 and $\bar{\psi}_0$. Let us proceed from the integration over Fermi fields to the integration over auxiliary Bose fields. This completes the analogy between the superconductivity and the superfluidity. To accomplish this idea, we insert the Gaussian integral

$$
\int \exp \left\{ \int d\tau \, d^3x(t')^{-1} \bar{c}(\mathbf{x}, \tau) c(\mathbf{x}, \tau) \right\} d\bar{c} \, dc
\tag{23.22}
$$

over an auxiliary Bose field $c(\mathbf{x}, \tau)$ into the integral over Fermi fields ψ_0 and $\bar{\psi}_0$. If the interaction potential $u(\mathbf{x} - \mathbf{y})$ is assumed to be attractive and the quantity t' negative, then the quadratic form in the integrand of (23.22) is negatively definite. After the translation of integration variables,

$$
c(\mathbf{x}, \tau) \rightarrow c(\mathbf{x}, \tau) + t' \psi_{0+}(\mathbf{x}, \tau) \psi_{0-}(\mathbf{x}, \tau);
$$

$$
\bar{c}(\mathbf{x}, \tau) \rightarrow \bar{c}(\mathbf{x}, \tau) + t' \bar{\psi}_{0-}(\mathbf{x}, \tau) \bar{\psi}_{0+}(\mathbf{x}, \tau)
\tag{23.23}
$$

the integrand of the integral over $c, \bar{c}$ transforms into $\exp S''$, where the functional S'' is defined by the relation

$$S'' = \int d\tau \, d^3x \left\{ \sum_s \left(\bar{\psi}_{0s} \partial_\tau \psi_{0s} - \frac{1}{2m} |(\nabla + ie\mathbf{A})\psi_{0s}|^2 + \right. \right.$$

$$\left. \left. + \left(\mu + ie\varphi + \frac{eHs}{2m} \right) \bar{\psi}_{0s}\psi_{0s} \right) + \psi_{0+}\psi_{0-}c + \bar{\psi}_{0-}\bar{\psi}_{0+}\bar{c} + \frac{1}{t'}|c|^2 \right\} \tag{23.24}$$

It is quadratic in ψ_{0s} and $\bar{\psi}_{0s}$. Now it is possible to calculate the Gaussian integral over ψ_{0s} and $\bar{\psi}_{0s}$. It is formally equal to the determinant of the operator

$$\hat{M} = \begin{pmatrix} \partial_\tau + \dfrac{eH}{2m} - \dfrac{1}{2m}(\nabla + ie\mathbf{A})^2 + \mu + ie\varphi, & -\bar{c} \\[2mm] -c, \ \partial_\tau + \dfrac{eH}{2m} + \dfrac{1}{2m}(\nabla - ie\mathbf{A})^2 - \mu - ie\varphi \end{pmatrix} \tag{23.25}$$

acting on the pair of Fermi fields $\psi_{0+}, \bar{\psi}_{0-}$. The operator depends on the functions $\varphi, \mathbf{A}, c$ and $\bar{c}$. Its determinant can be regularized by dividing it by the determinant of the operator M_0 corresponding to the zero values of the variables $\varphi, \mathbf{A}, c, \bar{c}$. Thus, we can write

$$\int \exp S'' \, d\bar{\psi}_0 \, d\psi_0 \, d\bar{c} \, dc$$

$$= \int \exp \left\{ \int (t')^{-1} |c|^2 \, d\tau \, dx + \ln \det \hat{M}/\hat{M}_0 \right\} d\bar{c} \, dc. \tag{23.26}$$

Let us evaluate $\ln \det \hat{M}/\hat{M}_0$ according to the formula

$$\ln \det \hat{M}/\hat{M}_0 = \ln \det \hat{M}/\hat{M}_1 + \ln \det \hat{M}_1/\hat{M}_0, \tag{23.27}$$

where $\hat{M}_1$ is the operator $\hat{M}$ (23.25) at $c = \bar{c} = 0$. The second term in (23.27) does not depend on the Bose field $c(\mathbf{x}, \tau)$; it will be investigated later. The first term can be expanded into the functional series c and $\bar{c}$:

$$\ln \det \hat{M}/\hat{M}_1 = \mathrm{Sp} \ln \hat{M}/\hat{M}_1 = \mathrm{Sp} \ln (I - \hat{G}\hat{u})$$

$$= - \sum_{n=1}^{\infty} \frac{1}{n} \int \prod_{i=1}^{n} d^3x_i \, d\tau_i \, \mathrm{tr} \, \hat{G}(\mathbf{x}_1, \tau_1; \mathbf{x}_2, \tau_2 | \varphi, \mathbf{A}) \times \tag{23.28}$$

$$\times \hat{u}(\mathbf{x}_2, \tau_2) G(\mathbf{x}_2, \tau_2; \mathbf{x}_3, \tau_3 | \varphi, \mathbf{A}) \hat{u}(\mathbf{x}_3, \tau_3), \dots, \hat{u}(\mathbf{x}_1, \tau_1).$$

Here, the symbol tr denotes the range-two matrix trace, $\hat{u}(\mathbf{x}, \tau)$ is the matrix

$$\hat{u}(\mathbf{x}, \tau) = \begin{pmatrix} 0 & c(\mathbf{x}, \tau) \\ \bar{c}(\mathbf{x}, \tau) & 0 \end{pmatrix}; \tag{23.29}$$

and $\hat{G}$ is the Green's function matrix

$$\hat{G} = \begin{pmatrix} G_+ & 0 \\ 0 & G_- \end{pmatrix}, \qquad (23.30)$$

in which $G_\pm$ are the Green's functions defined by the equations

$$\left[\partial_\tau + \frac{eH}{2m} \mp \left(\frac{1}{2m} (\nabla \pm ie\mathbf{A})^2 - \mu - ie\varphi \right) \right] \times$$
$$\times G_\mp(\mathbf{x}, \tau; \mathbf{y}, \tau_1 | \varphi, \mathbf{A}) = \delta(\mathbf{x} - \mathbf{y})\delta(\tau - \tau_1). \qquad (23.31)$$

Only the terms with even n contribute to expansion (23.28), as $\hat{G}$ is a diagonal matrix and the diagonal elements of $\hat{u}$ are equal to zero.

We restrict ourselves to terms of the second and fourth order in $c(\mathbf{x}, \tau)$ in expansion (23.38). This can be done near the phase transition point, where, as will be shown below, the leading contribution to the integral of $|c(\mathbf{x}, \tau)|^2$ originates from a neighbourhood of a positive constant ρ_0 which has the meaning of condensate density and vanishes in the limit $T \to T_c$.

We neglect the dependence of Green's functions $G_\mp$ on the fields φ and $\mathbf{A}$ in the fourth order term and assume that in the first approximation the arguments of all functions $\hat{u}$ are identical. This results in the following expression of the fourth-order term:

$$-\tfrac{1}{2} \int \prod_{i=1}^{4} d\tau_i \, dx_i |c(\mathbf{x}, \tau)|^4 G_+(1, 2) G_-(2, 3) G_+(3, 4) G_-(4, 1), \quad (23.32)$$

where $G_\pm(1, 2) = G_\pm(\mathbf{x}_1, \tau_1; \mathbf{x}_2, \tau_2)$.

Proceeding to the momentum-representation of Green's functions $G_\pm$, we are able to perform the integration over $(\mathbf{x}_2, \tau_2), (\mathbf{x}_3, \tau_3), (\mathbf{x}_4, \tau_4)$. This leads to the formula

$$-\frac{b}{2} \int |c(\mathbf{x}, \tau)|^4 \, d\tau \, d^3x. \qquad (23.33)$$

for (23.32). The coefficient b

$$b = \frac{T}{V} \sum_p (\omega^2 + \xi^2(\mathbf{k}))^{-2} \qquad (23.34)$$

can be easily calculated integrating (in the interval $(k - k_F) < k_0$) over $\xi(\mathbf{k}) = (k^2/2m) - \mu$:

$$b = mk_F \frac{T}{2\pi^2} \sum_\omega \int \frac{d\xi}{(\xi^2 + \omega^2)^2}$$

$$= mk_F \frac{T}{2\pi^2} \frac{\pi}{4} \sum_\omega \frac{1}{|\omega|^3} = \frac{7\zeta(3)mk_F}{32\pi^4 T^2},$$

(23.35)

where $\zeta(3)$ is the Riemann ζ-function for $s = 3$ and T in formula (23.35) should be put equal to the phase transition temperature T_c (22.31).

The second-order term in expansion (23.28) has the form

$$-\int dx\,dy\,d\tau_1\,G_+(\mathbf{x}, \tau; \mathbf{y}, \tau_1; \varphi, \mathbf{A}) \times$$

$$\times G_-(\mathbf{y}, \tau_1; \mathbf{x}, \tau; \varphi, \mathbf{A})\bar{c}(\mathbf{x}, \tau)c(\mathbf{y}, \tau_1).$$

(23.36)

In the sequel we use the approximation

$$G_\pm(\mathbf{x}, \tau; \mathbf{y}, \tau_1; \varphi, \mathbf{A})\exp\left\{\pm ie\int_\mathbf{x}^\mathbf{y} (\mathbf{A}\,d\mathbf{x} + \varphi\,d\tau)\right\}G_{0\pm}(\mathbf{x} - \mathbf{y}, \tau - \tau_1),$$

(23.37)

which takes into account the explicit dependence of $G_\pm$ on the fields φ and $\mathbf{A}$. This formula can be obtained from (13.14) for the Green's function of an electron in a slowly oscillating field using nonrelativistic approximation and performing the integral along the straight line connecting $(\mathbf{x}, \tau)$ and $(\mathbf{y}, \tau_1)$.

Employing the relation

$$c(\mathbf{y}, \tau_1) = \exp\left\{i\int_\mathbf{x}^\mathbf{y} (\nabla\,d\mathbf{x} + \partial_\tau\,d\tau)\right\}c(\mathbf{x}, \tau),$$

(23.38)

one can rewrite (23.36) in the form

$$-\int dx\,dy\,d\tau\,d\tau'\,G_{0+}(\mathbf{x} - \mathbf{y}, \tau - \tau')G_{0-}(\mathbf{y} - \mathbf{x}, \tau' - \tau)\bar{c}(\mathbf{x}, \tau) \times$$

$$\times \exp\left\{i\int_\mathbf{x}^\mathbf{y} ((\nabla + 2ie\mathbf{A})\,d\mathbf{x} + (\partial_\tau + 2ie\varphi)\,d\tau)\right\}c(\mathbf{x}, \tau).$$

(23.39)

From this formula it can be seen that, to calculate the value of this functional, one has first to calculate the functional for vanishing external fields and then to replace the derivatives of $c(\mathbf{x}, \tau)$ by the 'covariant derivatives' according to the rule

$$\nabla \to \nabla + 2ie\mathbf{A}; \qquad \partial_\tau \to \partial_\tau + 2ie\varphi.$$

(23.40)

Functional (23.36) can be for $\varphi = 0$ and $\mathbf{A} = 0$ most easily calculated in p representation, where it is equal to

$$- \sum_p A(p)c^+(p)c(p). \tag{23.41}$$

Here

$$A(p) = -\frac{T}{V} \sum_{p_1} [(i\omega_1 - \varepsilon(\mathbf{k}_1))(i\omega - i\omega_1 - \varepsilon(\mathbf{k} - \mathbf{k}_1))]^{-1}$$

$$\tag{23.42}$$

$$= -(2\pi)^{-3} \int_{|\mathbf{k}_1 - \mathbf{k}_F| < k_0} \frac{\tanh(\beta\varepsilon(\mathbf{k}_1)/2) + \tanh(\beta\varepsilon(\mathbf{k} - \mathbf{k}_1)/2)}{2(\varepsilon(\mathbf{k}_1) + \varepsilon(\mathbf{k} - \mathbf{k}_1) - i\omega)} \, d^3k_1,$$

where $\varepsilon(k) = (k^2/2m) - \mu$.

Let us extract the constant $A(0)$ from the $A(p)$ and add the contribution of this constant to $(t')^{-1} \int |c|^2 \, d\tau \, dx$. For $\varphi = 0$ and $\mathbf{A} = 0$ we obtain the expression

$$\sum_p (A(0) - A(p))c^+(p)c(p) + ((t')^{-1} - A(0)) \times$$

$$\times \int |c(\mathbf{x}, \tau)|^2 \, d\tau \, d^3x - \frac{b}{2} \int |c(\mathbf{x}, \tau)|^4 \, d\tau \, d^3x. \tag{23.43}$$

The functional resembles the action functional of a nonideal Bose gas. The first component is the kinetic term, and the third component describes the self-interaction of the Bose field $c(\mathbf{x}, \tau)$. Using (23.21), the coefficient of the second term can be written in the form

$$t_0^{-1} - \frac{mk_F}{2\pi^2}\left(2 - \ln\frac{2k_F}{k_0}\right) + (2\pi)^{-3} \int \frac{\tanh(\beta\varepsilon(\mathbf{k})/2)}{2\varepsilon(\mathbf{k})} \, d^3k$$

$$\tag{23.44}$$

$$= t_0^{-1} + \frac{mk_F}{2\pi^2}\left(-2 + \ln\frac{2k_F}{k_0} + \int_0^{k_F k_0/m} \frac{\tanh(\beta\xi/2)}{\xi} \, d\xi\right).$$

This expression vanishes at $T = T_c$ by virtue of (22.28). It can therefore be rewritten in the form

$$\frac{mk_F}{2\pi^2} \int_0^\infty \frac{\tanh(\beta\xi/2) - \tanh(\beta_c\xi/2)}{\xi} \, d\xi = \frac{mk_F}{2\pi^2} \ln\frac{T_c}{T} \approx \frac{mk_F}{2\pi^2}\frac{\Delta T}{T_c}, \tag{23.45}$$

where $\Delta T = T_c - T$. For $T < T_c$, coefficient (23.44) is positive and functional

(23.43) has a maximum at $|c(\mathbf{x}, \tau)|^2 = \rho_0 = \text{const}$. This constant quantity

$$\rho_0 = \frac{mk_F}{2\pi^2} \frac{\Delta T}{T_c b} = \frac{16 T_c \Delta T}{7\pi\zeta(3)} \tag{23.46}$$

has the meaning of condensate density.

Thus, in the formalism developed so far, the phase transition of a Fermi system can be described in the same way as in the theory of Bose gas. The main contribution to the integral over the auxiliary Bose field $c(\mathbf{x}, \tau)$ is given by functions with the modulus squared $|e|^2$ close to the condensate density ρ_0.

In the following, functional (23.43) will be applied only to the description of stationary effects in superconductors. To proceed, we integrate over the Fourier coefficients $c(p)$, $c^+(p)$ with $\omega \neq 0$. This yields a constant additional term to action (23.43) and leads to the renormalization of coefficients in the second- and fourth-order forms in (23.49). One can show that these corrections are small and can be neglected in the first approximation. Thus, up to an unessential constant term, the result of integration over $c(p)$, $c^+(p)$, $\omega \neq 0$ is that one can consider only τ-independent functions $c(\mathbf{x}, \tau)$, $c^+(\mathbf{x}, \tau)$, with Fourier coefficients having the frequency $\omega = 0$, in (23.43).

Finally, we obtain the functional

$$\beta \sum_{\mathbf{k}} (A(0) - A(\mathbf{k})) c^+(\mathbf{k}, 0) c(\mathbf{k}, 0) + \frac{\beta mk_F \Delta T}{2\pi^2 T_c} \times$$

$$\times \int d^3x |c(\mathbf{x})|^2 - \frac{b\beta}{2} \int |c(\mathbf{x}, \tau)|^4 d^3x. \tag{23.47}$$

instead of (23.43). For small $\mathbf{k}$ we have

$$A(0) - A(\mathbf{k}) = (2\pi)^{-3} \times$$

$$\times \int \left[\frac{\tanh(\beta/2)\varepsilon(\mathbf{k}_1 + (\mathbf{k}/2)) + \tanh(\beta/2)\varepsilon(\mathbf{k}_1 - (\mathbf{k}/2))}{2(\varepsilon(\mathbf{k}_1 + (\mathbf{k}/2)) + \varepsilon(\mathbf{k}_1 - (\mathbf{k}/2)))} - \right.$$

$$\left. - \frac{\tanh(\beta\varepsilon(\mathbf{k}_1)/2)}{2\varepsilon(\mathbf{k}_1)} \right] d^3k_1 \tag{23.48}$$

$$\approx -\frac{1}{(2\pi)^3} \int \frac{(\tanh(\beta/2)\xi)''}{4\xi} \left(\frac{\beta(\mathbf{k}_1 \mathbf{k})}{4m} \right)^2 d^3k_1$$

$$= -\frac{k^2 k_F^2}{384\pi^2 m T_c^2} \int (\tanh x)'' \frac{dx}{x},$$

where

$$\int (\tanh x)'' \frac{dx}{x} = -\frac{32}{\pi^2} \sum_{n=0}^{\infty} \frac{1}{(2n+1)^3} = -\frac{28}{\pi^2} \zeta(3). \tag{23.49}$$

Finally, the kinetic term in (23.47) takes the form

$$\beta \sum_{\mathbf{k}} (A(0) - A(\mathbf{k}, 0)) |c(\mathbf{k}, 0)|^2$$

$$= -\frac{\beta k_F \varepsilon_F 7\zeta(3)}{48\pi^4 T_c^2} \int |\nabla c(\mathbf{x}, \tau)|^2 \, d^3 x, \tag{23.50}$$

where $\varepsilon_F = \mu$ is the Fermi energy.

We now introduce the new function

$$\psi(\mathbf{x}) = \left(\frac{7mk_F \varepsilon_F \zeta(3)}{12\pi^4 T_c^2} \right)^{1/2} c(\mathbf{x}) \tag{23.51}$$

instead of $c(\mathbf{x})$ and write down functional (23.47), replacing the derivative ∇ in the kinetic term by the covariant derivative $\nabla + 2ie\mathbf{A}$ according to (23.40):

$$\beta \int \left(-\frac{1}{4m} |(\nabla + 2ie\mathbf{A})\psi(\mathbf{x})|^2 + \Lambda |\psi(\mathbf{x})|^2 - \frac{g}{2} |\psi(\mathbf{x})|^4 \right) d^3 x. \tag{23.52}$$

Here

$$\Lambda = \frac{6\pi^2 T_c \Delta T}{7\zeta(3)\varepsilon_F}; \qquad g = \frac{9\pi^4 T_c^2}{14\zeta(3)mk_F \varepsilon_F^2}. \tag{23.53}$$

The expression

$$-\frac{\beta}{2} \int (\operatorname{rot} \mathbf{A})^2 \, d^3 x, \tag{23.54}$$

describing the stationary magnetic field, as well as the contribution of $\ln \det \hat{M}_1/\hat{M}_0$ from (23.57), should be added to functional (23.52). This functional describes the magnetic polarization of a medium. For weak fields $\mathbf{A}(\mathbf{x})$ it differs from (23.54) by the polarization coefficient κ. The coefficient is of order $\alpha v_F/c$, where $\alpha = 1/137$ is the fine structure constant, v_F/c is the ratio of the Fermi-level velocity to the velocity of light. As $\kappa \ll 1$, one can neglect the contribution of $\ln \det \hat{M}_1/\hat{M}_0$ in the first approximation and consider the functional

$$S''' = -\beta \int d^3x \left(\frac{1}{2}(\text{rot}\,\mathbf{A})^2 + \frac{1}{4m}|(\nabla + 2ie\mathbf{A})\psi(\mathbf{x})|^2 - \right.$$

$$\left. - \Lambda|\psi(\mathbf{x})|^2 + \frac{g}{2}|\psi(\mathbf{x})|^4 \right). \tag{23.55}$$

The functional $\exp S''$ should be integrated over the complex functions $\psi(\mathbf{x})$, $\bar{\psi}(\mathbf{x})$ as well as over the vector potential $\mathbf{A}(x)$. The leading contribution to this integral is given by values close to the classical solutions. These solutions satisfy the system of equations

$$-\frac{1}{2m}(\nabla + 2ie\mathbf{A})^2\psi(\mathbf{x}) - \Lambda\psi(\mathbf{x}) + g|\psi(\mathbf{x})|^2\psi(\mathbf{x}) = 0;$$

$$\text{rot}\,\text{rot}\,\mathbf{A} = \mathbf{j} \equiv -\frac{ie}{2m}(\nabla\bar{\psi}\psi - \bar{\psi}\nabla\psi) - \frac{e^2}{m}\bar{\psi}\psi\mathbf{A}, \tag{23.56}$$

which are identical with the equations of Ginzburg and Landau.

The functional integration method employed in the derivation of these equations allows us, in principle, to also describe the fluctuations of the field $\psi, \bar{\psi}$ and $\mathbf{A}$ around the classical solutions. To account for these fluctuations, one has to shift by the solutions of classical equations

$$\psi \to \psi + \psi_0; \qquad \bar{\psi} \to \bar{\psi} + \bar{\psi}_0; \qquad \mathbf{A} \to \mathbf{A} + \mathbf{A}_0 \tag{23.57}$$

in the functional integral and to construct a perturbation theory where the quadratic form of the fields $\psi, \bar{\psi}$ and $\mathbf{A}$ represents the nonperturbed action and higher-order forms are regarded as a perturbation. This will be not be accomplished here.

As an example of the application of Ginzburg–Landau equations, we shall determine the vertex lattice structure near the phase transition and the value of the critical magnetic field H (usually denoted by H_{c_2}). The quantity H_{c_2} is the field which must be applied to a superconductor so as to cause its transition to the normal state. The right-hand side of the second equation and the term $g|\psi|^2\psi$ in the first equation can be neglected near the phase transition. The second equation has the solution $\mathbf{A} = \frac{1}{2}[\mathbf{Hx}]$.

The first equation

$$-\frac{1}{4m}(\nabla + ie[\mathbf{Hx}])^2\psi(\mathbf{x}) = \Lambda\psi(\mathbf{x}) \tag{23.58}$$

is the Schrödinger equation for a particle with charge $2e$ and mass $2m$ in a constant magnetic field $\mathbf{H}$. If the solution ψ does not depend on the component of $\mathbf{x}$ parallel to $\mathbf{H}$, the spectrum of possible values of Λ (i.e., the Landau levels) has the form

$$\Lambda = \frac{eH}{m}(n + \tfrac{1}{2}), \quad n = 0, 1, 2, \ldots \tag{23.59}$$

The smallest value of Λ ($n = 0$) corresponds to the highest value of the critical field

$$H = H_{c2} = \frac{2m\Lambda}{e} = \frac{12\pi^2 m T_c \Delta T}{7\zeta(3)e\varepsilon_F}. \tag{23.60}$$

Let us construct the solution of (23.58) describing a periodic structure. We take a particular solution $\varphi(\mathbf{x})$ corresponding to the lowest Landau level. The function $\varphi(\mathbf{x} + \mathbf{a})$ satisfies Equation (23.58) with $[\mathbf{Hx}]$ replaced by $[\mathbf{Hx}] + [\mathbf{Ha}]$, which corresponds to the gauge transformation of the vector potential. One can return to the previous gauge transforming $\varphi \to \varphi \exp(-ie[\mathbf{Ha}]\mathbf{x})$.

Thus, the function $\varphi(\mathbf{x} + \mathbf{a}) \exp(-ie[\mathbf{Ha}]\mathbf{x})$ satisfies Equation (23.58), too. Putting $\mathbf{a}$ equal to the lattice vector and performing the summation over all such vectors, we obtain the function

$$\psi(\mathbf{x}) = \sum_{\mathbf{a}} \exp(-ie[\mathbf{Ha}]\mathbf{x})\varphi(\mathbf{x} + \mathbf{a}), \tag{23.61}$$

describing a periodic structure. Recall that two arbitrary lattice vectors have to satisfy condition (23.11). Using this equation, we obtain the transformation law for $\psi(\mathbf{x})$ shifted by the lattice vector $\mathbf{b}$:

$$\psi(\mathbf{x} + \mathbf{b}) = \exp(ie[\mathbf{Hb}]\mathbf{x})\psi(\mathbf{x}). \tag{23.62}$$

It is suitable to take the function

$$\varphi(\mathbf{x}) = (x - iy)\exp\left(-\frac{eH}{2}(x^2 + y^2)\right), \tag{23.63}$$

where x, y are the coordinates in the plane perpendicular to H, as the particular solution corresponding to the lowest Landau level. The function (23.61) corresponding to this solution acquires zero values in the lattice sites and its phase acquires an increment 2π per passage around these points. One can therefore say that function (23.61) describes the quantum vortex lattice formed in a superconductor under the influence of a magnetic field less than the critical value H_{c_2} (23.60). One can show (considering also the nonlinear terms in Equations (23.56)) that the energy is minimal for a triangular lattice.

The periodic quantum vortex lattice also arises in a rotating superfluid Bose system, where the role of the magnetic field $\mathbf{H}$ is assumed by the angular velocity vector ω. The method of description of a vortex lattice developed in this paragraph also applies to Bose systems.

24. BOSE SPECTRUM OF SUPERFLUID FERMI GAS

It is natural to call the functional

$$
\int (\tilde{t}')^{-1} |C(\mathbf{x}, \tau)|^2 \, d\mathbf{x} \, d\tau + \ln \det \hat{M}/\hat{M}_0
$$
$$
= (\tilde{t}')^{-1} \sum_p C^+(p)C(p) + \ln \det \hat{M}/\hat{M}_0 \tag{24.1}
$$

introduced in Section 23 the functional of the hydrodynamical action of the superfluid Fermi system. Fields $C(\mathbf{x}, \tau)$, $\bar{C}(\mathbf{x}, \tau)$ describe the Cooper pairs of fermions. As already mentioned in Section 23, the phase transition in a Fermi system is nothing else than the Bose condensation of fields $C(\mathbf{x}, \tau)$, $\bar{C}(\mathbf{x}, \tau)$. Functional S_h contains complete information about collective Bose excitations in the Fermi system. Here we shall derive the spectrum of these excitations in two limiting cases:

(i) $|T - T_c| \ll T_c$, (ii) $T \ll T_c$.

In the case $|T - T_c| \ll T_c$ (the Ginzburg–Landau domain) the functional has the form

$$
\sum_p (A(0) - A(p))C^+(p)C(p) + \frac{mk_F(T_c - T)}{2\pi^2 T_c} \sum_p C^+(p)C(p) -
$$
$$
- \frac{b}{2\beta V} \sum_{p_1 + p_2 = p_3 + p_4} C^+(p_1)C^+(p_2)C(p_3)C(p_4). \tag{24.2}
$$

This formula is obtained from (23.43) replacing $t' - A(0)$ by the expression $mk_F \, \Delta T/2\pi^2 T_c$, and then by using the Fourier component $C(p), C^+(p)$ of fields $C(\mathbf{x}, \tau)$, $\bar{C}(\mathbf{x}, \tau)$.

(a) $T > T_c$. The Bose spectrum is determined by the quadratic part of S_h from the equation

$$
A(0) - A(p) + \frac{mk_F(T_c - T)}{2\pi^2 T_c} = 0 \tag{24.3}
$$

after the analytic continuation $i\omega \to E$. For small momenta k and the analytic continuation to a domain $|\omega| \ll T$ the expression

$$A(0) - A(p) = (2\pi)^{-3} \int d^3k_1 \times$$

$$\times \left(\frac{\tanh{(\beta/2)}\xi(\mathbf{k}_1 + (\mathbf{k}/2)) + \tanh{(\beta/2)}\xi(\mathbf{k}_1 - (\mathbf{k}/2))}{2(\zeta(\mathbf{k}_1 + (\mathbf{k}/2)) + \xi(\mathbf{k}_1 - (\mathbf{k}/2)) - i\omega)} - \frac{\tanh{(\beta/2)}\xi(\mathbf{k}_1)}{2\xi(\mathbf{k}_1)} \right),$$

(24.4)

has the form

$$-\frac{7\xi(3)k_F^3 k^2}{96\pi^4 m T^2} - \frac{mk_F |\omega|}{16\pi T},$$

(24.5)

where $\xi(\mathbf{k}) = (k^2 - k_F^2)/2m$. Replacing $T \to T_c, |\omega| \to \omega \to -iE$ in (24.5) and then substituting (24.5) into (24.3) we obtain the Bose spectrum of the form

$$E = -i\frac{7\xi(3)k_F^2 k^2}{6\pi^3 m^2 T_c} - i\frac{8}{\pi}(T - T_c)$$

(24.6)

(b) $T < T_c$. In action (24.2) we separate the condensate by the transformation

$$C(p) = a(p) + \sqrt{\beta V} \sqrt{\rho_0} \delta_{p,0}$$

(24.7)

where $\rho_0 = 16T_c \Delta T / 7\pi\xi(3)$ is the Bose condensate density. Consider the quadratic form of fields $a(p), a^+(p)$

$$-\sum_p A_1(p)a^+(p)a(p) -$$

$$-\frac{1}{2}\sum_p B_1(p)a^+(p)(a^+(-p) + a(p)a(-p)),$$

(24.8)

where

$$A_1(p) = A(p) - A(0) + \frac{mk_F(T - T_c)}{2\pi^2 T_c} + 2b\rho_0,$$

$$B_1(p) = b\rho_0.$$

(24.9)

The Bose spectrum for $T < T_c$ is determined in the first approximation by quadratic form (24.8) from the equation

$$\det\begin{pmatrix} A_1(p) & B_1(p) \\ B_1(p) & A_1(-p) \end{pmatrix} = A_1(p)A_1(-p) - B^2(p) = 0.$$

(24.10)

Since $A_1(p)$ is even in p for small $p = (\mathbf{k}, \omega)$ according to (24.3) and (24.5),

Equation (24.10) splits into two equations

$$A_1(p) - B_1(p) = 0, \qquad A_1(p) + B_1(p) = 0. \tag{24.11}$$

Now, by using (24.3) and (24.5) it is not difficult to find out two branches of the spectrum:

$$E_1(k) = -i\frac{7\zeta(3)k_F^2 k^2}{6\pi^3 m^2 T_c},$$

$$\tag{24.12}$$

$$E_2(k) = -i\frac{7\zeta(3)k_F^2 k^2}{6\pi^3 m^2 T_c} - i\frac{16(T_c - T)}{\pi}.$$

The first branch is determined by the equation $A_1 - B_1 = 0$ and begins from zero ($E(0) = 0$). The second one is determined by the equation $A_1 + B_1 = 0$ and differs from branch (24.6) appearing for $T > T_c$ by the replacement $T - T_c \to 2(T_c - T)$.

Let us consider now the case $T \ll T_c$. In order to evaluate the Bose spectrum, the condensate in action (24.11) will be separated by transformation (24.7). Variables $a(p)$, $a^+(p)$ describe fluctuations of fields C, C^+ around the condensate. The condensate density ρ_0 is determined from the condition for the maximum of the hydrodynamical action after the substitution $C = C^+ = \sqrt{\beta V \rho_0}\,\delta_{p,0}$. The equation for $\Delta = \sqrt{\rho_0}$ is nothing more than the equation for the gap (22.28). In the limit $T \to 0$ it has the solution

$$\Delta = 4k_F^2 m^{-1} \exp(-2 - \xi^{-1}) = \frac{\pi}{\gamma} T_c.$$

Consider the quadratic part from S_h of fields a, a^+. It is just this part that determines the energy spectrum in the first approximation. It has the form (24.8) with

$$A_1(p) = (\beta V)^{-1} \sum_{p_1} G_+(p_1) G_-(p + p_1) - (\tilde{t})^{-1},$$

$$\tag{24.13}$$

$$B_1(p) = (\beta V)^{-1} \sum_{p_1} G_1(p_1) G_1(p + p_1),$$

where

$$G_\pm(p) = \frac{i\omega \pm \xi(k)}{\omega^2 + \xi^2(k) + \Delta^2}, \qquad G_1(p) = \frac{\Delta}{\omega^2 + \xi^2(k) + \Delta^2}. \tag{24.14}$$

The equation for the spectrum has the form (24.10) and according to parity of $A_1(p)$ it splits into two equations (24.11) corresponding to two branches of the spectrum. The expression $A_1(0) - B_1(0)$ vanishes due to the equation determining gap Δ. Thus the branch of the spectrum determined by the first equation begins from zero. For its evaluation we rewrite the equation in the form

$$A_1(p) - B_1(p) - A_1(0) + B_1(0) = 0,$$

or

$$(\beta V)^{-1} \sum_{p_2 - p_1 = p} \left[\frac{(i\omega_1 - \xi_1)(i\omega_2 + \xi_2) - \Delta^2}{(\omega_1^2 + \xi_1^2 + \Delta^2)(\omega_2^2 + \xi_2^2 + \Delta^2)} + \frac{1}{\omega_1^2 + \xi_1^2 + \Delta^2} \right] = 0,$$

(24.15)

where $\xi_i = \xi(\mathbf{k}_i)$

For $T \to 0$ the sum in k, ω can be replaced by the integral and then we can introduce the integral over the neighbourhood of the Fermi sphere according to the rule

$$(\beta V)^{-1} \sum_p \to (2\pi)^{-4} \int d\omega \, d^3k \to$$

$$(2\pi)^{-4} \int d\omega k^2 \, dk \, d\Omega \to \frac{m k_F}{(2\pi)^4} \int d\xi \, d\omega \, d\Omega$$

where $d\Omega$ is an element of a solid angle. In order to evaluate the integral with respect to ω_1, ξ_1 it is useful to use the Feynman trick standard in relativistic quantum theory and based on the identity

$$(ab)^{-1} = \int_0^1 d\alpha [\alpha a + (1 - \alpha)b]^{-2}.$$

We substitute $a = \omega_1^2 + \xi_1^2 + \Delta^2$, $b = \omega_2^2 + \xi_2^2 + \Delta^2$ and make replacements

$$\omega_1 \to \omega_1 - \alpha\omega, \qquad \omega_2 \to \omega_1 + (1 - \alpha)\omega,$$

$$\xi_1 \to \xi_1 - \alpha C_F(\mathbf{nk}), \qquad \xi_2 \to \xi_1 - (1 - \alpha)C_F(\mathbf{nk})$$

where $C_F = k_F/m$ is a velocity on the Fermi surface, $\mathbf{n}$ a unit vector orthogonal to the Fermi surface, $(\mathbf{k}, \omega) = p$ an external four-momentum. After these substitutions the integral with respect to ω_1, ξ_1 can easily be evaluated and the left-hand side of Equation (24.15) takes the form

$$\frac{\pi m k_F}{(2\pi)^4} \int_0^1 d\alpha \int d\Omega \left[\ln(1 + \Delta^{-2}\alpha(1 - \alpha)(\omega^2 + C_F^2(\mathbf{nk})^2)) + \right.$$

(24.16)

$$\left. + \frac{2\alpha(1 - \alpha)(\omega^2 + C_F^2(\mathbf{nk})^2)}{\Delta^2 + \alpha(1 - \alpha)(\omega^2 + C_F^2(\mathbf{nk})^2)} \right]$$

For small ω, k we can expand the integrand function upto the second power of $q^2 = \omega^2 + C_F^2(\mathbf{nk})^2$ and obtain

$$\frac{mk_F}{8\pi^2\Delta^2}\left[\omega^2 + \frac{C_F^2 k^2}{3} - \frac{1}{6\Delta^2}(\omega^4 + \tfrac{2}{3}\omega^2 C_F^2 k^2 + \tfrac{1}{5}C_F^4 k^4)\right]. \quad (24.17)$$

Replacing $i\omega \to E$ in (24.17) and putting it equal to zero we find the spectrum

$$E = uk(1 - \gamma k^2), \quad u = \frac{C_F}{\sqrt{3}}, \quad \gamma = \frac{C_F^2}{45\Delta^2}. \quad (24.18)$$

This branch of the spectrum (the Bogoliubov sound) is linear in k for small k. Positivity of coefficient γ ('dispersion') means a stability of an excitation with respect to a decay into two (or more) excitations of the same type.

Let us consider excitations described by the equation $A_1(p) + B_1(p) = 0$. Let us write the equation in the form

$$(\beta V)^{-1} \sum_{p_2-p_1=p}\left[\frac{(i\omega_1 - \xi_1)(\omega_2 + \xi_2) + \Delta^2}{(\omega_1^2 + \xi_1^2 + \Delta^2)(\omega_2^2 + \xi_2^2 + \Delta^2)} + \frac{1}{\omega_1^2 + \xi_1^2 + \Delta^2}\right] = 0. \quad (24.19)$$

By using the Feynman method and by integrating with respect to ω_1, ξ_1 we write the left-hand side of (24.19) in the form

$$\frac{\pi m k_F}{(2\pi)^4}\int_0^1 d\alpha \int d\Omega \times$$

$$\times \left[\ln(1 + \Delta^{-2}\alpha(1-\alpha)(\omega^2 + C_F^2(\mathbf{nk})^2)) + 2\right]$$

$$= \frac{\pi m k_F}{2(2\pi)^4}\int d\Omega(\omega^2 + 4\Delta^2 + C_F^2(\mathbf{nk})^2)\int_0^1 d\alpha \times \quad (24.20)$$

$$\times [\Delta^2 + \alpha(1-\alpha)(\omega^2 + C_F^2(\mathbf{nk})^2]^{-1}.$$

Expression (24.20) is equal to zero when $\omega^2 \to -4\Delta^2$ and $\mathbf{k} = 0$. For E^2 near to $4\Delta^2$ and small k the internal integral with respect to α in (24.20) is reciprocal to the root $\sqrt{4\Delta^2 - E^2 + C_F^2(\mathbf{nk})^2}$. As a result equation (24.19) goes to

$$\int_0^1 dx\sqrt{x^2 + z^2} = 0, \quad (24.21)$$

where

$$z^2 = (E^2 - 4\Delta^2)/C_F^2 k^2.$$

(24.22)

By introducing the new variable

$$t = \ln \frac{\sqrt{1+z^2}+1}{\sqrt{1+z^2}-1},$$

(24.23)

the solution of Equation (24.21) reduces to finding nontrivial ($t \neq 0$) roots of the equation

$$\sinh t + t = 0.$$

(24.24)

The roots group to four ones $\pm a \pm ib$. For the smallest root with respect to its modulus in the first quadrant t_1 and for the asymptotics t_n for $n \to \infty$ we have

$$t_1 \approx 2.251 + i4.212,$$

$$t_n = \ln \pi(4n-1) + i\left(2\pi n - \frac{\pi}{2}\right) + O(1).$$

(24.25)

The sequence t_n determines the sequence of the roots of the equation $A_1(p) + B_1(p) = 0$

$$E_n = 2\Delta + k^2 C_F^2/4\Delta \sinh^2 \frac{t_n}{2}.$$

(24.26)

These roots lie in different places of a Riemannian surface and concentrate at $E = 2\Delta$ for $n \to \infty$. The immediate physical meaning has root E_1, the first one appearing when analytic continuation from the upper to the lower E halfplane is performed. It corresponds to oscillations of the system density which may be excited by acting on the system with the frequency near to 2Δ.

We found the Bose spectrum of a superfluid Fermi gas in the approximation in which quasiparticles of field $C, \bar{C}$ noninteract. Irrespective of this, in general, the spectrum proves to be complex.

In order to take into account excitations and, in particular, to construct the kinetic theory, the higher orders in expansion of field $a, \bar{a}$ in hydrodynamical action S_h are necessary. The sum of diagrams which reduces to solutions of the kinetic equations (analogously to the situation mentioned in section 19 for the Bose gas) must lead to the branch of the second sound for $0 < T < T_C$. For $T \to 0$ its velocity is

$$u_2 = \frac{C_F}{3}. \tag{24.27}$$

Let us still note that in order to construct a successive perturbation theory it is more convenient to use variables density and phase instead of variables $a, \bar{a}$ used here. They are defined by the formulae

$$C(\mathbf{x}, \tau) = \sqrt{\rho(\mathbf{x}, \tau)} \exp(i\varphi(\mathbf{x}, \tau),$$
$$\bar{C}(\mathbf{x}, \tau) = \sqrt{\rho(\mathbf{x}, \tau)} \exp(-i\varphi(\mathbf{x}, \tau)) \tag{24.28}$$

In the new variables the branch of the Bogoliubov sound (24.18) corresponds to oscillations of the phase and excitations (24.26) to oscillations of the density.

25. A SYSTEM OF THE He³ TYPE

The most complex and interesting of the superfluid systems is without any doubt that of He³. For the temperature of the order 10^{-3} K there might exist several superfluid phases in He³ which pass one into another when changing the external conditions – temperature, pressure, and magnetic field. The difficulty in constructing a complete microscopic theory of He³ makes it expedient to study simplified models analogous to that of the Bose gas for He⁴.

It is just such a simplified model of He³ that will be considered in the functional integral formalism here. With the aim to describe collective Bose excitations in the system we pass from the integral with respect to Fermi fields to the integral with respect to the auxiliary Bose field which corresponds to the Bose excitations. This method was already applied in the model of the superfluid Fermi gas with a pairing in the s state. The characteristic feature of He³ is the pairing in the p state. This necessarily leads not to a scalar but to a (3×3) matrix wave function of the superfluid state. It is just this fact that guarantees the possibility of the existence of a few superfluid phases and a rich spectrum of collective excitations. In the simplified model it is enough to use a local Bose field describing the collective excitations. In a general case, it is necessary to introduce a bilocal formalism, i.e., to introduce fields depending not on one but on two space-time arguments.

Here we restrict ourselves to deriving the functional of the hydrodynamical action S_h which describes collective excitations of the model system of the He³ type and then we shall study phonon branches of the Bose spectrum (such that $E(\mathbf{k}) \to 0$ for $\mathbf{k} \to 0$) in detail, in one of the superfluid phases of the model (the so-called B phase). A more detailed study of the other branches of the Bose spectrum in the B phase, as well as of the Bose excitations in the A phase of the model, was done by P. N. Brusov and V. N. Popov [167–169]. The same authors suggest a two-dimensional model [170] which describes a film of superfluid He^3, and in which the superfluid phases are found and the Bose spectrum of excitations evaluated.

The starting point of the construction of the He^3 model is the functional of the Fermi gas action

$$S = \int_0^\beta d\tau \int d\alpha \sum_s \bar{\chi}_s(\mathbf{x}, \tau) \partial_\tau \chi_s(\mathbf{x}, \tau) - \int_0^\beta H'(\tau) \, d\tau \qquad (25.1)$$

corresponding to the Hamiltonian

$$H'(\tau) = \int d\mathbf{x} \sum_s \left(\frac{1}{2m} \nabla \bar{\chi}_s(\mathbf{x}, \tau) \nabla \chi_s(\mathbf{x}, \tau) - (\lambda + s\mu_0 H) \bar{\chi}_s(\mathbf{x}, \tau) \chi_s(\mathbf{x}, \tau) \right) +$$
$$+ \frac{1}{2} \int d\mathbf{x} \, d\mathbf{y} \, u(\mathbf{x} - \mathbf{y}) \sum_{s, s'} \bar{\chi}_s(\mathbf{x}, \tau) \bar{\chi}_{s'}(\mathbf{y}, \tau) \chi_{s'}(\mathbf{y}, \tau) \chi_s(\mathbf{x}, \tau) \qquad (25.2)$$

in which λ is a chemical potential, μ_0 a magnetic moment of the fermion, H a magnetic field and $s = \pm\frac{1}{2}$ a spin index.

We integrate the functional $\exp S$ with respect to the 'rapid' Fermi fields $\chi_{1s}(\mathbf{x}, \tau)$, $\bar{\chi}_{1s}(\mathbf{x}, \tau)$, i.e., with respect to the expansion coefficients $a_s(\mathbf{k}, \omega)$ in

$$\chi_s(\mathbf{x}, \tau) = \frac{1}{\sqrt{\beta V}} \sum_{\mathbf{k}, \omega} a_s(\mathbf{k}, \omega) \exp(i(\omega \tau + \mathbf{k}\mathbf{x})) \qquad (25.3)$$

for which either $|k - k_F| > k_0$ or $|\omega| > \omega_0$. The result of the integration will be written in the form

$$\int \exp S \, d\bar{\chi}_{1s} \, d\chi_{1s} = \exp \tilde{S}[\chi_{0s}, \bar{\chi}_{0s}] \qquad (25.4)$$

in which $\tilde{S}$ has the meaning of the action of the slow fields $\chi_{0s}, \bar{\chi}_{0s}$ for which $|k - k_F| < k_0$ and $|\omega| < \omega_0$.

The auxiliary parameters k_0, ω_0, separating the rapid fields from the slow ones, are defined up to the order of their magnitude and the physical results must not depend on their concrete choice.

The general functional $\tilde{S}$ is the sum of the functionals of even orders with respect to fields $\chi_{0s}, \bar{\chi}_{0s}$

$$\tilde{S} = \sum_{n=0}^{\infty} \tilde{S}_{2n} \tag{25.5}$$

Neglecting the higher functionals $\tilde{S}_6, \tilde{S}_8, \ldots$ and omitting the constant S_0, nonessential in what follows, consider the forms S_2 and S_4. Form S_2 corresponds to noninteracting quasi-particles near the Fermi surface and has the form

$$\tilde{S}_2 = \sum_{\mathbf{k}, \omega, s} \varepsilon_s(\mathbf{k}, \omega) a_s^+(\mathbf{k}, \omega) a_s(\mathbf{k}, \omega) \tag{25.6}$$

where

$$\varepsilon_s(\mathbf{k}, \omega) \approx Z^{-1}(i\omega - C_F(k - k_F) + s\mu H) \tag{25.7}$$

and the sum in (25.6) is over $\mathbf{k}, \omega$ in the intervals $|k - k_F| < k_0$, $|\omega| < \omega_0$. Here, assuming that $\varepsilon_s(\omega = 0, k = k_F, H = 0) = 0$ we expand ε_s in a series of ω, $k - k_F, H$ keeping in it the linear terms. Coefficient C_F has a meaning of the velocity on the Fermi surface, μ is a magnetic momentum of the quasi-particle and Z is a normalization constant.

Form S_4 describes the interaction of quasi-particles and has the form

$$-\frac{1}{\beta V} \sum_{p_1 + p_2 = p_3 + p_4} t_0(p_1, p_2, p_3, p_4) a_+^+(p_1) a_-^+(p_2) a_-(p_4) a_+(p_3) -$$

$$-\frac{1}{2\beta V} \sum_{p_1 + p_2 = p_3 + p_4} t_1(p_1, p_2, p_3, p_4) [2a_+^+(p_1) a_-^+(p_2) a_-(p_4) a_+(p_3) +$$

$$+ a_+^+(p_1) a_+^+(p_2) a_+(p_4) a_+(p_3) + a_-^+(p_1) a_-^+(p_2) a_-(p_4) a_-(p_3)]. \tag{25.8}$$

Here $p = (\mathbf{k}, w)$ is a four momentum, $t_0(p_i)$ is a symmetric and $t_1(p_i)$ an antisymmetric scattering amplitude with respect to each of the permutations $p_1 \rightleftarrows p_2$, $p_3 \rightleftarrows p_4$. In the neighbourhood of the Fermi sphere we can put $\omega_i = 0$, $\mathbf{k}_i = k_F \mathbf{n}_i$ $(i = 1, 2, 3, 4)$ where $\mathbf{n}_i$ are unit vectors such that $\mathbf{n}_1 + \mathbf{n}_2 = \mathbf{n}_3 + \mathbf{n}_4$. The amplitudes t_0, t_1 can depend only on two invariants, e.g., on $(\mathbf{n}_1, \mathbf{n}_2)$ and $(\mathbf{n}_1 - \mathbf{n}_2, \mathbf{n}_3 - \mathbf{n}_4)$, and, at the same time, t_0 is even and t_1 is odd with respect to the second invariant.

Therefore, the formulae

$$t_0 = f((\mathbf{n}_1, \mathbf{n}_2), (\mathbf{n}_1 - \mathbf{n}_2, \mathbf{n}_3 - \mathbf{n}_4)),$$
$$t_1 = (\mathbf{n}_1 - \mathbf{n}_2, \mathbf{n}_3 - \mathbf{n}_4) g((\mathbf{n}_1, \mathbf{n}_2), (\mathbf{n}_1 - \mathbf{n}_2, \mathbf{n}_3 - \mathbf{n}_4)) \tag{25.9}$$

hold, expressing t_0 and t_1 in terms of the functions f and g which are even with respect to the second variable.

The functional $\tilde{S}_2 + \tilde{S}_4$ defined by formulae (25.6)–(25.9) represents the most general expression which describes fermionic quasi-particles and their pair interaction near the Fermi surface. The method, by the use of which this functional is obtained in the functional integral formalism, represents an approach alternative to that used in the Landau theory of the Fermi fluid.

Now we shall consider the functional $\tilde{S}_2 + \tilde{S}_4$ with

$$f = 0, \qquad g = \text{const} < 0 \tag{25.10}$$

as a simplified model of He^3 with the pairing in the p state. We shall use the idea of a new field describing the Coulomb pairs. The condition $g = \text{const}$ (like the condition $f = \text{const}$ in the model of the superfluid Fermi gas in the previous paragraph) allows us to use a local field without passing to a bi-local formalism. As a matter of fact, this can also be done in a general case when functions f, g in (25.9) depend only on the first variable.

In order to realize the idea mentioned we put into the integral with respect to the Fermi fields, the Gauss integral of $\exp(\bar{c}\hat{A}c)$ with respect to the auxiliary complex Bose field c, where $\bar{c}\hat{A}c$ is a quadratic form with an operator $\hat{A}$. Then, in order to eliminate the fourth degree form in S_n of the Fermi fields in the functional integral, we transform the Bose field into the quadratic form of the Fermi fields. After that the integral with respect to the Fermi field inverts to the Gaussian one and is equal to the determinant of operator $\hat{M}(c, \bar{c})$ depending on the Bose fields $c, \bar{c}$. We obtain the functional

$$S_h = \bar{c}\hat{A}c + \ln \det \hat{M}(c, \bar{c})/\hat{M}(0, 0) \tag{25.11}$$

in which $\ln \det$ is regularized dividing $\hat{M}(c, \bar{c})$ by $\hat{M}(0, 0) = \hat{M}(c, \bar{c})|_{c = \bar{c} = 0}$:

Functional S_h is called the functional of the hydrodynamical action. It determines the point of a phase transition of the initial Fermi system as a Bose condensation of fields $c, \bar{c}$ as well as the condensate density for $T < T_c$ and the spectrum of collective excitations.

For the case of the pairing in the s state (the model with $f = \text{const} < 0$, $g = 0$) it was sufficient to introduce a Gauss integral with respect to the scalar complex fields $c(\mathbf{x}, \tau)$, $\bar{c}(\mathbf{x}, \tau)$. In the case $f = 0$, $g = \text{const} < 0$ we have to integrate over a space of the complex functions $c_{ia}(\mathbf{x}, \tau)$, $\bar{c}_{ia}(\mathbf{x}, \tau)$ with a vector index i and a spin index a ($i, a = 1, 2, 3$). The Gauss integral inserted into the integral with respect to the Fermi fields has the form

$$\int d\bar{c}_{ia} \, dc_{ia} \exp\left(\frac{1}{g} \sum_{p,i,a} c_{ia}^+(p) c_{ia}(p)\right) \tag{25.12}$$

where g is constant (25.10). It is easy to check that the transformation

$$c_{i1}(p) \to c_{i1}(p) + \frac{g}{2(\beta V)^{1/2}} \sum_{p_1 + p_2 = p} (n_{1i} - n_{2i}) \times$$
$$\times [a_+(p_2)a_+(p_1) - a_-(p_2)a_-(p_1)],$$

$$c_{i2}(p) \to c_{i2}(p) + \frac{g_i}{2(\beta V)^{1/2}} \sum_{p_1 + p_2 = p} (n_{1i} - n_{2i}) \times \qquad (25.13)$$
$$\times [a_+(p_2)a_+(p_1) + a_-(p_2)a_-(p_1)],$$

$$c_{i3}(p) \to c_{i3}(p) + \frac{g}{(\beta V)^{1/2}} \sum_{p_1 + p_2 = p} (n_{1i} - n_{2i})a_-(p_2)a_+(p_1)$$

really eliminates form $\tilde{S}_4$.

In order to evaluate the Gauss integral with respect to the Fermi fields, we introduce a column $\psi_a(p)$ with the elements

$$\psi_1(p) = a_+(p), \qquad \psi_2(p) = -a_-(p),$$
$$\psi_3 = a_-^+(p), \qquad \psi_4 = a_+^+(p) \qquad (25.14)$$

and write the quadratic form of the Fermi fields in the form

$$K = \tfrac{1}{2} \sum_{p_1,p_2,a,b} \psi_a^+(p_1)M_{ab}(p_1,p_2)\psi_b(p_2). \qquad (25.15)$$

Matrix $M(p_1, p_2)$ of the fourth degree with elements $M_{ab}(p_1, p_2)$ has the form

$$M = \begin{pmatrix} Z^{-1}(i\omega - \xi + \mu\sigma_3)\delta_{p_1,p_2}, & (\beta V)^{-1/2}(n_{1i} - n_{2i})c_{ia}(p_1 + p_2)\sigma_a \\ -(\beta V)^{-1/2}(n_{1i} - n_{2i})c_{ia}^+(p_1 + p_2)\sigma_a, & Z^{-1}(-i\omega + \xi + \mu H\sigma_3)\delta_{p_1,p_2} \end{pmatrix},$$

$$(25.16)$$

where $\xi = c_F(k - k_F)$, $\sigma_a (a = 1, 2, 3)$ are the Pauli matrices of the second degree.

By integrating with respect to the Fermi fields

$$\int e^K \, d\bar{\chi}_{0s} \, d\chi_{0s} = (\det \hat{M})^{1/2} \qquad (25.17)$$

we obtain the functional of the hydrodynamical action

$$S_h = \frac{1}{g} \sum_{p,i,a} c_{ia}^+(p)c_{ia}(p) + \tfrac{1}{2}\ln \det \hat{M}(c, \bar{c})/\hat{M}(0, 0). \qquad (25.18)$$

Functional (25.18) contains the complete information on the physical properties of the model system. It depends on nine complex or 18 real functions, which correspond to 18 degrees of freedom (18 modes of the Bose spectrum).

We start to study functional (25.18) in the Ginzburg–Landau domain $|\Delta T| \ll T_c$, where we can expand ln det in powers of fields $c, \bar{c}$. Denoting

$$\hat{M}(0,0) = \hat{G}^{-1}, \qquad \hat{M}(c, \bar{c}) = \hat{G}^{-1} + \hat{u} \tag{25.19}$$

we keep the first two terms ($n = 1, 2$) in the expansion

$$\tfrac{1}{2} \ln \det \hat{M}(c, \bar{c})/\hat{M}(0, 0) = \tfrac{1}{2} \mathrm{Sp} \ln (1 + \hat{G}\hat{u})$$

$$= - \sum_{n=1}^{\infty} \frac{1}{4n} \mathrm{Sp}(\hat{G}\hat{u})^{2n}. \tag{25.20}$$

Let us consider the quadratic form of $\hat{u}$ in (25.20). If $H = 0$, the form is diagonal with respect to index a and has the form

$$- \sum_{p,i,a} A_{ij}(p)c_{ia}^{+}(p)c_{ja}(p)$$

$$= - \sum_{p,i,a} A_{ij}(0)c_{ia}^{+}(p)c_{ja}(p) - \sum_{p,i,a} (A_{ij}(p) - A_{ij}(0))c_{ia}^{+}(p)c_{ja}(p) \tag{25.21}$$

where

$$A_{ij}(0) = -\frac{4Z^2}{\beta V} \sum_{p_1} \frac{n_{1i}n_{ij}}{\omega_1^2 + \xi_1^2}$$

$$= -\delta_{ij} \frac{2Z^2 k_F^2}{3\pi^2} \int_0^{c_F k_0} \frac{\mathrm{d}\xi}{\xi} \tanh \frac{\beta\xi}{2} \tag{25.22}$$

$$= -\delta_{ij} A(0).$$

Here, when evaluating $A_{ij}(0)$ we pass to the integral over a neighbourhood of the Fermi sphere according to the rule

$$V^{-1} \sum_k \to (2\pi)^{-3} k_F^2 c_F^{-1} \int \mathrm{d}\xi \, \mathrm{d}\Omega.$$

Further we have

$$A_{ij}(p) - A_{ij}(0)$$

$$= \frac{Z^2}{2V} \sum_{|k-k_F| < k_0} (n_{1i} - n_{2i})(n_{1j} - n_{2j}) \times \tag{25.23}$$

$$\times \left[\frac{1}{i\omega - \xi_1 - \xi_2} \left(\tanh \frac{\beta\xi_1}{2} + \tanh \frac{\beta\xi_2}{2} \right) + \frac{1}{\xi_1} \tanh \frac{\beta\xi_1}{2} \right].$$

For small H ($\mu H \ll T$) it is sufficient to take into account the contribution from H to $A_{ij}(0)$ neglecting the H dependence of function $A_{ij}(p) - A_{ij}(0)$ for small p.

After summing up over frequencies, the following integrals arise

$$\int_{-c_F k_0}^{c_F k_0} \frac{d\xi}{\xi + \mu H} \tanh \frac{\beta}{2}(\xi \pm \mu H), \quad a = 1, 2,$$

(25.24)

$$\int_{-c_F k_0}^{c_F k_0} \frac{d\xi}{2\xi} \left(\tanh \frac{\beta}{2}(\xi + \mu H) + \tanh \frac{\beta}{2}(\xi - \mu H) \right), \quad a = 3.$$

Only the last of them depends on H. It obtains the supplement

$$\left(\frac{\beta}{2} \mu H \right)^2 \int_{-\infty}^{\infty} \frac{dx}{2x} (\tanh x)'' = -\frac{7\zeta(3)\mu^2 H^2}{2\pi^2 T^2}$$

(25.25)

proportional to H^2. As a result the supplement to S_h proportional to H^2 is equal to

$$\frac{-7\zeta(3)Z^2 k_F^2 \mu^2 H^2}{6\pi^4 T^2 c_F} \sum_{p,i} c_{i3}^+(p) c_{i3}(p)$$

(25.26)

We consider the components with small three-momenta of all fields $c_{ia}(p)$ ($c_F|k| \ll T$) in the form of the fourth order terms. These terms can be written as

$$-\frac{7\zeta(3)k_F^2 Z^4}{30\pi^4 T^2 c_F \beta V} \sum_{p_1 + p_2 = p_3 + p_4} [2c_{ia}^+(p_1)c_{jb}^+(p_2)c_{ia}(p_3)c_{jb}(p_4) +$$
$$+ 2c_{ia}^+(p_1)c_{ib}^+(p_2)c_{ja}(p_3)c_{jb}(p_4) +$$
$$+ 2c_{ia}^+(p_1)c_{jb}^+(p_2)c_{ja}(p_3)c_{ib}(p_4) -$$
$$- 2c_{ia}^+(p_1)c_{ja}(p_2)c_{ib}(p_3)c_{jb}(p_4) -$$
$$- c_{ia}^+(p_1)c_{ia}^+(p_2)c_{jb}(p_3)c_{jb}(p_4)].$$

(25.27)

We shall find the temperature of the Bose condensation T_c by putting the coefficient of the quadratic form at $c_{ia}^+(p)\, c_{ja}(p)$ equal to zero for $p = 0$. For $H = 0$ we have the equality

$$0 = \frac{1}{g} + A(0) = \frac{1}{g} + \frac{2Z^2 k_F^2}{3\pi^2 c_F} \int_0^{c_F k_0} \frac{d\xi}{\xi} \tanh \frac{\beta \xi}{2}.$$

(25.28)

The integral with respect to ξ in (25.28) depends logarithmically on k_0.

In order that T_c shall not depend on k_0, it is necessary that g^{-1} depend on k_0 according to the formula

$$g^{-1} = g_0^{-1} - \frac{2Z^2 k_F^2}{3\pi^2 c_F} \ln \frac{k_0}{k_F} \qquad (25.29)$$

where g_0 does not depend on $k_0 (g_0 < 0)$. Substituting formula (25.29) and the expression

$$\int_0^{c_F k_0} \frac{d\xi}{\xi} \tanh \frac{\beta \xi}{2} \approx \ln \frac{2\beta c_F k_0}{\pi} + C \qquad (25.30)$$

into (25.28) we find the temperature of the phase transition

$$T_c = \frac{2\gamma}{\pi} c_F k_F \exp \left(-\frac{3\pi^2 c_F}{2|g_0| Z^2 k_F^2} \right) \qquad (25.31)$$

where $\ln \gamma = C$ is the Euler constant.

In the domain $|\Delta T| \ll T_c$ we have

$$\frac{1}{g} + A(0) = \frac{2Z^2 k_F^2}{3\pi^2 c_F} \ln \frac{T_c}{T} \approx \frac{2Z^2 k_F^2}{3\pi^2 c_F} \frac{T_c - T}{T_c}. \qquad (25.32)$$

In order to find out the condensate density for $T < T_c$ we use the substitution

$$c_{ia}(p) \to (\beta V)^{1/2} \delta_{p0} a_{i0} \left(\frac{10 T_c |\Delta T|}{7\xi(3)} \right)^{1/2} \frac{\pi}{Z},$$

$$\qquad (25.33)$$

$$c_{ia}^+(p) \to (\beta V)^{1/2} \delta_{p0} \bar{a}_{ia} \left(\frac{10 T_c |\Delta T|}{7\xi(3)} \right)^{1/2} \frac{\pi}{Z},$$

which changes S_h into the expression

$$-\frac{20 k_F^2 (\Delta T)^2 \beta V}{21\xi(3) c_F} \pi. \qquad (25.34)$$

It depends on matrix A with the elements a_{ia} and also on its Hermitian conjugate A^+, transpose A^T and on its complex conjugate A^*:

$$\pi = -\operatorname{tr} AA^+ + v \operatorname{tr} A^+ AP + (\operatorname{tr} AA^+)^2 +$$
$$+ \operatorname{tr} AA^+ AA^+ + \operatorname{tr} AA^+ A^* A^T - \operatorname{tr} AA^T A^* A^+ - \qquad (25.35)$$
$$- \tfrac{1}{2} \operatorname{tr} AA^T \operatorname{tr} A^+ A^*$$

where

$$P = \begin{pmatrix} 0 & 0 & 0 \\ 0 & 0 & 0 \\ 0 & 0 & 1 \end{pmatrix} \tag{25.36}$$

is a projector on the third axis which is identical with the direction of the magnetic field

$$v = \frac{7\xi(3)\mu^2 H^2}{4\pi^2 T_c \Delta T}. \tag{25.37}$$

Notice the invariance of π under the transformations

$$A \rightarrow e^{i\alpha} U A V \tag{25.38}$$

where α is a real parameter, U a real orthogonal matrix, and V is a matrix of the form

$$V = \begin{pmatrix} U_2 & 0 \\ 0 & 1 \end{pmatrix}, \tag{25.39}$$

where U_2 is an orthogonal matrix of the second degree.

By minimizing π, we obtain a matrix which determines the density of the Bose condensate. The equation $\delta\pi = 0$, i.e.,

$$-A + vAP + 2(\operatorname{tr} AA^+)A + 2AA^+A +$$
$$+ 2A^*A^TA - 2AA^TA^* - A^* \operatorname{tr} AA^T = 0 \tag{25.40}$$

has a few nontrivial solutions corresponding to the different superfluid phases. Let us consider the following possibilities

$$A_1 = c_1 P, \quad A_2 = c_2 P_2, \quad A_3 = c_3' P + c_3'' P_2,$$
$$A_4 = c_4 \hat{C}_4, \quad A_5 = c_5 \hat{C}_5, \quad A_6 = c_6' \hat{C}_4 + c_6'' \hat{C}_5,$$
$$A_7 = c_7 \hat{C}_7. \tag{25.41}$$

Here P_2 is a projector on a two-dimensional subspace orthogonal to P; $\hat{C}_4, \hat{C}_5, \hat{C}_7$ are the matrices of the third degree

$$\hat{C}_4 = \begin{pmatrix} 1 & 0 & 0 \\ i & 0 & 0 \\ 0 & 0 & 0 \end{pmatrix}, \quad \hat{C}_5 = \begin{pmatrix} 0 & 0 & 0 \\ 0 & 0 & 0 \\ 1 & i & 0 \end{pmatrix}, \quad \hat{C}_7 = \begin{pmatrix} 1 & i & 0 \\ i & -1 & 0 \\ 0 & 0 & 0 \end{pmatrix}. \tag{25.42}$$

We shall write the moduli squared of the coefficients $|c_i|^2$ and the corresponding values of π_i:

$$|c_1|^2 = \tfrac{1}{3}(1 - v), \qquad |c_2|^2 = \tfrac{1}{4}, \qquad |c_3'|^2 = \tfrac{1}{5}(1 - 2v),$$
$$|c_3''|^2 = \tfrac{1}{10}(2 + v), \qquad |c_4|^2 = \tfrac{1}{4}, \qquad |c_5|^2 = \tfrac{1}{12}, \tag{25.43}$$
$$|c_6'|^2 = \tfrac{1}{8}(2 - 3v), \qquad |c_6''|^2 = \frac{v}{8}, \qquad |c_7|^2 = \tfrac{1}{16},$$

$$\pi_1 = -\tfrac{1}{6}(1 - v)^2, \qquad \pi_2 = \pi_4 = -\tfrac{1}{4},$$
$$\pi_3 = -\tfrac{1}{10}(3 - 2v + v^2), \qquad \pi_5 = -\tfrac{1}{12}, \tag{25.44}$$
$$\pi_6 = -\tfrac{1}{8}(2 - 4v + 3v^2), \qquad \pi_7 = -\tfrac{1}{8}.$$

For $H = 0$ ($v = 0$) the minimal of all possibilities, $\pi = -0.3$, is obtained for $A_3 = c_3\,I$. It is the symmetric Balyan–Werthammer phase (B phase). The next values (equal in magnitude), $\pi = -\tfrac{1}{4}$, are obtained on the flat 2D-phase with $A = c_2 P_2$ and for $A = c_4 \hat{C}_4$ which corresponds to the Anderson–Morela–Brinkman A phase. The remaining four phases A_1, A_5, A_6, A_7 are energetically disadvantageous compared with the B, A, 2D phases.

The quantity π for the A and 2D phases does not change when increasing the magnetic field. However, for the B phase π increases and the B phase gets deformed. For $v = \tfrac{1}{2}$ we have $A_2 = A_3$, $\pi_2 = \pi_3$. For $v > \tfrac{1}{2}$ the solution A_3 has no sense since $|c_3'|^2 = (1 - 2v)/5$ becomes negative. Thus, for

$$v = \tfrac{1}{2}, \qquad H^2 = H_c^2 = \frac{2\pi T_c \Delta T}{7\xi(3)\mu^2} \tag{25.45}$$

there is a phase transition of the B phase into the flat 2D phase the energy of which, in the first approximation, does not depend on H in the model. Continuity of the transition A_3 into A_2 indicates that this phase transition is of the second-order. Besides the second-order transition from the B to the 2D phase there is a competitive first-order transition from the B to the A phase for the same value of $v = \tfrac{1}{2}$ (since the energy of the A and the 2D phases are, in the first approximation, the same). In the real He^3 there is a phase transition of the first-order from the B into the A-phase.

Now we investigate the stability of various superfluid phases under small excitations. The necessary stability condition is a nonnegativity of the second variation $\delta^2 \pi$. We present the expressions $\delta^2 \pi_i$ for the most interesting cases $i = 2, 3, 4$.

$$\delta^2 \pi_2 = (v - \tfrac{1}{2})u_{33}^2 + (v + \tfrac{1}{2})v_{33}^2 + v(u_{13}^2 + u_{23}^2) + (v + 2)(v_{13}^2 + v_{23}^2) +$$
$$+ \tfrac{1}{2}(3u_{11}^2 + 3u_{22}^2 + 2u_{12}u_{22} + (u_{12} + u_{21})^2] +$$
$$+ \tfrac{1}{2}[3v_{11}^2 + 3v_{21}^2 - 2v_{12}v_{21} + (v_{11} - v_{22})^2];$$

$$\delta^2\pi_3 = \frac{v+2}{5}[3u_{11}^2 + 3u_{22}^2 + 2u_{11}u_{22} + (u_{12}+u_{21})^2 + u_{13}^2 + u_{23}^2] +$$

$$+ \frac{2(1-2v)}{5}(u_{31}^2 + u_{32}^2 + 3u_{33}^2) +$$

$$+ \frac{4}{5}\left(\frac{(1-2v)(2+v)}{2}\right)^{1/2}(u_{11}u_{33} + u_{22}u_{33} + u_{13}u_{31} + u_{23}u_{32} +$$

$$+ \frac{4-3v}{5}(v_{11}^2 + v_{22}^2) + \frac{8-v}{5}(v_{12}^2 + v_{21}^2) + \frac{2(2+v)}{5} \times$$

$$\times (v_{33}^2 - v_{11}v_{22} - v_{12}v_{21}) + \frac{8+9v}{5}(v_{13}^2 + v_{23}^2) \qquad (25.46)$$

$$+ \frac{8(1-2v)}{5}(v_{31}^2 + v_{32}^2) -$$

$$- \frac{4}{5}\left(\frac{(1-2v)(2+v)}{2}\right)^{1/2}(v_{11}v_{33} + v_{22}v_{33} + v_{13}v_{31} + v_{23}v_{32}),$$

$$\delta^2\pi_4 = v[u_{13}^2 + u_{23}^2 + u_{33}^2 + v_{13}^2 + v_{23}^2 + v_{33}^2] +$$
$$+ (u_{11} + v_{21})^2 + (u_{22} - v_{12})^2 + (u_{23} - v_{13})^2 +$$
$$+ \tfrac{1}{2}[(u_{11} - u_{21})^2 + (u_{21} + v_{11})^2 + (u_{12} - v_{22})^2 + (u_{22} + v_{12})^2 +$$
$$+ (u_{13} - v_{23})^2 + (u_{23} + v_{13})^2].$$

Here $u_{ia} = \operatorname{Re}\delta a_{ia}$ and $v_{ia} = \operatorname{Im}\delta a_{ia}$.

For $v < \tfrac{1}{2}$ the variation $\delta^2\pi_3$ is nonnegative and for $v > \tfrac{1}{2}$ that of $\delta^2\pi_2$. The variation $\delta^2\pi_4$ is nonnegative for all v. Thus, the B phase must pass either into the 2D phase or into the A phase for sufficiently large v.

The expressions $\delta^2\pi_1$, $\delta^2\pi_5$, $\delta^2\pi_7$ change the sign for all v. The corresponding phases are destroyed by small excitations and therefore are not realized in the model considered. The expression $\delta^2\pi_6$ changes the sign for all $v > 0$, and for $v = 0$ the corresponding phase coincides with the A phase.

Thus, in the model system considered, the stability condition is weakly satisfied only for the A phase, the B phase $(H \le H_c)$ and the 2D phase $(H \ge H_c)$.

The quadratic form $\delta^2\pi_3$ of 18 variables u_{ia}, v_{ia} has four zero eigenvectors for $v < \tfrac{1}{2}$; the form $\delta^2\pi_2$ has six zero eigenvectors for $v > \tfrac{1}{2}$, and $\delta^2\pi_4$ has nine zero eigenvectors for $v = 0$. This is connected with the existence of four phonon branches in the Bose spectrum in the B phase (for which $E(\mathbf{k} = 0) = 0$), of six such branches in the 2D phase and of nine in the A phase (for $H = 0$).

Let us consider the Bose spectrum of the system for $|\Delta T| \ll T_c$. For $T > T_c$ all branches of the spectrum are pure imaginary and $|E(k)| \ll T_c$. By analytic continuation of (25.23) in the domain $|\omega| \ll T$ we obtain

$$A_{ij}(p) - A_{ij}(0)$$
$$= \frac{Z^2 k_F^2}{12\pi T c_F} \left[\omega \delta_{ij} + \frac{7\xi(3)c_F^2}{10\pi^3 T}(\delta_{ij}k^2 + 2k_i k_j) \right] \tag{25.47}$$

for $\omega > 0$. By taking into account the component $g^{-1} + A(0)$ given in (25.32) and the magnetic supplement (25.26) we obtain the following branches of the spectrum for $T > T_c$:

$$E_{a,\parallel}(\mathbf{k}) = -i\left(\frac{21\xi(3)c_F^2 k^2}{10\pi^3 T_c} + \frac{8}{\pi}(T - T_c) + \delta_{a3}\frac{14\xi(3)\mu^2 H^2}{\pi^3 T_c} \right),$$

$$E_{a,\perp}(\mathbf{k}) = -i\left(\frac{7\xi(3)c_F^2 k^2}{10\pi^3 T_c} + \frac{8}{\pi}(T - T_c) + \delta_{a3}\frac{14\xi(3)\mu^2 H^2}{\pi^3 T_c} \right). \tag{25.48}$$

Here a is a spinorial index of the corresponding branch and the symbols $\parallel$ and $\perp$ indicate that the vector index is 'parallel' or 'perpendicular' to the direction of spreading.

For $T < T_c$ it is necessary to take into account the form of the fourth degree. It is not difficult to obtain the Bose spectrum after the shift transformation $c_{ia}(p) \to c_{ia}(p) + c_{ia}^{(0)}(p)$. All its branches have the form $E = -i\alpha k^2 - i\Gamma$; $\alpha > 0, \Gamma \ge 0$ with $\Gamma = 0$ for four branches in the B phase, six branches in the $2D$ phase and nine in the A phase. It is only these branches that go into the branches of the phonon spectrum for a decrease of the temperature.

Let us now consider the temperature domain $T_c - T \sim T_c$. In this domain we expand ln det in (25.18) in powers of the deviation $c_{ia}(p)$ from the condensate quantity $c_{ia}^{(0)}(p)$ different for different phases. We make the shift transformation

$$c_{ia}(p) = c_{ia}^{(0)}(p) + a_{ia}(p) \tag{25.49}$$

and extract the quadratic form of $a_{ia}(p)$, $a_{ia}^+(p)$ from S_h

$$\sum_p a_{ia}^+(p)a_{jb}(p)A_{ijab}(p) +$$
$$+ \frac{1}{2}\sum_p (a_{ia}(p)a_{jb}(-p) + a_{ia}^+(p)a_{jb}^+(-p))B_{ijab}(p). \tag{25.50}$$

This form determines the Bose spectrum in the first approximation by using the equation

$$\det Q = 0. \tag{25.51}$$

Here Q is a matrix of the quadratic form which is determined by the coefficient tensors A_{ijab}, B_{ijab} in (25.51). If $T \to 0$, the quantities A_{ijab}, B_{ijab} are proportional to the integrals of Green's function of fermions. These integrals are advantageous to calculate by applying the Feynman method. With its help it is easy to perform the integration with respect to variables ω and ξ and then with respect to angular variables and parameter α.

Here we shall restrict ourselves to an evaluation of the phonon branches of the Bose spectrum $E(\mathbf{k})$ (such that $E(0) = 0$) in the superfluid B phase model. In the B phase the condensate function $c_{ia}^{(0)}(p)$ has the form

$$c_{ia}^{(0)}(p) = (\beta V)^{1/2} c \delta_{p0} \delta_{ia}, \tag{25.52}$$

where constant c can be found from the equation

$$\frac{3}{g} + \frac{4Z^2}{\beta V} \sum_p (\omega^2 + \xi^2 + 4c^2 Z^2)^{-1} = 0. \tag{25.53}$$

Performing the shift transformation (25.49), we extract the quadratic form of the new variables from S_h

$$\frac{1}{g} \sum_{p,i,a} a_{ia}^+(p) a_{ia}(p) - \tfrac{1}{4} \operatorname{Sp}(\hat{G}\hat{u})^2 \tag{25.54}$$

where

$$u_{p_1 p_2} = (\beta V)^{-1/2} \begin{pmatrix} 0, & (n_1 - n_2)_i \sigma_a a_{ia}(p_1 + p_2) \\ -(n_1 - n_2)_i \sigma_a a_{ia}^+(p_1 + p_2), & 0 \end{pmatrix}, \tag{25.55}$$

$$G^{-1} = \begin{pmatrix} Z^{-1}(i\omega - \xi)\delta_{p_1 p_2}, & 2c(\mathbf{n}\boldsymbol{\sigma})\delta_{p_1 + p_2} \\ -2c(\mathbf{n}\boldsymbol{\sigma})\delta_{p_1 + p_2}, & Z^{-1}(-i\omega + \xi)\delta_{p_1 p_2} \end{pmatrix}. \tag{25.56}$$

Inverting G^{-1} we obtain

$$G = \frac{Z}{M} \begin{pmatrix} -(i\omega + \xi)\delta_{p_1 p_2}, & \Delta(\mathbf{n}\boldsymbol{\sigma})\delta_{p_1 + p_2} \\ -\Delta(\mathbf{n}\boldsymbol{\sigma})\delta_{p_1 p_2}, & (i\omega + \xi)\delta_{p_1 p_2} \end{pmatrix} \tag{25.57}$$

where

$$M = \omega^2 + \xi^2 + \Delta^2, \quad \Delta = 2cZ. \tag{25.58}$$

It follows from (25.55) and (25.57) that

$$\tfrac{1}{4}\mathrm{Sp}\,(\hat{G}\hat{u})^2 = \tfrac{1}{4}\sum_{p_1,p_2,p_3,p_4} \mathrm{tr}\,(G_{p_1p_2}u_{p_2p_3}G_{p_3p_4}u_{p_4p_1})$$

$$= \frac{Z^2}{4\beta V}\sum_p \left\{ 2a_{ia}^+(p)a_{jb}(p) \sum_{p_1+p_3=p}(M_1M_3)^{-1}(i\omega_1+\xi_1)(i\omega_3+\xi_3) \times \right.$$

$$\times\, \mathrm{tr}\,\sigma_a\sigma_b(n_1-n_3)_i(-n_1+n_3)_j + 4\Delta^2(a_{ia}(p)a_{jb}(-p)+a_{ia}^+(p)a_{jb}^+(-p)) \times$$

$$\left. \times \sum_{p_3-p_1=p}(M_1M_3)^{-1}(n_1+n_3)_i(n_1+n_3)_j\,\mathrm{tr}\,(n_1\sigma)\sigma_a(n_3\sigma)\sigma_b\right\} \qquad (25.59)$$

For small $p=(\mathbf{k},\omega)$ it is possible to replace $\mathbf{n}_3\to-\mathbf{n}_1$ if $p_1+p_3=p$, and $\mathbf{n}_3\to\mathbf{n}_1$ if $p_3-p_1=p$. Transforming in addition $p_1\to-p_1$ in the sum with $p_3-p_1=p$ and taking the trace, we obtain the quadratic part in the form (25.50) where

$$A_{ijab}(p)=\delta_{ab}A_{ij}(p)$$

$$= -\delta_{ab}\delta_{ij}g^{-1} - \frac{4\delta_{ab}}{\beta V}\sum_{p_1+p_2=p} n_{1i}n_{2j}\varepsilon(-p_1)\varepsilon(-p_2)G(p_1)G(p_2),$$

$$\qquad\qquad\qquad (25.60)$$

$$B_{ijab}(p)=\frac{4\Delta^2}{\beta V}\sum_{p_1+p_2=p} n_{1i}n_{2j}(2n_{1a}n_{1b}-\delta_{ab})G(p_1)G(p_2),$$

where

$$\varepsilon(p)=i\omega-\xi, \qquad G(p)=Z(\omega^2+\xi^2+\Delta^2)^{-1}. \qquad (25.61)$$

After the substitution

$$a_{ia}(p)=u_{ia}(p)+iv_{ia}(p), \qquad a_{ia}^+(p)=u_{ia}(p)-iv_{ia}(p) \qquad (25.62)$$

the quadratic form of a,a^+ splits into two independent forms, the first of which depends on u_{ia} and the second on v_{ia}:

$$-\sum_p (A_{ij}(p)u_{ia}(p)u_{ja}(p)+B_{ijab}(p)u_{ia}(p)u_{ja}(p)) -$$

$$-\sum_p (A_{ij}(p)v_{ia}(p)v_{ja}(p)-B_{ijab}(p)v_{ia}(p)v_{ja}(p)). \qquad (25.63)$$

The term with $p=0$ in (25.63) is equal to

$$-A_{ij}(0)(u_{ia}u_{ja}+v_{ia}v_{ja})-B_{ijab}(0)(u_{ia}u_{jb}-v_{ia}v_{jb}). \qquad (25.64)$$

$A_{ij}(0)$, $B_{ijab}(0)$ can easily be evaluated by passing on the integral near the

Fermi sphere

$$A_{ij}(0) = \delta_{ij}g^{-1} - \frac{4}{\beta V}\sum_p n_i n_j \varepsilon_+ \varepsilon_- G^2 = \frac{Z^2 k_F^2}{3\pi^2 c_F}\delta_{ij},$$

(25.65)

$$B_{ijab}(0) = \frac{4\Delta^2}{\beta V}\sum_p (2n_a n_b n_i n_j - \delta_{ab} n_i n_j)G^2$$

$$= \frac{Z^2 k_F}{15\pi^2 c_F}[2\delta_{ai}\delta_{bj} + 2\delta_{aj}\delta_{bi} - 3\delta_{ab}\delta_{ij}].$$

Substituting it in (25.64) we obtain

$$-\frac{2Z^2 k_F^2}{15\pi^2 c_F}[u_{ia}u_{ja} + u_{aa}u_{bb} + u_{ia}u_{ai} + 4v_{ia}v_{ia} - v_{aa}v_{bb} - v_{ia}v_{ai}].$$

(25.66)

The u-form appearing in (25.66) has three zero eigenvectors corresponding to the variables $u_{12} - u_{21}$, $u_{23} - u_{32}$, $u_{31} - u_{13}$ and the v-form has one zero eigenvector corresponding to the variable $v_{11} + v_{22} + v_{33}$. It is just these variables that are the phonon ones and the branches corresponding to them begin from zero.

Let us consider the difference $A_{ij}(p) - A_{ij}(0)$. By applying the Feynman method to the Green's functions $G(p_1)$, $G(p_2)$ in (25.61) and performing the integration near the Fermi sphere we obtain

$$A_{ij}(p) - A_{ij}(0) = \frac{4Z^2}{\beta V}\sum_{p_1+p_2=p} n_{1i}n_{1j}\int_0^1 d\alpha \times$$

$$\times \left\{ \frac{(\xi_1 + i\omega_1)(\xi_2 + i\omega_2)}{[\alpha(\omega_1^2 + \xi_1^2 + \Delta^2) + (1-\alpha)(\omega_2^2 + \xi_2^2 + \Delta^2]^2} - \right.$$

$$\left. - \frac{\omega_1^2 + \xi_1^2}{(\omega_1^2 + \xi_1^2 + \Delta^2)^2} \right\}$$

(25.67)

$$= -\frac{4Z^2 k_F^2}{(2\pi)^4 c_F}\int_0^1 d\alpha \int d\Omega_1\, d\omega_1\, d\xi_1\, n_{1i}n_{1j} \times$$

$$\times \left[\frac{\omega_1^2 + \xi_1^2 - \alpha(1-\alpha)q^2}{[\omega_1^2 + \xi_1^2 + \Delta^2 + \alpha(1-\alpha)q^2]^2} - \frac{\omega_1^2 + \xi_1^2}{(\omega_1^2 + \xi_1^2 + \Delta^2)^2} \right],$$

where $q^2 = w^2 + c_F^2(\mathbf{nk})^2$. By integrating with respect to the variable

$r_1^2 = \omega_1^2 + \xi_1^2$ we obtain

$$\frac{Z^2 k_F^2}{4\pi^3 c_F} \int_0^1 d\alpha \int d\Omega_1 \, n_{1i} n_{1j} \times$$
$$\times \left[\ln\left(1 + \frac{\alpha(1-\alpha)q^2}{\Delta^2}\right) + \frac{\alpha(1-\alpha)q^2}{\Delta^2 + \alpha(1-\alpha)q^2} \right]. \tag{25.68}$$

Decomposing the integrand in terms of a power series in a small parameter $\alpha(1-\alpha)q^2\Delta^{-2}$ and integrating with respect to α we find the formula

$$A_{ij}(p) - A_{ij}(0) \approx \frac{Z^2 k_F^2}{4\pi^3 c_F} \int d\Omega_1 \left(\frac{q^2}{3\Delta^2} - \frac{q^4}{20\Delta^4}\right) n_{1i} n_{1j}. \tag{25.69}$$

By performing the analogous steps for $B_{ijab}(p) - B_{ijab}(0)$ we obtain

$$B_{ijab}(p) - B_{ijab}(0) \approx \frac{Z^2 k_F^2}{4\pi^3 c_F} \int d\Omega_1 \left(\frac{q^2}{6\Delta^2} - \frac{q^4}{30\Delta^4}\right) \times$$
$$\times (\delta_{ab} - 2n_{1a}n_{1b})n_{1i}n_{1j}. \tag{25.70}$$

It is not difficult to evaluate the integrals with respect to the angular variable in (25.69) and (25.70). As a result we get for the quadratic form S_h the expression

$$-\frac{k_F^2 Z^2}{\pi^2 c_F} \sum_p \left\{ \left(\frac{1}{3} \pm \frac{1}{5}\right) w_{ia} w_{ia} \pm \frac{2}{15}(w_{aa} w_{bb} + w_{ia} w_{ai}) + \right.$$
$$+ w_{ia} w_{ja} \left[\frac{\delta_{ij}}{3}\left(\frac{\omega^2}{3\Delta^2} - \frac{\omega^4}{20\Delta^4}\right) + \frac{c_F^2}{15}(k^2\delta_{ij} + 2k_i k_j)\left(\frac{1}{3\Delta^2} - \frac{\omega^2}{10\Delta^4}\right) \right.$$
$$\left. - \frac{c_F^4}{700\Delta^4}(k^4\delta_{ij} + 4k^2 k_i k_j)\right] \pm w_{ia} w_{jb}\left(\delta_{ab}\left[\frac{\delta_{ij}}{3}\left(\frac{\omega^2}{6\Delta^2} - \frac{\omega^4}{30\Delta^4}\right) + \right.\right.$$
$$+ \frac{c_F^2}{15}(k^2\delta_{ij} + 2k_i k_j)\left(\frac{1}{6\Delta^2} - \frac{\omega^2}{15\Delta^4}\right) -$$
$$\left. - \frac{c_F^4}{1050\Delta^4}(k^4\delta_{ij} + 4k^2 k_i k_j)\right] + (\delta_{ab}\delta_{ij} + \delta_{ai}\delta_{bj} + \delta_{aj}\delta_{bi}) \times$$
$$\times \left[\frac{2}{15}\left(-\frac{\omega^2}{6\Delta^2} + \frac{\omega^4}{30\Delta^4}\right) + \frac{2k^2}{105}\left(-\frac{c_F^2}{6\Delta^2} + \frac{\omega^2 c_F^2}{15\Delta^4}\right) + \frac{c_F^4 k^4}{4725\Delta^4}\right] +$$

$$+\frac{4}{315}(\delta_{ij}k_ak_b+\delta_{ab}k_ik_j+\delta_{ai}k_bk_j+\delta_{bj}k_ak_i+\delta_{aj}k_bk_i+\delta_{bi}k_ak_j)\times$$

$$\times\left(-\frac{c_F^2}{2\Delta^2}+\frac{\omega^2c_F^2}{5\Delta^4}+\frac{c_F^4k^2}{15\Delta^4}\right)+\frac{8c_F^4}{4725\Delta^4}k_ak_bk_ik_j\right)\right\}. \tag{25.71}$$

Here we understand first to put $w_{ia}=u_{ia}$ and to take the upper sign in $\pm,\mp$ and then to put $w_{ia}=v_{ia}$ and to take the lower sign.

Expression (25.71) determines the phonon branch of the spectrum of the system. Consider now excitations spreading in a given direction, say, e.g., along the third axis. Putting $k_1=k_2=0$, $k_3=k$ in (25.71), it is not difficult to check that the quadratic form of variables w_{ia} decomposes into the sum of four, the first of which depends on w_{12}, w_{21}, the second on w_{13}, w_{23}, the third on w_{23}, w_{32} and the fourth on w_{11}, w_{22}, w_{33}.

Let us substitute $w_{ia}=v_{ia}$, $k_1=k_2=0$, $k_3=k$ into (25.71) and consider the quadratic form of the variables v_{11}, v_{22}, v_{33} which defined one of the phonon branches. This form is proportional to the expression

$$a(v_{11}^2+v_{22}^2)+bv_{33}^2+2cv_{11}v_{22}+2d(v_{11}+v_{22})v_{33} \tag{25.72}$$

with

$$a=\tfrac{4}{15}+a_1, \qquad b=\tfrac{4}{15}+b_1,$$
$$c=-\tfrac{2}{15}+c_1, \qquad d=-\tfrac{2}{15}+d_1,$$
$$a_1=\tfrac{11}{90}x^2+\tfrac{13}{630}y^2-\tfrac{17}{900}x^4-\tfrac{19}{3150}x^2y^2-\tfrac{1}{900}y^4,$$
$$b_1=\tfrac{11}{90}x^2+\tfrac{17}{210}y^2-\tfrac{17}{900}x^4-\tfrac{9}{350}x^2y^2-\tfrac{37}{3780}y^4, \tag{25.73}$$
$$c_1=\tfrac{1}{45}x^2+\tfrac{1}{315}y^2-\tfrac{1}{225}x^4-\tfrac{2}{1575}x^2y^2-\tfrac{1}{4725}y^4,$$
$$d_1=\tfrac{1}{45}x^2+\tfrac{1}{105}y^2-\tfrac{1}{225}x^4-\tfrac{2}{525}x^2y^2-\tfrac{1}{945}y^4,$$

where $x=\omega/\Delta$, $y=c_Fk/\Delta$.

The determinant of the third degree is equal to $(a-c)[b(a+c)-2d^2]$. Since the difference $(a-c)$ does not vanish for $\omega,\mathbf{k}\to0$ we obtain the equation

$$b(a+c)-2d^2=0 \tag{25.74}$$

or

$$\tfrac{2}{15}(b_1+2a_1+2c_1+4d_1)+b_1(a_1+c_1)-2d_1^2=0. \tag{25.75}$$

We have

$$b_1+2a_1+2c_1+4d_1=\tfrac{1}{2}x^2+\tfrac{1}{6}y^2-\tfrac{1}{12}x^4-\tfrac{1}{18}x^2y^2-\tfrac{1}{60}y^4,$$
$$b_1(a_1+c_1)-2d_1^2=\tfrac{1}{60}x^4+\tfrac{13}{945}x^2y^2+\tfrac{11}{6300}y^4.$$

Substitution into (25.75) leads to the equation

$$x^2 + \tfrac{1}{3}y^2 + \tfrac{1}{12}x^4 + \tfrac{2}{21}x^2 y^2 - \tfrac{1}{140}y^4 = 0. \tag{25.76}$$

Its solution after replacing $i\omega \to E$ and in terms of the variables ω, k gives the sound branch of the spectrum

$$E = \frac{c_F k}{\sqrt{3}} \left(1 - \frac{2 c_F^2 k^2}{45 \Delta^2} \right) \tag{25.77}$$

It is stable with respect to a decay of the excitation on two or several excitations of the same type. It is interesting that the dispersion coefficient $\gamma = 2 c_F^2 / 45 \Delta^2$ was obtained twice larger than that for the superfluid Fermi gas with a pairing in the s state (section 24).

Now we consider the form of variables u_{ia} splitting into four independent forms for $k_1 = k_2 = 0$. The phonon spectrum is determined by the forms of the variables (u_{12}, u_{21}), (u_{13}, u_{31}) which are proportional to the expressions

$$\tfrac{2}{15}(u_{12} + u_{21})^2 + (u_{12}^2 + u_{21}^2) \times \cdot$$
$$\times \left[\tfrac{13}{90}x^2 + \tfrac{19}{630}y^2 - \tfrac{7}{300}x^4 - \tfrac{31}{3150}x^2 y^2 - \tfrac{41}{18900}y^4 \right] -$$
$$- 2u_{12}u_{21}\left[\tfrac{1}{45}x^2 + \tfrac{1}{315}y^2 - \tfrac{1}{225}x^4 - \tfrac{2}{1575}x^2 y^2 - \tfrac{1}{4725}y^4 \right],$$
$$\tfrac{2}{15}(u_{13} + u_{31})^2 + u_{13}^2 \left(\tfrac{13}{90}x^2 + \tfrac{1}{42}y^2 - \tfrac{7}{300}x^4 - \tfrac{23}{3150}x^2 y^2 - \tfrac{1}{756}y^4 \right) + \tag{25.78}$$
$$+ u_{31}^2 \left(\tfrac{13}{90}x^2 + \tfrac{19}{210}y^2 - \tfrac{7}{300}x^4 - \tfrac{31}{1050}x^2 y^2 - \tfrac{41}{3780}y^4 \right) -$$
$$- 2u_{13}u_{31}\left(\tfrac{1}{45}x^2 + \tfrac{1}{105}y^2 - \tfrac{1}{225}x^4 - \tfrac{2}{525}x^2 y^2 - \tfrac{1}{945}y^4 \right)$$

and also by the form of the variables (u_{23}, u_{32}) which is obtained from the second form in (25.78) by replacing $(u_{13}, u_{31}) \to (u_{23}, u_{32})$. Putting the determinants of forms (25.78) equal to zero, we get the branch of longitudinal spin waves (the variables u_{12}, u_{21}))

$$E = \frac{c_F k}{\sqrt{5}} \left(1 - \frac{2 c_F^2 k^2}{105 \Delta^2} \right) \tag{25.79}$$

and the double degenerate branch of transversal spin waves (the variables (u_{13}, u_{31}), (u_{23}, u_{32}))

$$E = \sqrt{\frac{2}{5}} c_F k \left(1 - \frac{173 c_F^2 k^2}{3360 \Delta^2} \right). \tag{25.80}$$

Both these branches, like the sound branch (25.77), are stable under the decay of the excitation into two or several excitations of the same type.

CHAPTER 8

PLASMA THEORY

26. HYDRODYNAMICAL ACTION IN PLASMA THEORY

In this chapter some applications of the functional integral method to plasma physics, i.e., to the system of fermions with the Coulomb interaction, will be considered. For this system the standard perturbation theory gives rise to divergences analogous to the infrared divergenaces of quantum electrodynamics. By applying the method of integration with respect to the rapid and slow variables (see section 12), these divergences will be removed and the functional of the hydrodynamical action for plasma will be obtained. Then from a variety of possible applications of this functional, the problem of a plasma spectrum and its damping will be discussed.

For the sake of simplicity, a model of an electron system in a constant background potential of a compensating charge of opposite sign will be considered.

The starting action functional has the form

$$S = \int_0^\beta d\tau \int dx \, (\bar\psi(\mathbf{x},\tau)\partial_\tau\psi(\mathbf{x},\tau) - \frac{1}{2m}\nabla\bar\psi(\mathbf{x},\tau)\nabla\psi(\mathbf{x},\tau) +$$

$$+ \lambda\bar\psi(\mathbf{x},\tau)\psi(\mathbf{x},\tau)) - \frac{1}{2}\int_0^\beta d\tau \int dx\, dy\, \bar\psi(\mathbf{x},\tau)\bar\psi(\mathbf{y},\tau) \times \qquad (26.1)$$

$$\times \frac{e^2}{|\mathbf{x}-\mathbf{y}|}\psi(\mathbf{y},\tau)\psi(\mathbf{x},\tau).$$

Here, the first term describes the system of noninteracting electrons, and the second describes their Coulomb interaction.

The effect of the background is taken into account by skipping the term with $\mathbf{k}=0$ in the Fourier series of the potential

$$\frac{4\pi e^2}{V} \sum_{\mathbf{k}\neq 0} \exp[i\mathbf{k}(\mathbf{x}-\mathbf{y})]k^{-2}. \qquad (26.2)$$

As usual, Green's function is defined as an expectation value with the weight $\exp S$ of a product of a few Femi fields $\psi, \bar\psi$ of different space-time arguments. The standard perturbation technique may yield divergences since the expression corresponding to the vertex of the Coulomb interaction of fermions

$$4\pi e^2(|\mathbf{k}_1 - \mathbf{k}_3|^{-2} - |\mathbf{k}_1 - \mathbf{k}_4|^{-2}) \tag{26.3}$$

is infinite in the limit $\mathbf{k}_1 \to \mathbf{k}_3$ or $\mathbf{k}_1 - \mathbf{k}_4$. Thus, integrating products of type (26.3), we encounter singularities.

In order to build up a perturbation theory free of such singularities we introduce the following integral with respect to the auxiliary Bose field $\varphi(\mathbf{x}, \tau)$

$$\int d\varphi \exp\left(-\int d\tau\, d\mathbf{x}\frac{(\nabla\varphi(\mathbf{x},\tau))^2}{8\pi e^2}\right), \tag{26.4}$$

where $d\varphi$ is an integration measure, into the functional integral with respect to Fermi fields $\psi, \bar{\psi}$.

Let us perform the shift transformation

$$\varphi(\mathbf{x}, \tau) \to \varphi(\mathbf{x}, \tau) + ie^2 \int \frac{d^3 y}{|\mathbf{x} - \mathbf{y}|} \bar{\psi}(\mathbf{y}, \tau)\psi(\mathbf{y}, \tau) \tag{26.5}$$

in the integral with respect to $\varphi(\mathbf{x}, \tau)$ which cancels the Coulomb interaction term in the initial action (26.1). As a result, we obtain the action

$$\int d\tau\, d\mathbf{x}\left(\bar{\psi}\partial_\tau\psi - \frac{\nabla\bar{\psi}\nabla\psi}{2m} + \lambda\bar{\psi}\psi - i\varphi(\mathbf{x},\tau)\bar{\psi}(\mathbf{x},\tau)\psi(\mathbf{x},\tau) - \right.$$
$$\left. - (8\pi e^2)^{-1}(\nabla\varphi(\mathbf{x},\tau))^2 \right). \tag{26.6}$$

This expression describes a system of fermions interacting with the field of a scalar electric potential $\varphi(\mathbf{x}, \tau)$. Faraday's idea of a short-range interaction was essentially realized by replacing (26.1) with (26.6). For the system with action (26.6) it is not difficult to formulate a perturbation theory. Denoting field ψ by a directed line and field φ by a wavy line we have the following diagrammatic technique

$$\longrightarrow \qquad \left(i\omega - \frac{k^2}{2m} + \lambda\right)^{-1}$$

$$\sim\!\sim\!\sim\!\sim\!\sim \quad 4\pi e^2 k^{-2} \tag{26.7}$$

$i.$

This perturbation theory essentially coincides with the standard one with vertex (26.3) and thus does not lead to removal of divergences for small momenta. The required aim can be reached by applying the idea of successive integration over the rapid and slow variables (Section 12).

Let us define the slowly varying component of field $\varphi_0(\mathbf{x}, \tau)$ as the sum of terms in the expansion

$$\varphi(\mathbf{x}, \tau) = \frac{1}{\sqrt{\beta V}} \sum_{k,\omega} \exp[i(\omega\tau - \mathbf{kx})]\varphi(p) \qquad (26.8)$$

with momenta k smaller k_0. The difference $\varphi(\mathbf{x}, \tau) - \varphi_0(\mathbf{x}, \tau)$ is called the rapidly varying component $\varphi_1(\mathbf{x}, \tau)$.

A physical reason for decomposing field φ into the rapid and slow components is based on the fact that in order to describe a scattering of two electrons with a small momentum transfer, collective effects have to be taken into account, whereas pair collisions of electrons with a large momentum transfer can be studied by neglecting the presence of the other particles of the system.

It is useful to divide the electron field into the rapid and the slow components whenever electrons with a large momentum essentially determine the properties of plasma. For instance, the fast electrons can induce instability of the system in an external electric field.

By the hydrodynamical action of plasma we understand the action of fast electrons $\psi_1, \bar{\psi}_1$ in the slowly varying field φ_0. It is given by the formula

$$\exp \tilde{S}[\bar{\psi}_1, \psi_1, \varphi_0]$$
$$= \int \exp S[\bar{\psi}_0 + \bar{\psi}_1, \psi_0 + \psi_1, \varphi_0 + \varphi_1] \, \mathrm{d}\bar{\psi}_0 \, \mathrm{d}\psi_0 \, \mathrm{d}\varphi_1. \qquad (26.9)$$

If the external field is absent, there is no necessity to consider the fast electrons and thus we can put $\psi_1 = \bar{\psi}_1 = 0, \psi_0 = \psi, \bar{\psi}_0 = \bar{\psi}$. In this case the hydrodynamical action $S[\varphi_0]$ describes plasmons and their interactions. In order to obtain it we integrate the functional $\exp S$ with respect to the Fermi fields $\psi, \bar{\psi}$ and with respect to the rapid component φ_1 of the Bose field φ.

The integral with respect to the Fermi fields $\psi, \bar{\psi}$ is Gaussian and formally equal to the determinant of the operator

$$\hat{M} = \partial_\tau + \frac{\nabla^2}{2m} + \lambda + i\varphi(\mathbf{x}, \tau).$$

For regularization, we divide $\det \hat{M}$ by the determinant $\det \hat{M}_0$ of the operator obtained from $\hat{M}$ for $\varphi = 0$. In this way we can write

$$\int \exp S \, d\bar{\psi} \, d\psi = \exp \left\{ \ln \det \hat{M}/\hat{M}_0 - \frac{1}{8\pi e^2} \int (\nabla \varphi)^2 d\tau \, dx \right\}.$$
(26.10)

Before expanding $\ln \det \hat{M}/\hat{M}_0$ in a functional series of $\varphi(\mathbf{x}, \tau)$, it is useful to transform it into a form where its dependence on the slowly varying field $\varphi_0(\mathbf{x}, \tau)$ is apparent. We express $\varphi_0(\mathbf{x}, \tau)$ in the form of the sum

$$\varphi_0(\mathbf{x}, \tau) = \varphi_0(\mathbf{x}) + \tilde{\varphi}_0(\mathbf{x}, \tau)$$
(26.11)

consisting of $\varphi_0(\mathbf{x})$ which is constant with respect to 'time' τ, and components with frequencies $\omega \neq 0$:

$$\tilde{\varphi}_0(\mathbf{x}, \tau) = \frac{1}{\sqrt{\beta V}} \sum_{\substack{k < k_0 \\ \omega \neq 0}} \exp [i(\mathbf{kx} - \omega\tau)] \varphi(p).$$

In the integral with respect to $\psi, \bar{\psi}$ we perform the following change of variables

$$\psi(\mathbf{x}, \tau) \rightarrow \psi(\mathbf{x}, \tau) \exp i \int^\tau \tilde{\varphi}_0(\mathbf{x}, \tau) \, d\tau;$$

$$\bar{\psi}(\mathbf{x}, \tau) \rightarrow \bar{\psi}(\mathbf{x}, \tau) \exp -i \int^\tau \tilde{\varphi}(\mathbf{x}, \tau) \, d\tau,$$
(26.12)

where

$$\int^\tau \varphi_0(\mathbf{x}, \tau) \, d\tau = \frac{1}{\sqrt{\beta V}} \sum_{\substack{k < k_0 \\ \omega \neq 0}} \frac{1}{i\omega} \exp [i(\omega\tau - \mathbf{kx})] \varphi(p).$$
(26.13)

This transformation, which does not change the determined, maps the operator $\hat{M}$ into the operator

$$\hat{M}_1 \equiv \partial_\tau + \frac{1}{2m} \left(\nabla + i \int^\tau \nabla \bar{\varphi}_0(\mathbf{x}, \tau) \, d\tau \right)^2 +$$
$$+ \lambda - i(\varphi_0(\mathbf{x}) + \varphi_1(\mathbf{x}, \tau)).$$
(26.14)

Now we expand the functional $\ln \det \hat{M}_1/\hat{M}_0$ in a series with respect to variables φ_0, φ_1. In order to represent this expansion graphically we denote field φ_1 by a wavy line and the slowly varying field φ_0 by a double line.

In accordance with the structure of expression (26.14), we shall have two types of vertices describing the interaction of electrons with a slowly varying field

$$i \qquad \omega = 0;$$

$$\frac{(\mathbf{k}, \mathbf{k}_1)}{im\omega} \qquad \omega \neq 0; \tag{26.15}$$

$$-\frac{(\mathbf{k}_3, \mathbf{k}_4)}{m\omega_3\omega_4}.$$

In this notation the expansion

$$\ln \det \hat{M}_1/\hat{M}_0 = \quad \cdots \tag{26.16}$$

is represented by the sum of the circles of the electronic lines to which the interaction vertices with the slow and rapid components of potential φ are connected in all possible ways. Notice that the contribution of the circle with one leg is compensated by taking into account the background.

The integration with respect to φ_1 can be performed by using the usual diagrammatic technique (with a cutoff of the integrals over momenta smaller than the lower bound k_0). This corresponds to all possible interlocking of diagrams (26.16) by lines corresponding to field φ_1.

Thus we obtain the functional of the hydrodynamical action

$$S[\varphi_0] = C_0 - \frac{1}{8\pi e^2}\int(\nabla\varphi_0(x,\tau))^2\,d\tau\,dx + \quad \cdots \tag{26.17}$$

Here, C_0 is the sum of all vacuum diagrams that do not depend on field φ_0. The contributions to the quadratic part of the hydrodynamical action with respect to φ_0 originate from the integral of $(\nabla\varphi_0)^2$ and from all diagrams with two legs that describe field φ_0 (for instance the diagram a, b, c, d, e, f in (26.17)). The diagrams g, h are examples of terms of the third and fourth orders in $\tilde{S}[\varphi_0]$.

In the last step of functional integration with respect to field φ_0, the perturbation theory is determined by the structure of dynamical action (26.17). To diagram lines there corresponds the expression

$$G_0(p) = \left(\left(\frac{4\pi e^2}{k^2} \right)^{-1} - P \right)^{-1}, \tag{26.18}$$

reciprocal to the coefficient at $-\frac{1}{2}\varphi_0(p)\varphi_0(-p)$ in the quadratic form in $S[\varphi_0]$. Here $P = a + b + c + d + \cdots$ is the sum of contributions of all diagrams with two legs. Vertices of third and higher orders are determined by the forms with a degree higher than two in $S_0[\varphi]$.

The new perturbation theory in contrast to diagrammatic technique (26.6) has no divergences for small momenta. Indeed, the singular function $4\pi e^2 k^{-2}$ for $\mathbf{k} \to 0$ is replaced by expression (26.18). If $\omega = 0$, then P is finite and different from zero in the limit $\mathbf{k} \to 0$. The formula giving the first approximation of P is written below (26.23). As a result, function $G_0(\omega = 0, \mathbf{k})$ is finite for $\mathbf{k} \to 0$ and does not lead to any divergences when integrating with respect to $\mathbf{k}$. If $\omega \neq 0$, then $G_0(p)$ remains singular (proportional to k^{-2}) since $P \sim k^2$ due to formula (26.15) for the vertex of interaction of field φ_0 with an electron for $\omega \neq 0$. However, in the diagrams the line with $\omega \neq 0$ goes to two vertices, each of which contains a factor linear in $\mathbf{k}$ due to (26.15). This mechanism of singularities cancellation appeared earlier in the theory of bosonic systems (see Section 19).

Let us explain the physical meaning of the various terms in hydrodynamical action $S_0[\varphi_0]$. The integral of $(\nabla\varphi_0)^2$ together with one-loop diagrams a, b describes plasma oscillations and the Landau noncollision damping. The contribution of diagrams c, d, e, f is necessary when calculating dampings of plasma oscillations caused by collisions of charged particles and plasmon–plasmon interaction. Finally, the diagram g, h describe the interaction of long-wave plasmons (with small $\mathbf{k}$).

The problem of the plasma spectrum determined by action (26.17) is considered in the next section. It leads to the evaluation of poles of the correlation function

$$\langle \varphi_0(p)\varphi_0(-p)\rangle \tag{26.19}$$

for analytic continuation $i\omega \to E$.

Let us consider the correlation function (26.19) in the static case $\omega = 0$ when it has a meaning of an interaction potential of two electrons in a medium. In the first approximation, the correlation function coincides with the non-perturbed one (26.18) in which only the contribution of the simplest one-loop diagram b is taken into account (diagram a is equal to zero for $\omega = 0$). The expression corresponding to diagram b is given by

$$\frac{T}{V}\sum_{\mathbf{k}_1,\omega_1}(i\omega_1 - \varepsilon(\mathbf{k}_1))^{-1}(i\omega + i\omega_1 - \varepsilon(\mathbf{k}_1 + \mathbf{k}))^{-1}$$
$$= \frac{1}{V}\sum_{\mathbf{k}_1}\frac{n(\varepsilon(\mathbf{k}_1 + \mathbf{k})) - n(\varepsilon(\mathbf{k}_1))}{i\omega + \varepsilon(\mathbf{k}_1 + \mathbf{k}) - \varepsilon(\mathbf{k}_1)}, \tag{26.20}$$

where $n = (\exp(\beta\varepsilon) + 1)^{-1}$ is a distribution function of fermions. For the Boltzmann plasma

$$n(\varepsilon) \approx \exp(-\beta\varepsilon), \quad \varepsilon = \frac{k^2}{2m} - \lambda. \tag{26.21}$$

By putting $\omega = 0$ in (26.10) and assuming that the order of the external momentum $\mathbf{k}$ is smaller than that of $\sqrt{2mT}$, we can replace

$$\frac{n(\varepsilon(\mathbf{k}_1 + \mathbf{k})) - n(\varepsilon(\mathbf{k}_1))}{i\omega + \varepsilon(\mathbf{k}_1 + \mathbf{k}) - \varepsilon(\mathbf{k}_1)} \approx \frac{\partial n}{\partial\varepsilon} \approx -\beta n, \tag{26.22}$$

which allows us to express

$$P(\omega = 0, \mathbf{k}) \approx -\frac{1}{TV}\sum_{\mathbf{k}}n(\varepsilon(\mathbf{k})) = -\frac{\rho}{T} \tag{26.23}$$

in terms of the plasma density ρ and temperature T. Substituting (26.23) into (26.18) we obtain

$$G_0(\mathbf{k}, \omega = 0) = \frac{4\pi e^2}{k^2 + a^2}, \tag{26.24}$$

where

$$a = \left(\frac{4\pi e^2 \rho}{T}\right)^{1/2}. \tag{26.25}$$

The Fourier transform of (26.24)

$$(e^2/r)\exp(-ar) \tag{26.26}$$

is called the *Debye potential* and the quantity $r_D = a^{-1}$ is the *Debye radius*. The Debye potential is a screening potential of two interacting electrons in plasma and the Debye radius is a radius of screening.

In a sphere of radius r_D there have to be many particles, i.e., it has to be

$$\frac{4\pi}{3}r_D^3 \gg \rho^{-1}. \tag{26.27}$$

Due to the definition of r_D and (26.27) we obtain the inequality

$$e^2\rho^{1/3} \ll T \tag{26.28}$$

which means that the potential energy of interaction of two electrons at a distance $r \sim \rho^{-1/3}$ is much smaller than the average kinetic energy of the order T. The second condition for parameters of the system can be obtained by requiring that the contribution from diagram a is of an order smaller than that from diagram b. The contribution of d for $\omega = 0$

$$\begin{aligned}
d &= \frac{T^2}{V^2} \sum_{\substack{\mathbf{k}_1,\mathbf{k}_2 \\ \omega_1,\omega_2}} (i\omega_1 - \varepsilon(\mathbf{k}_1))^{-2}(i\omega_2 - \varepsilon(\mathbf{k}_2))^{-2}4\pi e^2|\mathbf{k}_1 - \mathbf{k}_2|^{-2} \\
&= \frac{1}{T^2}\frac{1}{V^2} \sum_{\mathbf{k}_1,\mathbf{k}_2} n(\varepsilon(\mathbf{k}_1))\frac{4\pi e^2}{|\mathbf{k}_1 - \mathbf{k}_2|}n(\varepsilon(\mathbf{k}_2))
\end{aligned} \tag{26.29}$$

is equal to the product of T^{-2} with the expression having the meaning of an average potential energy of the Coulomb interaction per unit volume. The condition $|d| \ll |b| \sim \rho T^{-1}$ indicates a smallness of the average potential energy in the system compared to the average kinetic energy. Since $|d| \sim e^2\rho^2 m^{-1}T^{-3}$ due to (26.29), the condition $|d| \ll |b|$ can be written in the form

$$\frac{e^2\rho}{m} \ll T^2 \quad \text{or} \quad a \ll \sqrt{mT}, \tag{26.30}$$

which implies that characteristic momentum a is small with respect to $\sqrt{mT}$.

It is not difficult to show that under conditions (26.28) and (26.30), the contribution of all remaining diagrams of the perturbation theory in $G(\omega = 0, \mathbf{k})$ is small compared to the contribution of the main diagram b.

27. DAMPING OF PLASMA OSCILLATIONS

We shall use hydrodynamical action (26.17) to evaluate the spectrum of plasma determined by the poles of correlation function (26.19) in analytic continuation $i\omega \to E$ [150].

Taking into account only the simplest one-loop diagrams a, b (26.17), we obtain the correlation function with poles $E = \pm E(k)$ where for $k \ll a \sim r_D^{-1}$

$$E(\mathbf{k}) = \omega_0 + \frac{3Tk^2}{2m\omega_0} - i\omega_0 \left(\frac{\pi}{8}\right)^{1/2} \left(\frac{a}{k}\right)^3 \exp(-a^2/2k^2) \qquad (27.1)$$

Here the first constant term ω_0 is the plasma frequency

$$\omega_0 = \left(\frac{4\pi e^2 \rho}{m}\right)^{1/2}. \qquad (27.2)$$

The second term in (27.1) determines the dispersion of the plasma spectrum and the third gives the known *Landau damping* [110].

Taking into account only diagram a equal to

$$-\frac{k^2}{m\omega^2}\frac{T}{V} \sum_{\mathbf{k}_1,\omega_1} \exp(i\omega_1 \varepsilon)(i\omega_1 - \varepsilon(\mathbf{k}_1))^{-1} = -\frac{k^2 \rho}{m\omega^2}, \qquad (27.3)$$

we obtain the correlation function

$$\frac{4\pi e^2}{k^2 \left(1 + \dfrac{4\pi e^2 \rho}{m\omega^2}\right)}.$$

Replacing $i\omega \to E$ we get the plasma spectrum $E = \pm \omega_0$ not depending on $\mathbf{k}$. In order to obtain dispersion and damping, we take into account diagram b which is equal to

$$-\frac{T}{V} \sum_{\mathbf{k},\omega_1} \left(\frac{\mathbf{k}\mathbf{k}_1}{im\omega}\right)^2 \left(i\omega_1 - \varepsilon\left(\mathbf{k}_1 - \frac{\mathbf{k}}{2}\right)\right)^{-1} \times$$

$$\times \left(i\omega + i\omega_1 - \varepsilon\left(\mathbf{k}_1 + \frac{\mathbf{k}}{2}\right)^{-1}\right)$$

$$= \frac{1}{V} \sum_{\mathbf{k}_1} \left(\frac{\mathbf{k}\mathbf{k}_1}{m\omega}\right)^2 \frac{[n(\varepsilon(\mathbf{k}_1 + (\mathbf{k}/2))) - n(\varepsilon(\mathbf{k}_1 - (\mathbf{k}/2)))] \times}{\omega^2 +}$$

$$\frac{\times [\varepsilon(\mathbf{k}_1 + (\mathbf{k}/2)) - \varepsilon(\mathbf{k}_1 - (\mathbf{k}/2))]}{+ (\varepsilon(\mathbf{k}_1 + (\mathbf{k}/2)) - \varepsilon(\mathbf{k}_1 - (\mathbf{k}/2)))^2} \qquad (27.4)$$

Putting $i\omega \to E$ and considering $\mathbf{k} \to 0$ this expression has the form

$$-(mE)^{-4}V^{-1}\sum_{\mathbf{k}_1}(\mathbf{k}\mathbf{k}_1)^4\frac{\partial n}{\partial\varepsilon}=\frac{3\rho Tk^4}{m^2E^4}. \tag{27.5}$$

The imaginary part of (27.4) is equal to

$$\frac{i\pi}{E^2V}\sum_{\mathbf{k}_1}\left(\frac{\mathbf{k}\mathbf{k}_1}{m}\right)^4\frac{\partial n}{\partial\varepsilon}\delta\left(E^2-\left(\frac{\mathbf{k}\mathbf{k}_1}{m}\right)^2\right)$$
$$=-i\frac{(2\pi m)^{1/2}\rho\omega_0}{3kT^{3/2}}\exp(-a^2/2k^2). \tag{27.6}$$

Adding (27.5) and (27.6) to (27.3) we obtain the spectrum (27.1). The Landau damping decreases for $\mathbf{k} \to 0$ faster than any power of k. This is explained by its collisionless mechanism which is connected with the transfer of plasmon energy $E(\mathbf{k})$ to electrons with velocities greater than the phase plasmon energy $E(\mathbf{k})/k \approx \omega_0/k$ (the inverse Vavilov–Cherenkov effect). The number of such electrons very rapidly decreases for $\mathbf{k} \to 0$ which leads to such a fast falling of the Landau damping.

There are other damping mechanisms besides collisionless one:

(a) *Collision damping* caused by collisions of charged particles. This mechanism gives damping proportional to k^2 for $\mathbf{k} \to 0$ in the considered model [111];

(b) *damping caused by plasmon–plasmon interaction.*

Both the collision and the plasmon–plasmon dampings exceed the Landau damping and are therefore essential.

Let us sketch a method of calculation of these quantities by means of the functional of the hydrodynamical action (26.17). Let momentum k_0, which divides the fast fields from the slow ones, be restricted by the inequalities

$$4\pi\rho e^2/T = r_D^{-2} \ll k_0^2 \ll k_T^2 = mT, \tag{27.7}$$

which are satisfied under the condition

$$T^2 \gg 4\pi\rho e^2/m = \omega_0^2, \tag{27.8}$$

It is true for nondegenerate plasma. The smallness of the phase volume with the small momenta domain gives us the possibility of restricting ourselves to the quadratic form of fields φ in functional (26.17).

Let us perform a partial summation in diagrams c, d, e, f by incorporating all the simplest proper energy contributions in the wavy line. Denoting the result of such a summation by a double wavy line we can write

$$(27.9)$$

The result of the summation (first done by Gell-Mann and Brueckner [145]) is given by formula (26.18), where $P(i\omega, \mathbf{k})$ has the form

$$\frac{1}{V} \sum_{\mathbf{k}_1} \frac{n(\varepsilon(\mathbf{k}_1 + (\mathbf{k}/2))) - n(\varepsilon(\mathbf{k}_1 - (\mathbf{k}/2)))}{i\omega + \varepsilon(\mathbf{k}_1 + (\mathbf{k}/2)) - \varepsilon(\mathbf{k}_1 - (\mathbf{k}/2))}. \qquad (27.10)$$

Let us stress that summation (27.9) was done here over the internal lines of the diagrams. Approximation (26.19) with P given by (27.10) for the whole correlation function is equivalent to taking into account only two diagrams a, b in (26.18) and was considered above.

The main result consists of the fact that the collision damping is determined by the imaginary parts of the diagrams

$$(27.11)$$

which are obtained from diagrams c, d, e (26.17) by replacing the single wavy line by double ones according to (27.9).

The plasmon–plasmon damping is determined by the diagram

$$(27.12)$$

The construction method of the hydrodynamical action functional and the damping theory of the long-wave plasma oscillations described in this chapter are generalized to a more realistic model of the two-component electron-ion plasma and to the case of plasma in a magnetic field [172] as well.

CHAPTER 9

THE ISING MODEL

28. THE STATISTICAL SUM OF THE ISING MODEL
AS A FUNCTIONAL INTEGRAL

The two-dimensional Ising model is one of the few nontrivial models of statistical physics having a second-order phase transition and, at the same time, an exact solution.

The first solution of the two-dimensional Ising model was obtained by Onsager [113] calculating the free energy of the system in the limit of an infinite volume. Onsager's result was then rederived by other authors by applying various methods [114–120]. All methods used are sufficiently clumsy and many cannot be treated as fully exact.

In this chapter we shall derive the formulae for the statistical sum and the correlation function of the finite two-dimensional Ising model by using the formalism of functional integration with respect to the anticommuting Fermi fields. We shall make use of some results of the work by A. A. Molokanov [121].

The functional integral was used for evaluation of the statistical sum in the limit of an infinite volume in the works by F. A. Berezin [122] and E. S. Fradkin [123].

Let us recall the definition of the Ising model. Consider a square lattice on a plane of dimension $L \times L$, the sites of which are numbered by indices (k, l), where k is a row index and l a column one, $1 \le k, l \le L$, and $L^2 = N$ is the number of sites of the lattice.

We associate the spin variables s_{kl} admitting two values ± 1 with the sites of the lattice. We put the periodic boundary condition

$$s_{kl} = s_{k+L,l} = s_{k,l+L},\tag{28.1}$$

which endows the lattice with the topology of a torus. The state $\{s\}$ of the spin system is defined by the set of values of N spin variables. We shall study a system with the Hamiltonian

$$H_N(s) = -g \sum_{k,l} (s_{kl} s_{k,l+1} + s_{kl} s_{k+1,l}),\tag{28.2}$$

It is advantageous to interpret this expression as the energy of the spin interaction. Note that only the nearest neighbours interact with the interaction constant g.

The statistical sum of the Ising model with Hamiltonian (28.2) is defined by the formula

$$Z_N = \sum_s \exp\left[-\beta H_N(s)\right]$$

$$= \sum \exp\left\{x \sum_{kl} (s_{kl}s_{k,l+1} + s_{kl}s_{k+1,l})\right\}, \qquad (28.3)$$

in which β is a reciprocal temperature; $x = \beta g$, $\sum_s$ denotes the sum over the 2^N spin states.

We rewrite expression (28.3) in the form

$$\sum_s \prod \exp(x s_{kl}s_{k'l'}), \qquad (28.4)$$

i.e., as the sum of the products, each factor of which corresponds to a definite pair of nearest neighbours or, as we shall say, to one bond. All-in-all, there are 2^N bonds in the lattice – N vertical ones (with the corresponding pairs $k, l; k, l+1$) and N horizontal ones (with the corresponding $k, l; k+1, l$).

We express the factor corresponding to one bond in the form

$$\exp(x s_{kl}s_{k'l'}) = \cosh(x s_{kl}s_{k'l'}) + \sinh(x s_{kl}s_{k'l'})$$

$$= \cosh x + s_{kl}s_{k'l'} \sinh x = \cosh x(1 + s_{kl}s_{k'l'} \tanh x). \qquad (28.5)$$

Putting $(\cosh \chi)^{2N}$ in front of the sum symbol we get

$$Z_N = (\cosh x)^{2N} \sum_s \prod(1 + s_{kl}s_{k'l'} \tanh x). \qquad (28.6)$$

Making the product over the bonds, we express the statistical sum in the form

$$(\cosh x)^{2N} \sum_{n_1,\ldots,n_{2N}=0} \prod_{j=1}^{2N} (s_{kl}s_{k'l'})^{n_j} (\tanh x)^{n_j}. \qquad (28.7)$$

Here every index $n_j (j = 1, \ldots, 2N)$ runs over two values: 0 and 1.

We associate with each term $\prod_{j=1}^{2N}(s_{kl}s_{k'l'})^{n_j}(\tanh x)^{n_j}$ a diagram on the lattice drawing the line along the bond if $n_j = 1$ and not drawing it if $n_j = 0$.

When summing up over the spin variables $s_{kl} = \pm 1$, summands in (28.7)

containing at least one variable s_{kl} in an odd power vanish and the terms
containing all spin variables in the even power are multiplied by the number
of states 2^N.

Thus

$$Z_N = 2^N(\cosh x)^{2N} S_N, \tag{28.8}$$

where S_N can be treated as a sum over all diagrams into each site of which
enters an even number of links (0, 2 or 4). The contribution of any such
diagram to S_N is $(\tanh x)^l$, where l is the number of diagram bonds.

Calculating the sum of the contributions of the mentioned diagrams is the
most difficult part of evaluating the statistical sum of the Ising model.

We shall begin by replacing each site of the lattice into which four links
enter by the three sites

$$(28.9)$$

There are no intersections of the links in sites a, b, whereas there is an
intersection of the vertical and horizontal links in site c. We may sum up
over the diagrams with sites a, b, c by taking sites a, b with a positive sign
and sites c with a negative one.

Each of the new diagrams consists of a few closed loops going through
the lattice bonds. The contribution of every diagram is

$$(-1)^n(\tanh x)^l, \tag{28.10}$$

where n is the number of diagram link intersections. It is clear that

$$n = \Gamma + \sum_\alpha \gamma_\alpha, \tag{28.11}$$

where γ_α is the number of self-interactions of the loops entering the diagram
and Γ is the number of intersections among the different loops. Γ is always
even for loops on the plane and $(-1)^\Gamma = 1$. That is not the case for loops
on a torus.

As a result, we obtain the expression for S_N

$$S_N = \sum(-1)^\Gamma \prod_\alpha (-1)^{\gamma_\alpha}(\tanh x)^{l_\alpha}, \tag{28.12}$$

where Σ is the sum over the diagrams, $\prod_\alpha$ is a product over the cycles
forming the diagrams, and l_α is the number of bonds forming loop α. Since

every diagram link is simple, each loop does not have a common bond with other loops nor with itself. This means that we need not take into account the diagrams

$$(28.13)$$

Now we insert the expressions for numbers γ_α, Γ into (28.12). It has been said that boundary condition (28.1) allows us to consider our lattice as a lattice on a torus. All closed loops on the torus are divided into homotopic classes which will be denoted by $[m, n]$. The loop of class $[m, n]$ admits m complete turns in a horizontal direction and n in a vertical one. The symbols $[m, n]$ and $[-m, -n]$ denote one and the same class for the nonoriented loops. The class $[0, 0]$ is formed by the loops which can be shrunk to a point by continuous deformation.

The number of pair intersections of a few loops is

$$\sum_{i \neq j} m_i n_j (\mathrm{mod}\, 2), \tag{28.14}$$

where m_i is the index of the ith loop and n_j that of the jth. Therefore,

$$(-1)^\Gamma = (-1)^{\sum_{i \neq j} m_i n_j}. \tag{28.15}$$

It is known that the full rotation angle of the tangent vector is equal to $2\pi (l + 1)$ when one winds once around the closed loop α on a plane (or the loop of class $[0, 0]$ on a torus). The integer number l has the same parity as the number of self-intersections γ_α of loop α. We associate the factor $\exp(i\varphi/2)$ (the Katz–Ward factor) with any site of loop α in which the loop rotates by angle $\varphi(\varphi = 0, \pm \pi/2)$. The product of these factors gives

$$\prod_\alpha \exp(i\varphi/2) = (-1)^{l+1} = (-1)^{\gamma_\alpha + 1} \tag{28.16}$$

after winding once around the loop so that

$$(-1)^{\gamma_\alpha} = -\prod_\alpha \exp(i\varphi/2). \tag{28.17}$$

The generalization of this formula for the loop of class $[m, n]$ has the form

$$(-1)^{\gamma_\alpha} = (-1)^{mn+m+n+1} \prod_\alpha \exp(i\varphi/2). \tag{28.18}$$

By virtue of (28.15) and (28.16), expression (26.12) for S_N can be written in the form

$$S_N = \sum (-1)^{(\Sigma_\alpha m_\alpha)(\Sigma_\alpha n_\alpha)} \prod_\alpha (-1)^{m_\alpha + n_\alpha + 1} \times$$

$$\times \left(\prod \exp(i\varphi/2) \right) (\tanh x)^{l_\alpha}. \tag{28.19}$$

We shall use the identity

$$(-1)^{mn} = \tfrac{1}{2}(1 + (-1)^m + (-1)^n - (-1)^{m+n}), \tag{28.20}$$

in which we insert $m = \Sigma_\alpha m_\alpha, n = \Sigma_\alpha n_\alpha$. This gives the possibility of expressing S_N in the form

$$S_N = \tfrac{1}{2}(S_N^{(1)} + S_N^{(2)} + S_N^{(3)} - S_N^{(4)}), \tag{28.21}$$

where

$$S_N^{(1)} = \sum \prod_\alpha (-1)^{m_\alpha + n_\alpha + 1} (\tanh x)^{l_\alpha} \prod \exp(i\varphi/2);$$

$$S_N^{(2)} = \sum \prod_\alpha (-1)^{n_\alpha + 1} (\tanh x)^{l_\alpha} \prod \exp(i\varphi/2);$$

$$\tag{28.22}$$

$$S_N^{(3)} = \sum \prod_\alpha (-1)^{m_\alpha + 1} (\tanh x)^{l_\alpha} \prod \exp(i\varphi/2);$$

$$S_N^{(4)} = \sum \prod_\alpha (-1)(\tanh x)^{l_\alpha} \prod \exp(i\varphi/2).$$

It appears to be useful if we represent any squared quantity $(S_N^{(i)})^2$ as a Gaussian functional integral with respect to anticommuting variables.

Taking $S_N^{(i)}$ squared can be interpreted as a passage from the nonoriented to the oriented loops. For instance,

$$(S_N^{(1)})^2 = \sum \prod_\alpha \left\{ (-1)^{m_\alpha + n_\alpha + 1} (\tanh x)^{l_\alpha} \prod \exp(i\varphi/2) \right\}, \tag{28.23}$$

where $\prod_\alpha$ is a product over all *oriented* loops having no common oriented bonds in the same sense. Indices m_α, n_α run over all positive and negative numbers.

We represent the sign factor $(-1)^{m_\alpha + n_\alpha}$ in the form

$$(-1)^{m_\alpha + n_\alpha} = \prod \exp[(i\pi/L)(\Delta x + \Delta y)], \tag{28.24}$$

where π is a product over the succeeding sites of the lattice when going round the oriented loop α; $\Delta x, \Delta y$ are increments of the coordinates x and

y expressed in length units of the lattice when passing from the centre of the site bond previous to the given one to the centre of the next site bond. For instance

$$\longrightarrow\!\!\bullet\!\!\longrightarrow \quad \Delta x = 1, \Delta y = 0; \qquad \rule{0.5pt}{20pt}\!\!\!\!\!\_\quad \Delta x = \tfrac{1}{2}, \Delta = \tfrac{1}{2} \qquad (28.25)$$

The analogous representation

$$(-1)^{n_\alpha} = \prod \exp\left[(i\pi/L)\Delta y\right];$$
$$(-1)^{m_\alpha} = \prod \exp\left[i\pi/L)\Delta x\right] \qquad (28.26)$$

exists for the sign factors in $S_N^{(2)}, S_N^{(3)}$.

Joining product (28.4) with $\pi \exp(i\varphi/2)$ we write down (28.23) in the form

$$(S_N^{(1)})^2 = \sum \prod_\alpha \left\{ (-1)(\tanh x)^{l_\alpha} \prod \exp\left(\frac{i\pi}{L}(\Delta x + \Delta y) + \frac{i\varphi}{2} \right) \right\}. \qquad (28.27)$$

This expression that can be written in the form of a functional integral. With every oriented bond we associate a pair of mutually-conjugated Fermi fields c, c^*:

$$\underset{\displaystyle \overset{mn}{\bullet\!\!\!\longrightarrow}}{} \qquad c_1(m,n), \qquad c_1^*(m,n);$$

$$\underset{\displaystyle \overset{mn}{\longrightarrow\!\!\!\bullet}}{} \qquad c_2(m,n), \qquad c_2^*(m,n);$$

$$\Big\downarrow mn \qquad c_3(m,n), \qquad c_3^*(m,n); \qquad (28.28)$$

$$\Big\uparrow mn \qquad c_4(m,n) \qquad c_4^*(m,n);$$

We shall show that expression (28.27) can be represented in the form of a Gaussian functional integral with respect to $8\,N$ variables (28.28):

$$\int \exp(-c^* \hat{C} c) \, dc^* \, dc. \qquad (28.29)$$

Here

$$dc^* \, dc = \prod_{m,n,i} dc_i^*(m,n) \, dc_i(m,n) \qquad (28.30)$$

is the measure of functional integration, and $c^*\hat{C}c$ a quadratic form given by

$$c^*\hat{C}c = \sum_{m,n} c_i^*(m, n)\hat{C}_{ij}c_j(m, n), \tag{28.31}$$

where $\hat{C}$ is a matrix operator

$$\hat{C} = I - \text{th}\,x\hat{A}, \tag{28.32}$$

Here I is a four-dimensional unit matrix and $\hat{A}$ an operator of the form

$$\hat{A} = \begin{pmatrix} \exp\left(\dfrac{i\pi}{L} + \dfrac{d}{dm}\right) & 0 & \exp\left(\dfrac{i\pi}{L} + \dfrac{i\pi}{4} + \dfrac{d}{dm}\right) & \exp\left(-\dfrac{i\pi}{4} + \dfrac{d}{dm}\right) \\ 0 & \exp\left(-\dfrac{i\pi}{L} - \dfrac{d}{dm}\right) & \exp\left(-\dfrac{i\pi}{4} - \dfrac{d}{dm}\right) & \exp\left(-\dfrac{i\pi}{L} + \dfrac{i\pi}{4} - \dfrac{d}{dm}\right) \\ \exp\left(\dfrac{i\pi}{L} - \dfrac{i\pi}{4} + \dfrac{d}{dm}\right) & \exp\left(\dfrac{i\pi}{4} + \dfrac{d}{dn}\right) & \exp\left(\dfrac{i\pi}{L} + \dfrac{d}{dn}\right) & 0 \\ \exp\left(\dfrac{i\pi}{4} - \dfrac{d}{dn}\right) & \exp\left(-\dfrac{i\pi}{L} - \dfrac{i\pi}{4} - \dfrac{d}{dn}\right) & 0 & \exp\left(-\dfrac{i\pi}{L} - \dfrac{d}{dn}\right) \end{pmatrix}$$

$$\tag{28.33}$$

where $\exp(\pm d/dm)$, $\exp(\pm d/dn)$ are the shift operators acting according to the rule

$$\exp\left(\pm\frac{d}{dm}\right)c_i(m, n) = c_i(m \pm 1, n);$$

$$\exp\left(\pm\frac{d}{dn}\right)c_i(m, n) = c_i(m, n \pm 1). \tag{28.34}$$

In order to prove that $(S_N^{(1)})^2$ is represented by functional integral (27.30), we first notice that the quadratic form

$$\tanh x\,c^*\hat{A}c = \tanh x \sum_{m,n,i,j} c_i^*(m, n)\hat{A}c_j(m, n) \tag{28.35}$$

can be graphically interpreted as a sum of the contributions over all possible ways of the passage from one directed bond to its neighbour:

$$(28.36)$$

The arrow going out of the site (m, n) is associated with the Fermi field $c_i^*(m, n)$ where the index $i = 1, 2, 3, 4$ is defined by the direction of the out-going arrow – left, right, up and down. To the arrow going into the site (m, n), there corresponds the field $c_j(m', n')$ where index j is determined by the direction of the ingoing arrow and (m', n') are coordinates of the site neighbouring (m, n) from which we can come into (m, n) moving along the arrow. The coefficient at $c_i^*(m, n) c_j(m', n')$

$$\tanh x \exp\left(\frac{i\varphi}{2} + \frac{i\pi}{L}(\Delta x + \Delta y)\right) \equiv z_{mnij} \qquad (28.37)$$

is equal to the product of $\tanh x$ with a factor appearing in the product inside formula (28.27).

Integral (28.29) is represented in the form

$$\int dc^* \, dc \exp(-c^*c) \prod_{mnij} \exp(z_{mnij} c_i^*(m, n) c_j(m', n'))$$

$$= \int dc^* \, dc \exp(-c^*c) \prod_{mnij} (1 + z_{mnij} c_i^*(m, n) c_j(m', n')), \qquad (28.38)$$

where

$$c^*c \equiv \sum_{m,n,i} c_i^*(m, n) c_i(m, n). \qquad (28.39)$$

Removing the product $\pi(1 + zc^*c)$ we reduce the problem to a calculation of the integral of the type

$$\int \exp(-c^*c)(c^*c)(c^*c), \ldots, (c^*c) \, dc^* \, dc \qquad (28.40)$$

of the product of pairs of the Fermi fields c^*c with the weight $\exp(-c^*c)$. If there are thought to be two identical ones among fields c or c^*, the integral is equal to zero. Integral (28.40) differs from zero if the graphical elements corresponding to the pairs c^*c, according to (28.36), form one or several closed loops from oriented bonds and if the loops have no common identically oriented bonds. In other words, we obtain the same loops as those taken in formula (28.27). Since there is the factor z_{mnij} (28.37) at each pair c^*c and integral (28.40) is equal to $(-1)^f$, where f is the number of independent loops, we obtain for integral (28.29) the expression

$$\sum \prod_\alpha \left\{ (-1) \prod z_{mnij} \right\}, \qquad (28.41)$$

which evidently coincides with (28.27). Thus, the representation of $(S_N^{(1)})^2$ by the functional integral (28.29) is proved.

In order to calculate integral (28.29), we change the variables c^*, c by their Fourier images according to the formulae

$$c_i(m, n) = \frac{1}{L} \sum_{1 \le p, q \le L} a_i(p, q) \exp\left(\frac{2\pi i}{L} (mp + nq) \right);$$

$$c_i^*(m, n) = \frac{1}{L} \sum_{1 \le p, q \le L} a_i^*(p, q) \exp\left(-\frac{2\pi i}{L} (mp + nq) \right), \qquad (28.42)$$

realizing a unitary transformation. At the same time

$$\sum_{m, n, i} c_i^*(m, n) c_i(m, n) = \sum_{p, q, i} a_i^*(p, q) a_i(p, q); \qquad (28.43)$$

and

$$\sum_{m, n, i, j} c_i^*(m, n) \hat{A}_{ij} c_j(m, n) = \sum_{p, q, i, j} a_i^*(p, q) A_{ij}(p, q) a_j(p, q), \qquad (28.44)$$

where $A_{ij}(p, q)$ forms a matrix obtained from the operator $\hat{A}$ by

$$\exp\left(\pm \frac{d}{dm} \right) \to \exp\left(\pm 2\pi i \frac{p}{L} \right);$$

$$\exp\left(\pm \frac{d}{dn} \right) \to \exp\left(\pm 2\pi i \frac{q}{L} \right). \qquad (28.45)$$

The quadratic form $c^* \hat{C} c$ of variables a, a^* is represented in the form of a sum of the quadratic forms of $a_i^*(p, q), a_j(p, q)$:

$$\sum_{p, q} \sum_{ij} (\delta_{ij} - \tanh x A_{ij}(p, q)) a_i^*(p, q) a_j(p, q). \qquad (28.46)$$

Integral (28.29) reduces to the product of the integrals

$$\prod_{p,q} \int \prod_i da_i^*(p,q)\, da_i(p,q) \times$$

$$\times \exp\left\{ -\sum_{ij} (\delta_{ij} - \tanh x A_{ij}(p,q)) a_i^*(p,q) a_i(p,q) \right\} \quad (28.47)$$

$$= \prod_{p,q} \det(I - \tanh x \hat{A}(p,q)),$$

where $\det(I - \tanh x \hat{A}(p,q))$ is a determinant of the four-dimensional matrix equal to

$$(1 + \tanh^2 x)^2 - 2\tanh x(1 - \tanh^2 x) \times$$
$$\times \left(\cos\frac{\pi}{L}(2p - q) + \cos\frac{\pi}{L}(2q - 1) \right). \quad (28.48)$$

The quantity

$$S_N^{(1)} = \prod_{p,q} \left[(1 + \tanh^2 x)^2 - 2\tanh x(1 - \tanh^2 x) \times \right.$$
$$\left. \times \left(\cos\frac{\pi}{L}(2p - 1) + \cos\frac{\pi}{L}(2q - 1) \right) \right]^{1/2} \quad (28.49)$$

is obtained by taking the square root of (28.47) in such a way that we get an expression which goes to 1 in the limit $x \to 0$. Analogously we obtain the formulae

$$S_N^{(2)} = S_N^{(3)} = \prod_{p,q} \left[(1 + \tanh^2 x)^2 - \right.$$
$$\left. - 2\tanh x(1 - \tanh^2 x)\left(\cos\frac{\pi}{L}(2p - 1) + \cos\frac{2\pi q}{L} \right) \right]^{1/2};$$

$$S_N^{(4)} = \prod_{p,q} \left[(1 + \tanh^2 x)^2 - 2\tanh x(1 - \tanh^2 x) \times \right.$$
$$\left. \times \left(\cos\frac{2\pi}{L}p + \cos\frac{2\pi}{L}a \right) \right]^{1/2}. \quad (28.50)$$

The final answer for the statistical sum of the finite two-dimensional Ising model has the form

$$Z_N = 2^{N-1}(\cosh x)^{2N}(S_N^{(1)} + S_N^{(2)} + S_N^{(3)} - S_N^{(4)}). \quad (28.51)$$

We shall consider expression (28.51) in the limit $N \to \infty$. For this aim we write quantity $(S_N^{(1)})^2$ in the form

$$(S_N^{(1)})^2 = \exp\left\{ - \frac{N}{2\pi^2} \oint \frac{dz_1}{z_1} \frac{1}{1+z_1^L} \oint \frac{dz_2}{z_2} \frac{\ln f(z_1, z_2)}{1+z_2^L} \right\};$$

$$S_N^{(2)})^2 = (S_N^{(3)})^2 = \exp\left\{ - \frac{N}{4\pi^2} \oint \frac{dz_1}{z_1} \frac{1}{1+z_1^L} \oint \frac{dz_2}{z_2} \frac{\ln f(z_1, z_2)}{1+z_2^L} \right\};$$

$$(S_N^{(4)})^2 = \exp\left\{ - \frac{N}{4\pi^2} \oint \frac{dz_1}{z_1} \frac{1}{1-z_1^L} \oint \frac{dz_2}{z_2} \frac{\ln f(z_1, z_2)}{1-z_2^L} \right\}, \qquad (28.52)$$

where

$$f(z_1, z_2) = (1 + \tanh^2 x)^2 - 2 \tanh x (1 - \tanh^2 x) \times$$
$$\times (z_1 + z_1^{-1} + z_1^{-1} + z_2 + z_2^{-1}), \qquad (28.53)$$

and contour Γ consists of two circles $|z_i| = 1 \pm \delta (i = 1, 2)$ with a sufficiently small $\delta > 0$. One winds the external circle round in a positive direction and the internal circle in a negative direction. The integrals along Γ reduce to a sum of values of polynomials $1 \pm z_1^L, 1 \pm z_2^L$ in roots. This returns us to (28.49) and (28.50) and thus proves formulae (28.52).

It is clear from the expression

$$\left. \begin{array}{l} (1 \pm z^L)^{-1} = 1 \mp z^L + z^{2L} \mp z^{3L} + \ldots |z| < 1; \\ (1 \pm z^L)^{-1} = \pm z^{-L} - z^{-2L} \pm z^{-3L} + \ldots |z| > 1 \end{array} \right\} \qquad (28.54)$$

that for $L \gg 1$ the main contribution to the integrals along Γ gives 1 in the first expansion (28.52) and the rest is a quantity of the order $L^2 \exp(-\alpha L)$. This can be seen by enlarging the big circle and shrinking the small one up to their intersection with the singularities of the function $\ln f(z_1, z_2)$. Quantity α is defined by the distance from the unit circle to the singularities of function $\ln f(z_1, z_2)$. Function $f(z_1, z_2)$ is a positive definite for all $|z_1| = |z_2| = 1$ for any x except $|x| = x_c$ where

$$\tanh x_c = \sqrt{2} - 1. \qquad (28.55)$$

For $x = x_c$ and $x = -x_c$, function $f(z_1, z_2)$ vanishes for $z_1 = z_2 = 1$ and $z_1 = z_2 = -1$ respectively.

Our discussion leads to the following asymptotic formula

$$(S_N^{(i)})^2 = \exp\left\{ \frac{N}{4\pi^2} \int_0^{2\pi} dx_1 \int_0^{2\pi} dx_2 \ln f(\exp(ix_1), \exp(ix_2)) + \right.$$

$$\left. + O(L^2 \exp(-\alpha L)) \right\}, \qquad (28.56)$$

which holds for all x except $|x| = x_c$ and the O-estimate is uniform for $||x| - x_c| > \varepsilon$.

If constant g in Hamiltonian (28.2) is positive (the case of attraction), the physical domain of definition of the parameter $x = \beta g$ is the positive semi-axis $x > 0$. For $x = x_c$ the expression $S_N^{(4)}$ changes the sign since the cofactor in $\Pi_{p,q}$ with $p = q = 0$ has the form

$$[(1 + \tanh^2 x)^2 - 4\tanh x(1 - \tanh^2 x)]^{1/2}$$
$$= (\sqrt{2} + 1 + \tanh x)(\sqrt{2} - 1 - \tanh x). \tag{28.57}$$

The remaining quantities $S_N^{(i)}, i = 1, 2, 3$, remain positive for all $x > 0$. We, therefore, obtain the following asymptotic formulae

$$Z_N(x < x_c) = Z_{\text{Ons}}(1 + O(1));$$
$$Z_N(x > x_c) = 2Z_{\text{Ons}}(1 + O(1)), \tag{28.58}$$

where the expression

$$Z_{\text{Ons}} = 2^N(\cosh x)^{2N} \times \tag{28.59}$$

$$\times \exp\left\{ -\frac{N}{8\pi^2} \int_0^{2\pi} \int_0^{2\pi} dx_1\, dx_2\, \ln f(\exp(ix_1), \exp(ix_2)) \right\}$$

is usually called the Onsager statistical sum. The expression for the limiting free energy for one spin

$$F = -\frac{1}{\beta} \lim_{N \to \infty} \frac{\ln Z_N}{N}$$

$$= -\frac{1}{\beta}\left\{ \ln 2 + 2\ln\cosh x + \frac{1}{8\pi^2} \int_0^{2\pi} \int_0^{2\pi} dx_1\, dx_2 \times \right. \tag{28.60}$$

$$\left. \times \ln\left[(1 + \tanh^2 x)^2 - 2\tanh x(1 - \tanh^2 x)(\cos x_1 + \cos x_2)\right] \right\}$$

follows from (28.58) and (28.59) and was first derived by Onsager [113].

For $x = x_c$, expression (28.60) has a nonanalytic singularity due to the vanishing of the expression in the logarithm for $x_1 = x_2 = 0$. The corresponding temperature

$$T_c = \frac{1}{\beta_c} = \frac{g}{x_c}$$

is interpreted as the phase transition temperature.

Let us remark that factor 2 appears in the point of Z_{Ons} in asymptotic formulae (28.58) for $T < T_c$. This corresponds to the fact that the system has a tendency to be in one of two possible states with minimum energy – in the state with the parallel spins for ferromagnetic system ($g > 0$) and with the antiparallel ones for antiferromagnetic system ($g < 0$). Factor 2 has no effect on the thermodynamical limit (28.60) for free energy F.

Finally, notice that the method discussed in this section can be applied to arbitrary plane lattices without intersecting bonds.

29. THE CORRELATION FUNCTION OF THE ISING MODEL

The method previously developed for calculating the statistical sum of a two-dimensional Ising model can be used for finding the correlation function as well. For simplicity, we consider a pair correlation function of two spins appearing in one row of the lattice:

$$G_N(r) = Z_N^{-1} \sum_s s_{kl} s_{k,l+r} \exp\left\{ x \sum_{kl} (s_{kl} s_{k+1,l} + s_{kl} s_{k,l+1}) \right\}. \quad (29.1)$$

Here Z_N is the statistical sum evaluated above.

By using formula (28.5) and putting $(\cosh \chi)^{2N}$ under the summation sign, the problem is reduced to calculating the expression

$$\sum_s s_{kl} s_{k,l+2} \prod (1 + \tanh x s_{kl} s_{k'l'}), \quad (29.2)$$

in which the product is taken over all lattice bonds. Performing the product and summing up over spin variables s_{kl}, (29.2) will be represented as a product of the number of states 2^N with the sum of the contributions of the diagrams which differ from those for the statistical sum by the fact that their sites $(k, l), (k, l+r)$ contain an odd number of links (1 or 3). This leads to the formula

$$G_N(r) = S_N^{-1} g_N(r), \quad (29.3)$$

where S_N is determined above (28.8) and $g_N(r)$ is a sum of the contributions of the indicated diagrams, each of which is $(\tanh x)^l$ where l is the number of bonds forming the diagram.

The next step is a passage to the connected diagrams. With this aim we replace each site containing four links by three sites according to scheme (28.9). The site containing three links will be replaced by three sites according to the scheme

$$\top \;\longrightarrow\; \overset{a}{\top}\; + \;\overset{b}{\top}\; + \;\overset{c}{\top} \tag{29.4}$$

We associated with site a, b factor l and with site c factor $-l$. As a result, quantity $g_N(r)$ is represented as a sum of the contributions of the diagrams each of which consists of a few closed loops and unclosed links connecting sites (k, l) and $(k, l + r)$.

The contribution of any diagram is

$$(-1)^n (\tanh x)^l, \tag{29.5}$$

where n is the number of link intersections of the diagram. We have

$$n = \Gamma - \gamma + \sum_\alpha \gamma_\alpha, \tag{29.6}$$

where γ_α is the number of self-interactions of loop α entering the diagram; γ is the number of self-interactions of the unclosed link; and Γ is the number of self-interactions among different bond components.

Connecting chosen sites (k, l), $(k, l + r)$ by the link

$$\overset{\mu}{\underbrace{\subset \cdots\cdots \supset}} \tag{29.7}$$

we express the sign factors

$$\begin{aligned}
(-1)^{\gamma_\alpha} &= (-1)^{m_\alpha n_\alpha + m_\alpha + n_\alpha} \prod \exp(i\varphi/2); \\
(-1)^{\gamma} &= (-1)^{mn + m + n} \prod \exp(i\varphi/2); \\
(-1)^{\Gamma} &= (-1)^{\Sigma_{l+j} m_l n_j + \Sigma_\alpha v_\alpha} \prod \exp(i\varphi/2)
\end{aligned} \tag{29.8}$$

by means of v which is the number of intersections of unclosed link λ with link μ, of v_α, the number of intersections of loop α with link μ, and of indices m_α, n_α, m, n indicating the number of full turnings along the closed loops and the unclosed link around a torus in horizontal and vertical directions. By virtue of (28.8) we obtain

$$\begin{aligned}
g_N(r) = \sum (-1)^{(\Sigma_\alpha m_\alpha + m)(\Sigma_\alpha n_\alpha + n)} &[(-1)^{m+n+1+v}(\tanh x)^{l_\lambda} \times \\
&\times \prod \exp(i\varphi/2)] \prod_\alpha (-1)^{m_\alpha + n_\alpha + 1 + v_\alpha}(\tanh x)^{l_\alpha} \times \quad (29.9) \\
&\times \prod \exp(i\varphi/2).
\end{aligned}$$

Here Σ is the sum over all diagrams of the structure shown; $\prod \exp(i\varphi/2)$

is a product of the Katz–Ward factors for the closed loops α or the unclosed link λ. By using identity (28.20), in which we replace $m \to m + \Sigma_\alpha m_\alpha$, $n \to n + \Sigma_\alpha n_\alpha$, $g_N(r)$ will be expressed in the form

$$g_N(r) = \tfrac{1}{2}(g_N^{(1)}(r) + g_N^{(2)}(r) + g_N^{(3)}(r) - g_N^{(4)}(r)), \tag{29.10}$$

where

$$g_N^{(1)}(r) = \sum (-1)^{m+n+1+v}(\tanh x)^{l_\lambda} \prod \exp(i\varphi/2) \times$$
$$\times \prod_\alpha (-1)^{m_\alpha + n_\alpha + 1 + v_\alpha}(\tanh x)^{l_\alpha} \prod \exp(i\varphi/2);$$

$$g_N^{(2)}(r) = \sum [(-1)^{n+v+1}(\tanh x)^{l_\lambda} \prod \exp(i\varphi/2)] \times$$
$$\times \prod_\alpha (-1)^{n_\alpha + 1 + v_\alpha}(\tanh x)^{l_\alpha} \prod \exp(i\varphi/2); \tag{29.11}$$

$$g_N^{(3)}(r) = \sum [(-1)^{m+1+v}(\tanh x)^{l_\lambda} \prod \exp(i\varphi/2)] \times$$
$$\times \prod_\alpha (-1)^{m_\alpha + 1 + v_\alpha}(\tanh x)^{l_\alpha} \prod \exp(i\varphi/2);$$

$$g_N^{(4)}(r) = \sum [(-1)^{v+1}(\tanh x)^{l_\lambda} \prod \exp(i\varphi/2)] \times$$
$$\times \prod_\alpha (-1)^{1 + v_\alpha}(\tanh x)^{l_\alpha} \prod \exp(i\varphi/2).$$

The next step consists of writing the expression for squares $(g_N^{(i)}(r))^2$ and interpreting them in terms of oriented diagrams. The product of the contributions of unclosed links can be compared with the closed link going through two chosen points of the lattice (k, l) and $(k, l+r)$.

We introduce the expression for $(g_N^{(1)}(r))^2$

$$(g_N^{(1)}(r))^2 = \sum [(-1)^{m+n+1+v}(\tanh x)^l \prod \exp(i\varphi/2)] \times$$
$$\times \prod_\alpha (-1)^{m_\alpha + n_\alpha + 1 + v_\alpha}(\tanh x)^{l_\alpha} \prod \exp(i\varphi/2). \tag{29.12}$$

Here Σ is the summation over all oriented loops, one of which goes through the two chosen points of the lattice. The formulae for $(g_N^{(i)})^2$, $i = 2, 3, 4$, can be obtained from (29.12) by replacing the sign factor $(-1)^{m+n}$ by $(-1)^n$, $(-1)^m$ or (1) and factor $(-1)^{m_\alpha + n_\alpha}$ by $(-1)^{n_\alpha}$, $(-1)^{m_\alpha}$ or 1.

Expression (29.12) can be written in the form of a functional integral with respect to the Fermi fields $c_i(m, n)$, $c_i^*(m, n)$ defined by (28.28), which differs from the Gaussian one (28.29) in two respects:

(1) Besides the exponential of the quadratic form, there will be a product of two quadratic forms of the Fermi fields, corresponding to two lattice sites, under the integral sign.

(2) The quadratic form itself differs from that defined by formulae (28.31)–(28.33). The difference is explained through the existence of additional sign factors $(-1)^v$, $(-1)^{v_\alpha}$ counting intersections of the links of the diagram with the above-specified link μ (29.7). The inclusion of these factors can be reached by replacing

$$\hat{A} \to \hat{B} \equiv \hat{A} - 2\hat{A}\hat{P} \tag{29.13}$$

where $\hat{A}$ is a matrix operator defined by (28.33) and diagonal with respect to indices (m, n); $\hat{P}$ is a projector operator on $r+1$ bonds intersected by link μ.

Representation of $(g_N^{(1)}(r))^2$ by a functional integrals has the form

$$(g_N^{(1)}(r))^2 = \int \exp(-c^*\hat{C}c)\, dc^*\, dc \times$$
$$\times \{\langle D(k, l)D(k, l+r)\rangle - \langle D(k, l)\rangle\langle D(k, l+r)\rangle\}. \tag{29.14}$$

Here $dc^*\, dc$ is integration measure (28.30),

$$\hat{C} = I - \tanh x\hat{B} \tag{29.15}$$

is an operator of the quadratic form and the sign $\langle \ldots \rangle$ denotes the functional mean value:

$$\langle X \rangle = \frac{\int X \exp(-c^*\hat{C}c))\, dc^*\, dc}{\int \exp(-(c^*\hat{C}c))\, dc^*\, dc} \tag{29.16}$$

with the weight $\exp(-c^*\hat{C}c)$; $D(k, l)$ is the quadratic form

$$D(k, l) = \sum_{ij} c_i^*(k, l)\hat{D}_{ij}c_j(k, l) \tag{29.17}$$

with the matrix operator

$$\hat{D} = \begin{pmatrix} \exp\left(\dfrac{d}{dm}\right) & 0 & \exp\left(\dfrac{i\pi}{4}+\dfrac{d}{dm}\right) & \exp\left(-\dfrac{i\pi}{4}+\dfrac{d}{dm}\right) \\ 0 & \exp\left(\dfrac{d}{dm}\right) & \exp\left(-\dfrac{i\pi}{4}-\dfrac{d}{dn}\right) & \exp\left(\dfrac{i\pi}{4}-\dfrac{d}{dm}\right) \\ \exp\left(-\dfrac{i\pi}{4}+\dfrac{d}{dn}\right) & \exp\left(\dfrac{i\pi}{4}+\dfrac{d}{dn}\right) & \exp\left(\dfrac{d}{dn}\right) & 0 \\ \exp\left(\dfrac{i\pi}{4}-\dfrac{d}{dn}\right) & \exp\left(-\dfrac{i\pi}{4}-\dfrac{d}{dn}\right) & 0 & \exp\left(-\dfrac{d}{dn}\right) \end{pmatrix} \tag{29.18}$$

in which $\exp(\pm d/dm)$, $\exp(\pm d/dn)$ are shift operators (28.34).

In order to prove representation (29.14), we consider the expression

$$\int \exp(-c^* \hat{C} c) \, dc^* \, dc \, \langle D(k, l) D(k, l + r) \rangle$$

(29.19)

$$= \int D(k, l) D(k, l + r) \exp(-c^* \hat{C} c) \, dc^* \, dc.$$

The functional integral on the right-hand side can be calculated by means of the perturbation theory taking the operator $-\tanh x \hat{B}$ in (29.15) as a correction to the unit operator. The obvious graphical representation arising gives diagrams of two types. The first type consists of a few closed loops and one loop going through chosen points (k, l), $(k, l + r)$. It is just these diagrams that contribute in the expression $(g_N^{(1)}(r))^2$. The second type consists of a few closed loops, one loop going through point (k, l) and one loop going through point $(k, l + r)$. It is not difficult to see that the contribution of the diagrams of the second type is given by the expression

$$\langle D(k, l) \rangle \langle D(k, l + r) \rangle \int \exp - (c^* \hat{C} c) \, dc^* \, dc,$$

(29.20)

entering (29.14) with a negative sign. This means that the diagrams of the second type do not contribute in formula (29.14). Thus representation (29.14) can be considered as proved.

The expression appearing in (29.14) can be calculated by means of the formulae

$$\int \exp(-c^* \hat{C} c) \, dc^* \, dc = \det \hat{C};$$

(29.21)

and

$$\langle D(k, l) D(k, l + r) \rangle - \langle D(k, l) \rangle \langle D(k, l + r) \rangle$$
$$= \operatorname{tr} \hat{D} \hat{C}^{-1}(k, l; k, l + r) \hat{D} \hat{C}^{-1}(k, l + r; k, l).$$

(29.22)

In the first one $\det \hat{C}$ is the determinant of the operator acting in a $4N$-dimensional space. In the second formula, the sign tr denotes the trace with respect to the matrix indices i, j running over four values, $i, j = 1, 2, 3, 4$; $C^{-1}(i, k, l; j, k, l + r)$ is the matrix element of the operator $\hat{C}^{-1}$ inverse to $\hat{C}$.

First of all we find determinant (29.21). We have

$$\det \hat{C} = \det \hat{C} \det(\hat{C}^{-1}\hat{C}) \tag{29.23}$$

$$= \det(I - \tanh x\hat{A}) \det\left(I - \frac{2\tanh x\hat{A}}{I - \tanh x\hat{A}}\hat{P}\right).$$

Here the first factor is nothing more than the quantity $(S_N^{(1)})^2$ which was evaluated in the previous section (see formula (28.47)):

$$\det(I - \tanh x\hat{A}) = (S_N^{(1)})^2. \tag{29.24}$$

The second factor can be represented as the determinant of the operator

$$\hat{K} = \hat{P}\frac{I + \tanh x\hat{A}}{I - \tanh x\hat{A}}\hat{P}, \tag{29.25}$$

acting in a $2(r + 1)$-dimensional proper subspace of the projector $\hat{P}$. Operator $\hat{K}$ can be written in the form

$$\hat{K} = \begin{pmatrix} D, & -F \\ F, & D \end{pmatrix}, \tag{29.26}$$

where D, F are submatrices of the $(r + 1)$th order with the elements:

$$D_{mn} = L^{-2} \sum_{k_1,k_2 = 1}^{L} \exp\left[-\frac{i\pi}{L}(2k_1 + 1)(m - n)\right] \times$$

$$\times \frac{(1 - y^4 - 2y(1 + y^2)\cos(\pi/L)(2k_1 + 1) + 2iy(1 - y)\sin(\pi/L)(2k_2 + 1))}{(1 + y^2)^2 - 2y(1 - y^2)(\cos(\pi/L)(2k_1 + 1) + \cos(\pi/L)(2k_2 + 1)},$$

$$\tag{29.27}$$

$$F_{mn} = L^{-2} \sum_{k_1,k_2 = 1}^{L} \exp\left[-\frac{i\pi}{L}(2k_1 + 1)(m - n)\right] \times$$

$$\times \frac{4y^2 \sin(\pi/L)(2k_1 + 1)}{(1 + y^2)^2 - 2y(1 - y)^2(\cos(\pi/L)(2k_1 + 1) + \cos(\pi/L)(2k_2 + 1))},$$

where $y \equiv \tanh x$.

Determinant $\det \hat{K}$ is calculated according to the formula

$$\det \hat{K} = \det(D + iF)\det(D - iF). \tag{29.28}$$

Let us consider expression (29.22). The appearance of operator $\hat{D}$ under tr allows us to replace operator $\hat{C}^{-1}$ by $\hat{P}\hat{C}^{-1}\hat{P}$ and then reduce expression

(29.22) to

$$(\Phi_N^{(1)}(r))^2, \tag{29.29}$$

where

$$\Phi_N^{(1)}(r) = (\det \hat{K})^{-1} \det \begin{pmatrix} 0...01 & 0...0,ik \\ \hline \bar{D} & -\bar{E} \\ \hline F & D \end{pmatrix}, \tag{29.30}$$

and

$$k = \frac{1+y^2}{2y}, \tag{29.31}$$

the line above matrices D, F denotes that the corresponding matrix has no first row. The line to the right of the matrix denotes that the matrix has no first column on the right.

Evaluating the determinant in (29.30) and taking into account that matrices $D + iF$ and $D - iF$ are mutually transposed and therefore have the same determinants, expression $(g_N^{(1)}(r))^2$ will be of the form

$$(S_N^{(1)})^2 \{\cosh^2 x^* \Delta_r^{(1)} - \sinh^2 x^* \Delta_r^{(1)}\}^2, \tag{29.32}$$

where

$$\cosh x^* = \frac{k+1}{2} = \frac{(1+y)^2}{4y}, \qquad \sinh^2 x^* = \frac{k-1}{2} = \frac{(1-y)^2}{4y}; \tag{29.33}$$

and $\Delta_r^{(1)}, \Delta_{-r}^{(1)}$ are the Teplitz determinants of the order r written down below. Analogously for any $i = 1, 2, 3, 4$ we get

$$(g_N^{(i)}(r))^2 = (S_N^{(i)})^2 \cosh^2 x^* \Delta_r^{(i)} - \sinh^2 x^* \Delta_{-r}^{(i)}\}^2. \tag{29.34}$$

In order to derive the expression for $\Delta_{\pm r}^{(i)}$, we transform first formulae (29.27) for matrix elements D_{mn}, F_{mn} by summing up with respect to k_2:

$$\sum_{k_2=1}^{L} \frac{\sin(\pi/L)(2k_2+1)}{(1+y^2)^2 - 2y(1-y^2)(\cos(\pi/L)(2k_1+1) + \cos(\pi/L)(2k_2+1))} = 0; \tag{29.35}$$

$$\sum_{k_2=1}^{L} [(1+y^2)^2 - 2y(1-y)^2(\cos(\pi/L)(2k_1+1) + \cos(\pi/L)(2k_2+1))]^{-1}$$

$$= \frac{1}{b} \frac{1}{2\pi i} \oint_\Gamma \frac{dz}{(1+z^L)(z-\xi)(z-\xi^{-1})}, \tag{29.36}$$

where

$$b = y(1 - y^2); \qquad \xi_1 = \frac{a + \sqrt{a^2 - 4b^2}}{2b} \geq 1;$$

$$a = (1 + y^2)^2 - 2y(1 - y^2)\cos\frac{\pi}{L}(2k_1 + 1). \tag{29.37}$$

Contour Γ in (29.36) consists of two circles $|z| = 1 \pm \delta$ and the larger of them is wound round in a positive direction and the smaller one in a negative direction.

Calculating the integral, we shall write the elements of the matrices in the form of the simple sum:

$$(D + iF)_{mn} = \frac{1}{L}\sum_{k=1}^{L}\exp\left[-\frac{i\pi}{L}(2k + 1)(m - n)\right] \times$$

$$\times \left\{\frac{(1 - yy^*\exp[(i\pi/L)(2k + 1)])(y^*\exp[(i\pi/L)(2k + 1)] - y)}{(\exp[(i\pi/L)(2k + 1)] - yy^*)(y^* - y\exp[(i\pi/L)(2k + 1)])}\right\}^{1/2} \times$$

$$\times \frac{\xi_1^L(k) - 1}{\xi_1^L(k) + 1}; \tag{29.38}$$

$$(D - iF)_{mn} = \frac{1}{L}\sum_{k=1}^{L}\exp\left[-\frac{i\pi}{L}(2k + 1)(m - n)\right] \times$$

$$\times \left\{\frac{(\exp[(i\pi/L)(2k + 1)] - yy^*)(y^* - y\exp[(i\pi/L)(2k + 1)])}{(1 - yy^*\exp[(i\pi/L)(2k + 1)])(y^*\exp[(i\pi/L)(2k + 1)] - y)}\right\}^{1/2} \times$$

$$\times \frac{\xi_1^L(k) - 1}{\xi_1^L(k) + 1},$$

where

$$y^* = \frac{1 - y}{1 + y} = \tanh x^*; \quad 0 \leq m; n \leq r. \tag{29.39}$$

The Toeplitz determinants

$$\Delta_r^{(i)} = \begin{vmatrix} \Sigma_1^{(i)} & \Sigma_0^{(i)} & \dots & \Sigma_{2-r}^{(i)} \\ \Sigma_2^{(i)} & \Sigma_1^{(i)} & \dots & \Sigma_{3-r}^{(i)} \\ \dots \dots \dots \dots \\ \Sigma_r^{(i)} & \Sigma_{r-1}^{(i)} & \dots & \Sigma_1^{(i)} \end{vmatrix}; \qquad \Delta_{-r}^{(i)} = \begin{vmatrix} \Sigma_{-1}^{(i)} & \Sigma_{-2}^{(i)} & \dots & \Sigma_{-r}^{(i)} \\ \Sigma_0^{(i)} & \Sigma_{-1}^{(i)} & \dots & \Sigma_{1-r}^{(i)} \\ \dots \dots \dots \dots \\ \Sigma_{r-2}^{(i)} & \Sigma_{r-3}^{(i)} & \dots & \Sigma_{-1}^{(i)} \end{vmatrix} \tag{29.40}$$

have as elements the expressions:

$$\Sigma_n^{(1)} = \frac{1}{L} \sum_{k=1}^{L} \exp\left[-\frac{i\pi}{L}(2k+1) \right] \times$$

$$\times \left[\frac{(1 - yy^* \exp\left[(i\pi/L)(2k+1)\right])(y^* \exp\left[(i\pi/L)(2k+1)\right] - y)}{(\exp\left[(i\pi/L)(2k+1)\right] - yy^*)(y^* - y\exp\left[(i\pi/L)(2k+1)\right])} \right]^{1/2} \times$$

$$\times \frac{\xi_1^L(k) - 1}{\xi_1^L(k) + 1};$$

$$\Sigma_n^{(2)} = \frac{1}{L} \sum_{k=1}^{L} \exp\left[-\frac{i\pi}{L}(2k+1) \right] \times$$

$$\times \left[\frac{(1 - yy^* \exp\left[(i\pi/L)(2k+1)\right])(y^* \exp\left[(i\pi/L)(2k+1)\right] - y)}{(\exp\left[(i\pi/L)(2k+1)\right] - yy^*)(y^* - y\exp\left[(i\pi/L)(2k+1)\right])} \right]^{1/2} \times$$

$$\times \frac{\xi_2^L(k) - 1}{\xi_2^L(k) + 1}; \tag{29.41}$$

$$\Sigma_n^{(3)} = \frac{1}{L} \sum_{k=1}^{L} \exp\left(-i2\pi k/L\right) \times$$

$$\times \left[\frac{(1 - yy^* \exp(i2\pi k/L))(y^* \exp(i2\pi k/L) - y)}{(\exp(i2\pi k/L) - yy^*)(y^* - y\exp(i2\pi k/L))} \right]^{1/2} \times$$

$$\times \frac{\xi_1^L(k) - 1}{\xi_1^L(k) + 1};$$

$$\Sigma_n^{(4)} = \frac{1}{L} \sum_{k=1}^{L} \exp\left(-i2\pi k/L\right) \times$$

$$\times \left[\frac{(1 - yy^* \exp(i2\pi k/L))(y^* \exp(i2\pi k/L) - y)}{(\exp(i2\pi k/L) - yy^*)(y^* - y\exp(i2\pi k/L))} \right]^{1/2} \times$$

$$\times \frac{\xi_2^L(k) - 1}{\xi_2^L(k) + 1},$$

where

$$\xi_2(k) = \frac{a_2 + \sqrt{a_2^2 - 4b^2}}{2b}; \tag{29.42}$$

$$a_2 = (1 + y^2)^2 - 2y(1 - y^2)\cos\frac{2\pi k}{L}.$$

Note that the equality $\xi_2(k) = 1$ holds only for $y = y_c$, i.e., at the point of phase transition and for $k = 0$.

Taking the square root of expression (29.34), we get the formula

$$g_N^{(i)}(r) = (-1)^r S_N^{(i)} \{\cosh^2 x^* \Delta_r^{(i)} - \sinh^2 x^* \Delta_{-r}^{(i)}\}. \qquad (29.43)$$

The existence of the sign factor $(-1)^r$ can easily be checked by taking the limit $y = \tanh x \to 0$.

Formulae (29.3), (29.10), (29.34) and (29.43) lead to the final answer concerning the correlation of two spins in one row

$$G_N(r) = (-1)^r \left\{ \cosh^2 x^* \frac{\Delta_r^{(1)} S_N^{(1)} + \Delta_r^{(2)} S_N^{(2)} + \Delta_r^{(3)} S_N^{(3)} - \Delta_r^{(4)} S_N^{(4)}}{S_N^{(1)} + S_N^{(2)} + S_N^{(3)} + S_N^{(4)}} - \right.$$

$$\left. - \sinh^2 x^* \frac{\Delta_{-r}^{(1)} S_N^{(1)} + \Delta_{-r}^{(2)} S_N^{(2)} + \Delta_{-r}^{(3)} S_N^{(3)} - \Delta_{-r}^{(4)} S_N^{(4)}}{S_N^{(1)} + S_N^{(2)} + S_N^{(3)} - S_N^{(4)}} \right\}. \qquad (29.44)$$

This expression coincides in form with the Kaufman–Onsager formula for an infinite lattice [117] and goes to it in the limit $L \to \infty$.

The passage to the limit is realized in the same way as in the case of the statistical sum and leads to the following asymptotic formula:

$$G_N(r) = (-1)^r \{\cosh^2 x^* \Delta_r - \sinh^2 x^* \Delta_{-r}\} + O(\exp(-\alpha L)). \qquad (29.45)$$

Here Δ_r, Δ_{-r} are the Toeplitz determinants (19.40) with the elements

$$\sum_N = \frac{1}{2\pi} \int_0^{2\pi} d\omega \exp(-in\omega) \left\{ \frac{(1 - yy^* \exp(i\omega))(y^* \exp(i\omega) - y)}{(\exp(i\omega) - yy^*)(y^* - \exp(i\omega)y)} \right\}^{1/2}, \qquad (29.46)$$

and the O-estimate in (29.45) is uniform for $\left||x| - x_c\right| > \varepsilon$.

By taking the limit $L \to \infty$, we obtain the Kaufman–Onsager formula:

$$G(r) = \lim_{N \to \infty} G_N(r) = (-1)^r \{\cosh^2 x^* \Delta_r - \sinh^2 x^* \Delta_{-r}\}. \qquad (29.47)$$

CHAPTER 10

PHASE TRANSITIONS

30. SPECIAL ROLE OF DIMENSION $d = 4$

In this chapter the method of functional integration is applied to the theory of second-order transitions. The characteristic feature of phase transition is the appearance of *long-range correlations*. This means that the correlation function, which at temperature values higher than those of the transition temperature decreases exponentially, decreases more slowly at the point of the phase transition.

The most essential appears to be the assumption of a power decrease of the corresponding correlation. The power function $g(r) = cr^{-a}$ under the change $r \to \lambda r$ (the similarity transformation) is multiplied by a constant factor

$$g(r) \to g(\lambda r) = \lambda^{-a} g(r), \tag{30.1}$$

i.e., it also undergoes a similarity transformation. Thus, the assumption of a power decrease of the correlation in the phase transition point is usually called the *similarity hypothesis* or the *scaling hypothesis*. The power exponent of the correlation decrease is said to be the *critical index*. The other critical indices are defined as exponents of the powers determining the behaviour of thermodynamical functions near the phase transition as functions of differences $T - T_c$ of the form

$$|T - T_c|^b. \tag{30.2}$$

We shall give the definition of the most useful critical indices. Let a 'long-range' correlator in the phase transition point be the correlator

$$G(\mathbf{x} - \mathbf{y}, \tau - \tau_1) = \langle \varphi(\mathbf{x}, \tau) \varphi(\mathbf{y}, \tau^1) \rangle. \tag{30.3}$$

Here, $\langle \ldots \rangle$ denotes, as usual, the mean in the sense of a functional integral with the weight $\exp S$. Assuming a power decrease of correlator (30.3), we can write its asymptotics for $r = |\mathbf{x} - \mathbf{y}| \to \infty$ in the form

$$G \approx ar^{2-d-\eta}, \tag{30.4}$$

where d is the dimension of the space and η is a critical index

262

determining the decrease law of the correlator in the phase transition point.

Let the correlator decrease exponentially for $T > T_c$:

$$G \sim \exp\left(-\frac{r}{\xi(T)}\right). \tag{30.5}$$

Here we write down the most rapidly decreasing part of the asymptotics for $r \to \infty$ and the part before the exponent is omitted. Quantity $\xi(T)$ determining a decrease of the correlator above the phase transition temperature, is said to be a *correlation length*. The appearance of long-range correlations for $T \to T_c + 0$ indicates that $\xi(T)$ increases unboundedly in this limit. By assuming that the increase is given according to the formula

$$\xi(T) \sim |T - T_c|^{-\nu}, \tag{30.6}$$

we obtain the second critical index ν.

The critical index α determines a singularity of the thermal capacity (at the constant volume) in the limit $T \to T_c$:

$$C_v \sim |T - T_c|^{-\alpha}. \tag{30.7}$$

The critical index β gives a behaviour of the anomalous mean value

$$\langle \varphi(\mathbf{x}, \tau) \rangle \sim (T_c - T)^{\beta}, \tag{30.8}$$

existing for $T < T_c$ in the limit $T \to T_c - 0$.

The next critical index δ is connected with the behaviour of receptivity χ in the limit $T \to T_c + 0$:

$$\chi \sim (T - T_c)^{-\nu}. \tag{30.9}$$

Receptivity χ is a coefficient at ξ in the mean value $\langle \varphi(x, t) \rangle$ calculated in the presence of the contribution

$$\zeta \int dy \, d\tau' \varphi(\mathbf{y}, \tau') \tag{30.10}$$

to action S (in the limit $\xi \to 0$). From the given definition follows the formula

$$\chi = \int G(\mathbf{x} - \mathbf{y}, \tau - \tau') \, d(\mathbf{x} - \mathbf{y}) \, d(\tau - \tau'), \tag{30.11}$$

expressing χ as an integral of correlator (30.3) with respect to space and time $\tau - \tau'$.

Finally, the critical index δ describes the dependence of $p - p_c$ on $\rho - \rho_c$ for $T = T_c$:

$$p - p_c \sim (\rho - \rho_c)^\delta. \tag{30.12}$$

In this section we explain the method of calculating the critical indices mostly developed by Wilson [124]. Let us mention its main idea.

As was shown in the examples from quantum field theory and statistical physics above, in order to calculate the infrared asymptotics of the correlation function by the functional integral method, it is useful to integrate first with respect to the rapid and then to the slow variables applying in these two steps the different schemes of perturbation theory. Let us recall that the rapid variables are the components of the field with large wave vectors $\mathbf{k}(k > k_0)$ and the slow ones, with small $\mathbf{k}(k < k_0)$.

In the phase transition problem, such an approach is not sufficient. The functional ('hydrodynamical action') obtained after integration with respect to the rapid variables does not allow us to find out the correlation functions in the long-range limit we are interested in or to construct a perturbation theory for their evaluation with respect to a small parameter.

The method suggusted by Wilson reduces successive integration to an approximative multiple with respect to the fields with a diminishing lower bound k_0. This approach leads to a definite qualitative picture of the system's behaviour near the phase transition and gives the approximative method for calculating the critical indices. Moreover, it clarifies the exceptional role of the dimensionality of space $d = 4$. For $d > 4$ it turns out to be possible to use the usual perturbation theory. The case $d = 4$ is studied in detail in the work of A. I. Larkin and D. E. Khmelnitski [125]. They selected the main sequence of the perturbation theory diagrams, which was possible to sum up by a 'parquet' method. This allows us to find the asymptotics of the correlation function in the neighbourhood of the phase transition and to determine the character of nonanalyticity of the thermodynamical function near $T \sim T_c$ for $d = 4$. This helped Wilson and Fisher [126] to suggest the idea of ε-*expansion* with respect to $\varepsilon = 4 - d$ i.e. to the difference of the critical dimension $d_c = 4$ and the dimension of the space d. This idea leads to the method of calculating the critical indices by means of the power series in parameter ε. When applying it to the real system, it is necessary to put $\varepsilon = 1$ $(d = 3)$ or $\varepsilon = 2(d = 2)$. The calculation shows that the coefficients of ε-series first decrease and then start to increase rapidly. This can be taken as an indication that the ε-series is the asymptotic one. Nevertheless, by restricting ourselves to two-three terms it is possible to get reasonable values for the critical indices which agree with calculations via the perturbation theory for lattice systems.

The foundation problem for the Wilson approach is far from a satisfactory solution. Moreover, the method of calculation of the critical indices when applied to the real systems with a dimension $d = 2$ or $d = 3$ is basically approximative and the evaluation of corrections appears to be unrealistic at the present time. The similarity hypothesis itself has not been proved. In this situation there is a possibility that the correlation function in the phase transition point has the asymptotic

$$\exp\{-(r/r_0)^\gamma\} \tag{30.13}$$

with $0 < \gamma < 1$. Function (30.13) decreases slower than exponent $\exp -(r/r_0)$ with arbitrary $r_0 > 0$ but not faster than any negative power r^{-a}.

In this section we explain the method of approximative successive integration in the example of the model of the real scalar field φ with self-interaction proportional to φ^4. In the next section it will be shown how to make the method asymptotically exact in the limit $\varepsilon = 4 - d \to 0$ and to get the Wilson ε-expansion for the critical indices.

Consider a system of statistical mechanics with the action functional:

$$S = -\int d\tau\, dx \left(\frac{1}{2}(\partial_\tau\varphi)^2 + \frac{1}{2}(\nabla\varphi)^2 - \frac{\lambda}{2}\varphi^2 + \frac{g}{4!}\varphi^4\right); \tag{30.14}$$

where $x \in V$, $0 \le \tau \le \beta$. This functional describes the relativistic real scalar field φ for finite temperature T. The temperature diagrammatic technique has as elements vertices and lines

$$\underline{\hspace{3cm}} \quad (\omega^2 + k^2 - \lambda)^{-1},$$

$$\tag{30.15}$$

g.

If coefficient λ is negative in (30.14), the functional is negative definite and the phase transition is absent in the system. If $\lambda > 0$, the system may manifest the phase transition below which the anomalous mean value appears

$$\langle \varphi(\mathbf{x}, \tau)\rangle \ne 0. \tag{30.16}$$

In order to describe the system in the neighbourhood of the phase transition, we integrate functional $\exp S$ with respect to the fast variables which will be the Fourier coefficient $\tilde{\varphi}(\mathbf{k}, \omega)$ with $\omega \neq 0$ and $\omega = 0, k > k_0$ in the expansion:

$$\varphi(\mathbf{x}, \tau) = (\beta V)^{-1/2} \sum_{\mathbf{k}, \omega} \exp\left[i(\omega\tau - \mathbf{kx})\right]\tilde{\varphi}(\mathbf{k}, \omega). \tag{30.17}$$

Functional S_0, defined by the formula

$$\int \exp S \prod_{\substack{\omega \neq 0 \\ \mathbf{k}}} \prod_{\substack{\omega = 0 \\ \mathbf{k} > \mathbf{k}_0}} d\tilde{\varphi}(\mathbf{k}, \omega) = \exp S_0, \tag{30.18}$$

depends on components $\tilde{\varphi}(\mathbf{k}, 0)$ which will be denoted further by $\varphi(\mathbf{k})$. The general form of functional S_0

$$S_0 = C_0 - \frac{1}{2} \int_{k < k_0} u_2(\mathbf{k})\varphi(\mathbf{k})\varphi(-\mathbf{k}) \, d\mathbf{k} -$$

$$- \sum_{n=2}^{\infty} \frac{1}{(2n)!} \int_{k_i < k_0} u_{2n}(\mathbf{k}_i)\delta\left(\sum_{i=1}^{2n} \mathbf{k}_i\right) \prod_{i=1}^{2n} \varphi(\mathbf{k}_i) \, dk_i \tag{30.19}$$

is the sum of functionals of power $2n(n = 0, 1, 2, \ldots)$ in which the integrals with respect to $\mathbf{k}_i$ are cut on the upper bound k_0 and δ-function $\delta(\Sigma\mathbf{k}_i)$ guarantees translation invariance. Constant C_0 has no influence on further integration and therefore is inessential.

We consider the coefficient function $u_2(\mathbf{k})$. Let there exist the expansion of $u_2(\mathbf{k})$

$$u_2(\mathbf{k}) = u_{20} + u_{22}k^2 + \ldots \tag{30.20}$$

in even degrees k at the neighbourhood $\mathbf{k} = 0$. Let us make the similarity transformation

$$\varphi(\mathbf{k}) \to \zeta\varphi(\mathbf{k}), \tag{30.21}$$

by taking its coefficient ξ from the condition

$$\zeta^2 u_{22} = 1. \tag{30.22}$$

As a result, we obtain functional S_0 of form (30.19) in which

$$u^2(\mathbf{k}) = r + k^2 + \ldots \tag{30.23}$$

We now integrate functional $\exp S_0$ with respect to the components $\varphi(\mathbf{k})$ with $k_0/2 < k < k_0$.

Functional S_1 determined by the formula

$$\int \exp S_0 \prod_{k_0/2 \,<\, k \,<\, k_0} d\varphi(\mathbf{k}) = \exp S_1 \tag{30.24}$$

differs from S_0 by the value of constant C_0 and by the form of the coefficient functions $u_{2n}(\mathbf{k}_i)$. Integrals with respect to $\mathbf{k}_i$ in S_1 are cut on upper bound $k_0/2$ (besides k_0 in S_0). Now we perform the transformation

$$\mathbf{k} \to 2\mathbf{k}, \qquad \varphi(\mathbf{k}) \to \varphi(2\mathbf{k}), \tag{30.25}$$

which replaces the domain of momenta $k < k_0/2$ by the domain $k < k_0$ which we have met in functional S_0. We then perform transformation (30.21) in order that the new coefficient function of the quadratic form in S_1 (denoted by $u_2^{(1)}(\mathbf{k})$) may have the expression

$$u_2^{(1)}(\mathbf{k}) = r_1 + k^2 + \ldots, \tag{30.26}$$

analogous (30.23), i.e., with the coefficient at k^2 equal to one.

Thus functional integration (30.24) with a successive replacement of variables (30.21) and (30.25) changes functional $S_0 - C_0$ into functional $S_1 - C_1$ of the same form but with other coefficient functions $u_{2n}^{(1)}(\mathbf{k}_i)$. We can speak about the nonlinear transformation

$$\hat{M}u = u^{(1)} \tag{30.27}$$

of the sequence of the coefficient function

$$u = \{u_2(\mathbf{k}), u_4(\mathbf{k}_i), u_6(\mathbf{k}_i), \ldots\} \tag{30.28}$$

into the sequence

$$u^{(1)} = \{u_2^{(1)}(\mathbf{k}), u_4^{(1)}(\mathbf{k}_i), u_6^{(1)}(\mathbf{k}_i), \ldots\}. \tag{30.29}$$

By applying the same method to $u(1)$ as that to u we obtain

$$u^{(2)} = \hat{M}(u^{(1)}) = \hat{M}^2 u. \tag{30.30}$$

By repeating we can consider the functional integration of fields with smaller and smaller momenta as a successive application of the $\hat{M}$ transformation to the initial sequence u.

The similarity hypothesis for the phase transitions gives the following assumption.

On the phase transition line the successive application of the M transformation to the sequence has as a limit the 'stationary' sequence

$$u_0 = \hat{M}u_0. \tag{30.31}$$

Actually, if we have

$$u^{(n+1)} \equiv \hat{M}u^{(n)} \approx u^{(n)},$$

for a sufficiently large n, then the double change of the momentum scale in the k-space reduces to the similarity transformation $\mathbf{k} \to 2\mathbf{k}$, $\varphi \to \xi\varphi$.

We show that correlator (30.3) has a power asymptotics given by (30.4) in the limit $r \to \infty$; the exponent of the power is simply expressed in terms of parameter ξ of transformation (30.1). We consider correlator (30.3) in the $\mathbf{k}$ space

$$\langle \varphi(\mathbf{k})\varphi(\mathbf{k}')\rangle = D(\mathbf{k})\delta(\mathbf{k} + \mathbf{k}'). \tag{30.32}$$

Let the stationarity condition (30.32) hold for $n \geq n_0$, i.e., in the momentum domain

$$k < 2^{-n_0}k_0 . \tag{30.33}$$

We now take the arbitrarily small $\mathbf{k}$ and an integer n so that momentum $\mathbf{k}_1 = 2^n\mathbf{k}$ lies in the domain

$$2^{-n_0-1}k_0 < k_1 < 2^{-n_0}k_0. \tag{30.34}$$

Then, due to stationarity condition (30.32), we shall have

$$\langle \varphi(\mathbf{k})\varphi(\mathbf{k}')\rangle = \zeta^{2n}\langle \varphi(\mathbf{k}_1)\varphi(\mathbf{k}_1')\rangle \tag{30.35}$$

or

$$D(\mathbf{k}) = \zeta^{2n}2^{-nd}D(\mathbf{k}_1). \tag{30.36}$$

Rewriting this equality in the form

$$D(\mathbf{k})k^{(2\ln\zeta/\ln 2)-d} = D(\mathbf{k}_1)k_1^{(2\ln\zeta/\ln 2)-d}, \tag{30.37}$$

we obtain the asymptotics

$$D(\mathbf{k}) = ck^{d-(2\ln\zeta/\ln 2)}, \tag{30.38}$$

which is true in the limit $\mathbf{k} \to 0$. Taking the Fourier representation, we obtain that correlator (30.3) behaves as the power

$$r^{(2\ln\zeta/\ln 2)-2d}, \tag{30.39}$$

Comparing this with (30.4) we get the formula for the critical index

$$\eta = 2 + d - \frac{2\ln\zeta}{\ln 2}. \tag{30.40}$$

We derive the formula for correlation length $\xi(T)$ in the neighbourhood of the phase transition. A deviation from the phase transition line has the consequence that the successive application of the $\hat{M}$ transformation has no stationary sequence as a limit and after all that it brings u_n out of the u_0 neighbourhood. The nearer to the phase transition, the more steps, $\tilde{n}$, need to be done in order to take out u_n from the fixed u_0 neighbourhood. For the estimate of number $\tilde{n}$ we linearize the $\hat{M}$ transformation in the neighbourhood

$$u_{n+1} - u_0 = \hat{A}(u_n - u_0),\tag{30.41}$$

where $\hat{A}$ is a linear operator. Supposing that the system is near the phase transition, we can assume that the difference $u_{n_0} - u_0$ is proportional to the difference $T - T_c$ for a sufficiently big u_0. Then by virtue of (30.41), the quantity $u_{n+n_0} - u_0$ has the order

$$\lambda_1^n (T - T_c),\tag{30.42}$$

where λ_1 is the highest eigenvalue of linear operator A. We estimate quantity $\tilde{n}$ by demanding that expression (30.42) should be of the order T_c for $n \sim \tilde{n}$. This leads to the expression

$$\tilde{n} \sim \frac{\ln(T_c/T - T_c)}{\ln \lambda_1}.\tag{30.43}$$

One may consider that applying the n multiple $\hat{M}$ transformation to the system with functional S_0, we bring out the system from the neighbourhood of the phase transition. The correlation function of the new system decreases exponentially and has the asymptotics

$$\exp\left(-\frac{r}{r_0}\right)\tag{30.44}$$

(up to the pre-exponential term) in the limit $r \to \infty$ and r_0 may be considered of an order equal to k_0^{-1}. However, the new system is nothing more than the initial one for the enlarged scale by $2^{\tilde{n}}$ times. Thus, the correlator of the initial system has the asymptotics

$$\exp\left(-\frac{r}{2^{\tilde{n}} r_0}\right).\tag{30.45}$$

for $r \to \infty$. From here and (30.43) follows the required formula for the correlation length:

$$\xi(T) \sim 2^{\tilde{n}} \sim (T - T_c)^{-(\ln 2/\ln \lambda_1)}.\tag{30.46}$$

Recalling definition (30.6) of the critical index v, we obtain for it the expression

$$v = \frac{\ln 2}{\ln \lambda_1}. \tag{30.47}$$

We consider still receptivity χ in the limit $T \to T_c$. Definitions (30.11) and (30.32) yield the formula

$$\chi \sim D(\mathbf{k} = 0). \tag{30.48}$$

Reasonings analogous to those used above for deriving the formulae for indices η, v lead to the relation

$$D(\mathbf{k} = 0) \sim \zeta^{2\tilde{n}} 2^{-\tilde{n}d}, \tag{30.49}$$

where $\tilde{n}$ is the number of steps of the $\hat{M}$ transformation necessary for the system to leave the fixed neighbourhood of the phase transition. From (30.43) for $T \to T_c + 0$ (30.48) and (30.49) follows the asymptotics

$$\chi \sim (T - T_c)^{(\ln 2d/\ln \lambda_1) - (2\ln \zeta/\ln \lambda_1)}, \tag{30.50}$$

and the expression for the corresponding critical index γ

$$\gamma = \frac{2\ln \zeta}{\ln \lambda_1} - d\frac{\ln 2}{\ln \lambda_1}, \tag{30.51}$$

The critical indices η, v, γ obtained here are coupled by the relation

$$\gamma = (2 - \eta)v. \tag{30.52}$$

Notice that replacing sign $=$ by sign $\leq$ in (30.52), this equality turns into one of the Griffits inequalities [147].

Thus, the method of functional integration together with the Wilson hypothesis on connection of the phase transition with the unmovable points of the $\hat{M}$ transformation, allows us to express the critical indices by means of the parameters of this transformation and to derive the exact expression (30.52). It is, therefore, natural to call the Wilson hypothesis the *generalized similarity hypothesis*.

The approximative methods are necessary to calculate the critical indices remaining independent. One of them, suggested by Wilson, consists of the following. Neglecting all higher coefficient functions starting with u_6, the approximation

$$u_2^{(n)} = k^2 + r_n; \qquad u_4^{(n)} = u_n. \tag{30.53}$$

is used for $u_2^{(n)}$ and $u_4^{(n)}$. The transition formulae from $(u_2^{(n)}, u_4^{(n)})$ to $(u_2^{(n+1)}, u_4^{(n+1)})$ can be rewritten in the form

$$u_2^{(n+1)}(k) = \zeta^2 2^{-d}\left(u_2^{(n)}\left(\frac{k}{2}\right) + \overset{a}{\underset{}{\bigcirc}} + \cdots\right);$$

$$u_4^{(n+1)}(k_i) \approx u_4^{(n+1)}(0) = \zeta^4 2^{-3d}\left(u_4^{(n)} + \overset{b}{\bigcirc\!\!\!\bigcirc} + \cdots\right). \tag{30.54}$$

By applying the similarity transformations (30.21) and (30.25), we restrict ourselves to diagrams a, b introduced in (30.54). Diagram a in the approximation where $u_4^{(n)}$ is a constant, does not depend on the external momentum. By requiring the coefficient to be equal to 1 for k^2 in $u_2^{(n+1)}(k)$ we find:

$$\zeta = 2^{1+d/2}. \tag{30.55}$$

After this, formulae (30.54) take the form

$$u_2^{(n+1)}(\mathbf{k}) = k^2 + r_{n+1}$$

$$= k^2 + 4\left(r_n + \frac{u_n}{2(2\pi)^d}\int_{(k_0/2)<k<k_0}\frac{dk}{k^2 + r_n}\right);$$

$$u_4^{(n+1)}(k)(\mathbf{k}_i) \approx u_{n+1} \tag{30.56}$$

$$= 2^{4-d}\left(u_n - \frac{3}{2}\frac{u_n^2}{(2\pi)^d}\int_{(k_0/2)<k<k_0}\frac{dk}{(k^2 + r_n)^2}\right).$$

The multiplier 3 in front of the integral in the second of these formulae is obtained by taking into account three diagrams

$$\overset{1}{\underset{2}{\bigcirc\!\!\!\bigcirc}}\overset{3}{\underset{4}{}} + \overset{1}{\underset{3}{\bigcirc\!\!\!\bigcirc}}\overset{2}{\underset{4}{}} + \overset{1}{\underset{4}{\bigcirc\!\!\!\bigcirc}}\overset{2}{\underset{3}{}} \tag{30.57}$$

and the multiplier $\frac{1}{2}$ is a multiplier of the diagram symmetry.

Thus the $\hat{M}$ transformation reduces into a nonlinear transformation of pair (r_n, u_n) into (r_{n+1}, u_{n+1}) according to formulae (30.56). The second of them shows the exceptional role of dimensionality $d = 4$.

For $d > 4$ it follows from formula (30.56) that

$$0 < u_{n+1} < u_n 2^{4-d}, \tag{30.58}$$

only if initial u_0 is sufficiently small and positive. Hence $u_n \to 0$ and u must be zero in the limiting pair (u, r) and consequently also $r = 0$. As a result action S_n is the action of the free field

$$-\frac{1}{2}\int k^4 \varphi(\mathbf{k})\varphi(-\mathbf{k})\, dk. \tag{30.59}$$

in the limit $n \to \infty$. This implies that both the correlation function and the thermodynamical functions behave in the neighbourhood of the phase transition just like the corresponding quantities in the theory of a free field. The critical indices coincide with the critical indices of the free-field theory. In particular

$$\eta = 0, \qquad \nu = \tfrac{1}{2}, \qquad \gamma = 1. \tag{30.60}$$

For $d = 4$, the $\hat{M}$ transformation (30.56) also goes to the limit $r = 0$, $u = 0$ and the critical indices coincide with their values in the free-field theory. More detailed information on the behaviour of the system for the limiting dimensionality $d = 4$ can be found in the mentioned work by A. I. Larkin and D. E. Khmelnitski [125].

In the case $d < 4$, the situation is quite different. Starting from the pair (r_0, u_0) with sufficiently small r_0, u_0 $(u_0 > 0)$ we necessarily obtain then nontrivial solution (r, u) of the stationarity equation with $u \neq 0$, namely

$$u = (1 - 2^{d-4})\left[\frac{3}{2(2\pi)^d}\int\frac{dk}{(k^2 + r)^2}\right]^{-1}, \tag{30.61}$$

where r is determined from the equation

$$r = \frac{4}{9}(2^{d-4} - 1)\left[\int\frac{dk}{k^2 + r}\bigg/\int\frac{dk}{(k^2 + r)^2}\right], \tag{30.62}$$

and the integrals in (30.61) and (30.62) are considered in the domain $k_0/2 < k < k_0$.

Linearized transformation (30.56) in the neighbourhood of the stationary point takes the form

$$(r_{n+1} - r) = (r_n - r)\left[4 - \frac{2u}{(2\pi)^d}\int\frac{dk}{(k^2 + r)}\right] +$$
$$+ (u_n - u)\frac{2}{(2\pi)^d}\int\frac{dk}{k^2 + r}; \tag{30.63}$$

$$(u_{n+1} - u) = (r_n - r)\frac{2^{4-d}3u^2}{(2\pi)^d}\int\frac{dk}{(k^2 + r)^3} +$$

$$+ (u_n - u)2^{4-d}\left(1 - \frac{3u}{(2\pi)^d}\int\frac{dk}{(k^2 + r)^2}\right).$$

In this case operator $\hat{A}$ is the matrix of the second order

$$\begin{pmatrix} 4 - \dfrac{2u}{(2\pi)^d}\displaystyle\int\frac{dk}{(k^2 + r)^2}, & \dfrac{2}{(2\pi)^d}\displaystyle\int\frac{dk}{k^2 + r} \\[2ex] \dfrac{2^{4-d}3u^2}{(2\pi)^d}\displaystyle\int\frac{dk}{(k^2 + r)^3}, & 2^{4-d}\left(1 - \dfrac{3u}{(2\pi)^d}\displaystyle\int\frac{dk}{(k^2 + r)^2}\right) \end{pmatrix}. \tag{30.64}$$

Quantity λ determined above is the biggest eigenvalue of this matrix. For $d \geq 4$ we have $u = r = 0$ and we obtain $\lambda_1 = 4$, the result of the free-field theory. For $d < 4$ λ_1 differs from 4 and gives, in accordance with (30.47), critical index ν which differs from the value $\frac{1}{2}$ in the free-field theory.

Recall that approximation (30.55) for ξ coincides with the result of the free-field theory and gives the null critical index η.

Thus the approximative method for calculating the critical indices was obtained. The conclusion derived in this approach of the exceptional role of dimensionality $d = 4$ remains true in a more detailed study of the system with 'a dimensionality near to 4' done in the next section.

31. CALCULATION OF CRITICAL INDICES AND THE WILSON EXPANSION

The exceptional role of dimensionality $d = 4$ lead Wilson and Fisher [126] to the idea of making an expansion with respect to parameter $\varepsilon = 4 - d$. Such an expansion takes into consideration systems with a 'noninteger' dimensionality, besides the usual systems of statistical mechanics in spaces with an integer dimensionality. This means extending the diagrammatic technique of the perturbation theory in which the expressions for the diagrams are given by multiple integrals over a space of dimensionality d to the case when d is no more an integer. We explain it on the case of the diagrammatic technique with the elements.

$$G(\mathbf{k}) = -(k^2 + r)^{-1}, \qquad V_4 = g, \tag{31.1}$$

where $G(\mathbf{k})$ is the expression corresponding to the line and V_4 corresponds to the vertex of the fourth order.

The problem here consists in an extension of the integral of the product of functions

$$\prod_{i=1}^{l} (k_i^2 + r)^{-1}, \tag{31.2}$$

which corresponds to the lines of the diagram, onto the case of noninteger dimensionality d. By using the Feynman representation

$$\prod_{i=1}^{l} (k_i^2 + r)^{-1} = (l-1)! \int \prod_{i=1}^{l} d\alpha_i \delta\left(\sum_{i=1}^{l} \alpha_i - 1\right)\left(\sum_{i=1}^{l} (k_i^2 + r)\alpha_i\right)^{-l}, \tag{31.3}$$

we transfer the problem to that of continuation onto an arbitrary value d of the integral

$$\int \prod_{\alpha=1}^{c} dk_\alpha \left(\sum_{i=1}^{l} (k_i^2 + r)\alpha_i\right)^{-l} \tag{31.4}$$

with respect to the diagram momenta k_d the number of which is equal to the number of independent contours.

By means of the linear transformation

$$k_\alpha \rightarrow \sum_{\beta=1}^{c} a_{\alpha\beta} k_\beta + k_{0\alpha} \tag{31.5}$$

the expression

$$\sum_{i=1}^{l} \alpha_i (k_i^2 + r) \tag{31.6}$$

can be brought to the form

$$\sum_{\alpha=1}^{c} k_\alpha^2 + D, \tag{31.7}$$

where D is a function of external diagram momenta and of parameters α_i of the Feynman representation. Elements $a_{\alpha\beta}$ of the transformation matrix (31.5) do not depend on dimensionality d of the k space and are obtained under the transformations of the volume element

$$\prod_{\alpha=1}^{c} dk_\alpha \rightarrow (\det a_{\alpha\beta})^d \prod_{\alpha=1}^{c} dk_\alpha \tag{31.8}$$

Multiplier $(\det a_{\alpha\beta})^d$ is taken in front of the integral sign with respect to

k_α. We get the integral

$$\int \prod_{\alpha=1}^{c} dk_\alpha (\sum k_\alpha^2 + D)^{-l},$$ (31.9)

equal to the integral

$$\int \frac{dk}{(k^2 + D)^l}$$ (31.10)

over the k space of dimensionality cd. This integral is a product of the one-dimensional integral

$$\int_0^\infty \frac{k^{cd-1} \, dk}{(k^2 + D)^l}$$ (31.11)

with the surface element of the unit sphere in the cd-dimensional space

$$\Omega(cd) = \frac{2\pi^{cd}}{\Gamma\left(\dfrac{cd}{2}\right)}.$$ (31.12)

Expressions (31.11) and (31.12) are also defined for noninteger d. Thus the extension of the diagrammatic technique with elements (31.1) onto noninteger d does not cause difficulties.

We consider the method of a successive functional integration for the systems of dimensionality $d = 4 - \varepsilon$ with small ε. For such systems we can define the $\hat{M}$ transformation as done for the systems with integer d. It is natural to assume that analogously to integer d, the phase transition is connected with the fact that the $\hat{M}$ transformation when applied many times onto the initial action S_0, gives action S invariant under the $\hat{M}$ transformation. The formulae of the $\hat{M}$ transformation have the form

$$u_2^{(n+1)}(\mathbf{k}) = \zeta^2 2^{-d}\left(u_2^{(n)}\left(\frac{k}{2}\right) + \text{⎯⎯◯⎯⎯} + \text{⎯◯⎯◯⎯} + \cdots \right)$$

$$u_4^{(n+1)}(\mathbf{k}_i) = \zeta^4 2^{-3d}\left(u_4^{(n)}\left(\frac{k_i}{2}\right) + \text{⋈⋈} + \cdots \right)$$

$$u_6^{(n+1)}(\mathbf{k}_i) = \zeta^6 2^{-5d}\left(u_6^{(n)}\left(\frac{k_i}{2}\right) + \text{⎯◯⎯◯⎯} + \cdots \right)$$ (31.13)

$$u_8^{(n+1)}(\mathbf{k}_i) = \zeta^8 2^{-7d}\left(u_8^{(n)}\left(\frac{k_i}{2}\right) + \text{⎯◯⎯◯⎯◯⎯} + \cdots \right)$$

. . . .

We shall show that the stationarity condition $\hat{M}u = u$ admits solutions in the form of power series in ε in the limit $\varepsilon \to 0$, The ε-expansion of functions u_{2n}, obtained in this way, leads to the ε-expansion of the critical indices.

The solution of the stationarity equation of the $\hat{M}$ transformation (31.13) has the form

$$u_2(\mathbf{k}) = \mathbf{k}^2 + O(k_0^2\varepsilon); \qquad u_4(\mathbf{k}_i) = c\varepsilon + O(\varepsilon^2), \tag{31.14}$$

where c is a constant not depending on ε. The smallness of the interaction vertex allows us to restrict ourselves to the first few diagrams of the perturbation theory. If functions $u_2(\mathbf{k})$, $u_4(\mathbf{k}_i)$ have form (31.14), the first diagram introduced in (31.13) for $u_2(\mathbf{k})$ is a sum of he summand not depending on $\mathbf{k}$ (of the order εk_0^2) and of the quantity the order of which exceeds $k_0^2\varepsilon^2$. The other diagrams also give contributions of an order not exceeding $k_0^2\varepsilon^2$. Requiring the coefficient at k^2 in $u_2(\mathbf{k})$ to be equal to 1 we obtain the equality

$$1 = \zeta^2 2^{-2-d}(1 + O(\varepsilon^2)), \tag{31.15}$$

from which

$$\zeta = 2^{1+d/2}(1 + O(\varepsilon^2)). \tag{31.16}$$

This shows that formula (30.55) for ξ holds up to $O(\varepsilon^2)$.

We insert the formula for ξ (31.16) into the second equation (31.13) in which we put $\mathbf{k}_i = 0$ ($i = 1, 2, 3, 4$). By denoting

$$u_4(\mathbf{k}_i = 0) = u, \tag{31.17}$$

we obtain the equation

$$u = (1 + O(\varepsilon^2))2^{4-d}\left(u - \frac{3}{2(2\pi)^d}\int \frac{(u + O(\varepsilon^2))^2\,d\mathbf{k}}{(k^2 + O(\varepsilon^2 k_0^2))^2} + O(\varepsilon^2)\right), \tag{31.18}$$

in which the integral is taken over the domain $k_0/2 < k < k_0$. Since the d-dimensional integral with a relative error of order ε is equal to the four-dimensional one, the equation can be rewritten in the form

$$u = 2^\varepsilon\left(u - \frac{3u^2}{2(2\pi)^4}\int_{k_0/2}^{k_0} \frac{d^4k}{k^4} + O(\varepsilon^2)\right)$$

$$= 2^\varepsilon\left(u - \frac{3u^2}{16\pi^2}\ln 2 + O(\varepsilon^2)\right). \tag{31.19}$$

The solution of this equation is given by

$$u = \frac{16\pi^2}{3} \frac{(1 - 2^{-\varepsilon})}{\ln 2} + O(\varepsilon^2) = \frac{16\pi^2\varepsilon}{3} + O(\varepsilon^2). \tag{31.20}$$

This at the same time gives for constant c in (31.14) the value:

$$c = \frac{16\pi^2}{3}. \tag{31.21}$$

Knowing that $u_4(k_i) \approx u$ in the first approximation we find a complement to k^2 in $u_2(k)$. This correction does not depend on k in the first approximation. We denote it by r. The equation for r

$$r = A\left(r + \frac{u}{2(2\pi)^4} \int \frac{d^4k}{k^2} + O(\varepsilon^2)\right) \tag{31.22}$$

has the solution

$$r = -\frac{4}{3} \frac{k_0^2 u}{32\pi^2} = -\frac{2}{9} k_0^2 \varepsilon. \tag{31.23}$$

Thus

$$u_2(k) = k^2 - \tfrac{2}{9} k_0^2 \varepsilon + O(k_0^2 \varepsilon^2); \qquad u_4 = \frac{16\pi^2}{3}\varepsilon + O(\varepsilon^2). \tag{31.24}$$

Let us remark now that coefficients $\xi^{2n} 2^{d(1-2n)}$ in front of $u_{2n}(k_i/2)$ in the stationarity equations for higher coefficient functions $u_{2n}(k_i)$ starting from u_6, are smaller than one in the approximation (31.55) for ξ and go to zero (as 2^{-d}) for $n \to \infty$. The diagrams entering these equations are of higher orders with respect to ε (ε^2 in u_6, ε^3 in u_8, and so on). This implies that the higher coefficient functions are not essential in the limit $\varepsilon \to 0$ and that they must be taken into account only in higher orders with respect to ε.

In the lower orders with respect to ε, the $\hat{M}$ transformation reduces to the transformation of u_2 and u_4 according to formulae (31.56). In other words, the approximative method described in the previous section becomes exact in the limit $\varepsilon \to 0$.

This idea allows us to find out the biggest eigenvalue λ_1 of the linearized operator $\hat{A}$ of the $\hat{M}$ transformation as the biggest eigenvalue of matrix (31.64). With accuracy $O(\varepsilon^2)$, λ_1 is equal to element A_{11} of this matrix

$$\lambda_1 = 4 - \frac{2u}{(2\pi)^d} \int \frac{dk}{(k^2 + r)^2} + O(\varepsilon^2) = 4 - \frac{u \ln 2}{4\pi^2} + O(\varepsilon^2)$$

$$= 4 - \tfrac{4}{3}\varepsilon \ln 2 + O(\varepsilon^2). \tag{31.25}$$

Putting λ_1 in (31.47) we obtain the critical index

$$v = \frac{1}{2} + \frac{\varepsilon}{12} + O(\varepsilon^2). \tag{31.26}$$

In order to find the critical index with a higher degree of accuracy, we have to know exactly the coefficient ξ which, by virtue of (31.40), defines critical index η. This can be done by taking into account the second diagram described in Equation (31.13) for $u_2(k)$. We obtain the expression for η in another way, and accompanied by smaller calculation difficulties. The critical index $\eta = O(\varepsilon^2)$ is connected with the asymptotics of the function determined by relation (31.32)

$$D(k) \sim k^{\eta-2}. \tag{31.27}$$

Function $D(k)$ up to the sign is equal to the Green's function of the field φ in the k space. It is expressed via the self-energy part $\Sigma(k)$ by the formula

$$D(k) = (u_2(k) + \Sigma(k))^{-1}, \tag{31.28}$$

where

$$\Sigma(k) = \quad + \quad + \cdots \tag{31.29}$$

is a sum over the diagrams of the perturbation theory in which functions $u_2^{-1}(k)$ correspond to the lines and $u_{2n}(k_i)$ with $n \geq 2$ to the vertices. In these diagrams the integrals with respect to the momenta of the internal lines are taken over the whole domain $k < k_0$ in contradistinction to the domain $k_0/2 < k < k_0$ used in the diagrams describing the $\hat{M}$ transformation.

We calculate critical index η by comparing the formula

$$D(k) \sim k^{(-2+\eta)} \approx (k^2 - k^2\eta \ln(k/k_0))^{-1}, \tag{31.30}$$

which follows from (31.27) in the domain $k \ll k_0, |\eta \ln k_0/k)| \ll 1$, with the formula (31.28) in the same domain. We must obtain the asymptotic formula

$$k^2 - k^2\eta \ln(k/k_0) \approx u_2(k) + \Sigma(k). \tag{31.31}$$

We subtract at the right-hand side the value of this formula at $k = 0$, i.e. $u_2(0) + \Sigma(0)$, which is obviously equal to zero by virtue of (31.27).

Since

$$u_2(\mathbf{k}) - u_2(0) = k^2 + O(k^4), \tag{31.32}$$

we get the relation

$$\Sigma(\mathbf{k}) - \Sigma(0) \approx -\eta k^2 \ln(k/k_0). \tag{31.33}$$

in the domain indicated above.

The main contribution to the difference $\Sigma(\mathbf{k}) - \Sigma(0)$ is brought by diagram b (31.31) since diagram a gives a constant contribution. In the expression

$$\Sigma(\mathbf{k}) - \Sigma(0) = \frac{1}{3!(2\pi)^d} \int u^2 \, dk_1 \, dk_2 \times \tag{31.34}$$

$$\times \{u_2^{-1}(\mathbf{k}_1)u_2^{-1}(\mathbf{k}_2)[u_2^{-1}(\mathbf{k}_1 + \mathbf{k}_2) - u_2^{-1}(\mathbf{k}_1 + \mathbf{k}_2 + \mathbf{k})]\}$$

we can put $u_2(\mathbf{k}) = k^2, u_4 = u, d = 4$ in the first approximation so that

$$\Sigma(\mathbf{k}) - \Sigma(0) = \frac{v^2}{3!(2\pi)^8} \int \{dk_1 \, dk_2 k_1^{-2} k_2^{-2} \times \tag{31.35}$$

$$\times [(\mathbf{k}_1 + \mathbf{k}_2)^{-2} - (\mathbf{k}_1 + \mathbf{k}_2 + \mathbf{k})^{-2}]\}.$$

We apply the four-dimensional Laplace operator squared Δ_4^2 to this expression. By using the formula

$$\Delta_4 k^{-2} = -4\pi^2 \delta(\mathbf{k}). \tag{31.36}$$

the Laplace operator can be brought under the sign of the integral which produces

$$\Delta_4^2(\Sigma(\mathbf{k}) - \Sigma(0)) = \Delta_4 \frac{u^2}{3!(2\pi)^3} \times$$

$$\times \int dk_1 \, dk_2 k_1^{-2} k_2^{-2} 4\pi^2 \delta(\mathbf{k}_1 + \mathbf{k}_2 + \mathbf{k})$$

$$= \frac{u^2 4\pi^2}{3!(2\pi)^8} \int dk_1 k_1^{-2} \Delta_4(\mathbf{k}_1 + \mathbf{k})^{-2} \tag{31.37}$$

$$= -\frac{u^2}{3!(2\pi)^4} k^{-2} = -\frac{8\varepsilon^1}{27k^2}.$$

From here, it is not difficult to evaluate $\Sigma(\mathbf{k}) - \Sigma(0)$:

$$\Sigma(\mathbf{k}) - \Sigma(0) = -\frac{\varepsilon^2 k^2}{54}(\ln(k/k_0) + \text{const}). \tag{31.38}$$

By comparing this formula with (31.33) we obtain

$$\eta = \varepsilon^2/54. \tag{31.39}$$

Investigating the stationarity equation of the $\hat{M}$ transformation with great accuracy, it is not hard to get the next terms of the expansion of critical indices v and η.

Finally, we introduce the results of calculation [124] of the critical indices v, η, α, β, γ, δ for the systems with the initial action functional:

$$S = -\int \left\{ \frac{1}{2} \sum_{i=1}^{n} ((\partial_\tau \varphi_i)^2 + (\nabla \varphi_i)^2 - \lambda \varphi_i^2) \right.$$
$$\left. - \frac{g}{4!} \left(\sum_{i=1}^{n} \varphi_i^2 \right)^2 \right\} d\tau \, dx, \tag{31.40}$$

It generalizes expression (31.14) for the case of many-components field $\varphi_i(x, \tau)$, $i = 1, \ldots, n$. In this chapter was considered the system with $n = 1$. The case $n = 2$ corresponds to the complex scalar field.

The formulae for the critical indices are:

$$v = \frac{1}{2} + \frac{(n+2)}{4(n+8)} \varepsilon + \frac{n+2}{8} \frac{n^2 + 23n + 60}{(n+8)^3} \varepsilon^2 + \cdots ;$$

$$\eta = \frac{n+2}{2(n+8)^2} \varepsilon^2 + \frac{n+2}{2(n+8)^2} \left[\frac{6(3n+14)}{(n+8)^2} - \frac{1}{4} \right] \varepsilon^2 + \cdots ;$$

$$\alpha = \frac{(4-n)\varepsilon}{2(n+8)} - \frac{(n+2)^2(2n+1)}{2(n+8)^3} \varepsilon^2 + \cdots ; \tag{31.41}$$

$$\beta = \frac{1}{2} - \frac{3\varepsilon}{2(n+8)} + \frac{(n+2)(2n+1)}{2(n+8)^3} \varepsilon^2 + \cdots ;$$

$$\gamma = 1 + \frac{(n+2)}{2(n+8)} \varepsilon + \frac{(n+2)(n^2 + 22n + 52)}{4(n+8)^3} \varepsilon^2 + \cdots ;$$

$$\delta = 3 + \varepsilon + \left(\frac{1}{2} - \frac{(n+2)}{(n+8)^2} \right) \varepsilon^2 + \cdots$$

where n denotes the number of field components.

CHAPTER 11

VORTEX-LIKE EXCITATIONS IN RELATIVISTIC FIELD THEORY

32. VORTICES IN THE RELATIVISTIC GOLDSTONE MODEL

Recently the concept of quantum vortex excitations which originated in superconductive and superfluid theory, has been introduced into relativistic quantum field theory. It is based on the hypothesis that the strongly interacting particles (if not all, then at least some of them) are vortex-like excitations. This hypothesis allows us to reduce the number of fundamental fields. The necessity of doing this has been seen recently when the number of discovered strongly interacting particles together with resonances was found to be of the order 100–200. In this situation the standard scheme of field theory, in which to every particle there corresponds a fundamental field, becomes clumsy, ineffective for practical use, and unattractive from the point of view of the beauty of the theory.

Vortex-like excitations exist in exactly solvable models where they are called 'solitons', and in other theories as well. We shall consider one of the simplest models of relativistic field theory in which the vortex-like excitations are possible – the Goldstone model in one temporal and two spatial dimensions [127]. In the next section the possibility of the existence of vortex-like excitations in four-dimensional space-time is considered in some field theoretical models.

The method of functional integration is the most suitable one in the region where the excitations are essentially collective and originated from many identical particles.

Let us consider now the relativistic Goldstone model. The action functional of this model can advantageously be expressed in terms of the Euclidean variables

$$S = -\int \left(|\nabla \psi|^2 - \lambda |\psi|^2 + \frac{g}{4} |\psi|^4 \right) dx. \tag{32.1}$$

Functional (32.1) describes a complex scalar field self-interacting with the coupling constant $g > 0$. The coefficient λ is positive. This corresponds to the fact that there are particles with a negative value of the mass squared (tachyons) when the interaction is switched off ($g = 0$).

Goldstone observed that there is a Bose condensation in the system whenever the interaction is switched on [127]. As a result, particles appear with zero and finite masses (with a positive square). The particles with finite masses are unstable and have a finite lifetime.

The form of the action functional (32.1) reminds us of the corresponding functional for a nonideal Bose gas. It is natural to suppose that excitations can exist of the quantum vortex type in the Goldstone model which are characteristic of superfluid Bose systems.

Before we consider quantum vortices we shall construct the perturbation theory for the Goldstone model analogous to that developed in Section 16 for Bose gas. Such a perturbation theory does not contain infrared divergences and appears useful, in particular for evaluating the lifetime of unstable particles. We shall not yet specify the dimensionality of the model n.

The density of the condensate is determined in the first approximation by the condition for the maximum of the expression

$$\lambda |\psi|^2 - \frac{g}{4} |\psi|^4 \tag{32.2}$$

and is equal to

$$\rho_0 = |\psi_0|^2 = \frac{2\lambda}{g}. \tag{32.3}$$

In action (32.1) we introduce polar coordinates $\psi = \sqrt{\rho} \exp(i\varphi), \bar{\psi} = -\sqrt{\rho} \exp(-i\varphi)$ and then we use $\pi = \rho - \rho_0$ instead of ρ. In the variables φ, π the action has the form

$$S = -\int \left((\rho_0 + \pi)(\nabla \varphi)^2 + \frac{(\nabla \pi)^2}{4(\rho_0 + \pi)} + \frac{g}{4} \pi^2 \right) dx + \frac{\lambda^2}{g} \int dx. \tag{32.4}$$

Here the expression $(\lambda^2/g) \int dx$ is the contribution of the Bose condensate.

We shall make the transformation

$$\varphi \to \frac{1}{\sqrt{2\rho_0}} \varphi, \quad \pi \to \sqrt{2\rho_0} \pi, \tag{32.5}$$

in the functional integral with respect to variables $\varphi(x), \pi(x)$ which changes expression (30.4) into

$$S = - \int \left(\frac{1}{2}(\nabla \varphi)^2 + \frac{1}{2}\sqrt{\frac{2}{\rho_0}} \pi (\nabla \varphi)^2 + \right.$$
$$\left. + \frac{1}{2}\left(\frac{(\nabla \pi)^2}{1 + \sqrt{2/\rho_0}\, \pi} + 2\lambda \pi^2 \right) \right) dx + \frac{\lambda^2}{g} \int dx. \tag{32.6}$$

The constant $\sqrt{2/\rho_0}$ determines the strength of the interaction of φ and π fields as well as the self-interaction of the field π.

Notice that the unstability of the system of interacting tachyons with respect to an arbitrary small, weak stability excitation follows from formula (32.6). Really, if we consider the limit $g \to +0$ in (32.6) with λ fixed, we find that the second term in (32.6) goes to infinity and the first one inverts to the quadratic form describing noninteracting massless particles with a positive value of the mass squared $m^2 = 2\lambda$. The effect of this interaction leads to the unstability of the massive particle.

We construct the perturbation theory by means of variables φ, π. Elements of the diagrams of the perturbation theory will be two lines corresponding to fields φ and π, one vertex corresponding to the $\varphi-\pi$ interactions, and an infinite number of vertices corresponding to the $\pi-\pi$ interaction. We shall write down the expressions for the lines, the vertex of the $\varphi-\pi$ interaction, and the first of the vertices of the $\pi-\pi$ interaction:

$$\underline{\hspace{3cm}} -k^{-2}; \quad \underline{\hspace{2cm}} -(k^2 + 2\lambda)^{-1};$$

$$k_3 \prec^{k_1}_{k_2} \sqrt{\frac{2}{\rho_0}}(k_1 k_2); \quad k_3 \prec\!\!\!<^{k_1}_{k_2} \quad -\sqrt{\frac{2}{\rho_0}}(k_1 k_2) \tag{32.7}$$

We denote field φ by a single line and field π by a double line. In the $\pi-\pi$ vertex the strokes denote out-lines of the vertex carrying momenta k_1 and k_2. It is these momenta of the scalar product which occur in the expression corresponding to this vertex.

The expression corresponding to the perturbation theory diagram can be obtained by integrating the product of the expressions corresponding to the elements of the diagram with respect to the independent momenta and by multiplying the result by the factor

$$\frac{1}{r}\left(\frac{-1}{(2\pi)^n}\right)^c, \tag{32.8}$$

where r is the order of the symmetry group and c is the number of independent contour lines of the diagram. Since the contribution of the theory is via the Euclidean variables, it is necessary to continue the expressions corresponding to the diagrams into the domain of physical energy and momenta in order to get the physical results.

The perturbation theory obtained brings to mind that constructed in Section 20 for Bose gas theory. It does not contain infrared divergences but diverges for large momenta and is formally non-renormalizable. It would be, therefore, more logical first to integrate with respect to the rapidly varying components of fields $\psi, \bar{\psi}$ and then to use the polar coordinates in the integral of the slowly varying fields. By integrating functional $\exp S$ with respect to the rapidly varying components of fields $\psi, \bar{\psi}$, we obtain functional $\exp \tilde{S}$ containing only the slowly varying components of fields $\psi, \bar{\psi}$. The expression for $\tilde{S}$ is, in the first approximation, equal to the expression for S and differs from it by corrections which eliminate divergences from integration with respect to the slowly varying fields. These corrections will not be considered here. Momentum k_0, dividing the large and the small momenta, can be estimated by an order of magnitude analogous to the Bose gas theory. The result of the estimate is formulated in the form of the inequalities

$$\sqrt{\lambda} \ll k_0 \ll \sqrt{\frac{\lambda}{g}} \qquad \text{for } n = 4,$$

$$\tag{32.9}$$

$$\sqrt{\lambda} \ll k_0 \ll \frac{\lambda}{g} \qquad \text{for } n = 3,$$

which may be fulfilled whenever the coupling constant g is small.

As an example of the application of the perturbation theory with elements (32.7), we calculate the lifetime of the massive particle determined by the second-order diagram

$$\Sigma_2 = \quad \tag{32.10}$$

The diagrams arising due to the $\pi-\pi$ interaction do not contribute to the imaginary part. Physically, this means that the decay of the massive particle into two particles of the same mass is impossible. The expression corresponding to (32.10) has the form

$$-\frac{1}{2}\frac{2}{\rho_0}\frac{1}{(2\pi)^n}\int d^n k_1 \frac{(k_1 k_2)}{k_1^2 k_2^2}$$

$$= -\frac{g}{2\lambda(2\pi)^n}\int d^n k_1 \frac{(k_1 k_2)^2}{k_1^2 k_2^2}. \tag{32.11}$$

We consider the imaginary part of this expression for $k^2 = -2\lambda$. Due to the equality $k = k_1 + k_2$ we have

$$k_1 k_2 = \tfrac{1}{2}((k_1 + k_2)^2 - k_1^2 - k_2^2) = -\lambda - \frac{k_1^2 + k_2^2}{2},$$

$$(k_1 k_2)^2 = \lambda^2 + \lambda(k_1^2 + k_2^2) + \tfrac{1}{4}(k_1^2 + k_2^2)^2.$$

Only the first term λ^2 on the right-hand side contributes to the imaginary part. The remaining terms yield real though formally divergent integrals of the type

$$\int \frac{d^n k_1}{k_1^2}; \qquad \int d^n k_1; \qquad \int \frac{k_2^2}{k_1^2} d^n k_1.$$

In this way we have

$$\mathrm{Im}\Sigma = -\mathrm{Im}\frac{\lambda g}{2(2\pi)^n}\int \frac{d^n k_1}{k_1^2 k_2^2}. \tag{32.12}$$

The integral in the formula converges for the case $n = 3$ and diverges logarithmically for $n = 4$. The imaginary part of the integral is finite for both $n = 3$ and $n = 4$. After making the analytic continuation $k^2 \to -2\lambda + i0$ we obtain

$$\mathrm{Im}\Sigma = \frac{g\sqrt{2\lambda}}{2^5} \text{ for } n = 3; \qquad \mathrm{Im}\Sigma = \frac{2\lambda}{2^5\pi} \text{ for } n = 4. \tag{32.13}$$

The corresponding formulae for the lifetime of a π particle have the form

$$\tau = 2^6 g^{-1} \quad \text{for } n = 3;$$
$$\tau = 2^7 \pi g^{-1}(2\lambda)^{-1/2} \quad \text{for } n = 4 \tag{32.14}$$

Modification of the functional integral which is necessary in order to take into account quantum vortices, is analogous to that made in Section

21 for the Bose system. Here we shall consider the case $n = 3$ (one-dimensional time and two-dimensional space). There are lines in three-dimensional space (x_0, x_1, x_2) corresponding to quantum vortices on which functions $\psi, \bar{\psi}$ vanish and with respect to which we integrate. Phase φ of function ψ increases by $2\pi n$ (n-integer) for each turn around the line. We shall consider only the vortices with an increment of the phase by $\pm\pi$ ($|n| = 1$). The states with $|n| > 1$ are unstable and decay on vortices with $|n| > 1$.

To each vortex there corresponds the solution of the equation

$$-\Delta\psi - \lambda\psi + \frac{g}{2}\bar{\psi}\psi\psi = 0, \tag{32.15}$$

obtained by variation of action S (32.1) with respect to $\bar{\psi}$. It depends on the plane variables orthogonal to the world line and has the form $f(r)$ $\exp(i\theta)$ where θ is a polar angle and $f(r)$ is a real function of distance r from the vortex axis. Equation (32.15) reduces to the ordinary differential equation for function $f(r)$:

$$f'' + \frac{1}{r}f' - \frac{f}{r^2} + \lambda f - \frac{3}{2}f^3 = 0. \tag{32.16}$$

It coincides with the corresponding equation in the Bose gas theory studied by L. P. Pitayevski [106]. The solution of the equation vanishes for $r = 0$ and goes to $\sqrt{\rho_0}$ for $r \to \infty$. The characteristic length $\lambda^{-1/2}$ is naturally called a radius of the vortex stem.

In order to describe the situation with nonlocal vortices, we close each world line in a tube of the radius r_0 much bigger than the radius of the vortex stem $\lambda^{-1/2}$. The sum of the integrals over the vortex tubes in the action functional S can be expressed in the form

$$-\sum_i m_B(r_0) \int ds_i. \tag{32.17}$$

in the first approximation. Here ds_i is an element of the length of the ith vortex line and $m_B(r_0)$ is the mass (energy) of the vortex closed in the tube. The second quantity depends logarithmically on r_0 and is given by the formula

$$m_B(r_0) = 2\pi\rho_0 \ln(r_0/a), \tag{32.18}$$

here a is a parameter of the order of the radius of the vortex stem. Formulae

(32.17) and (32.18) are obtained analogously to the corresponding formulae (21.8) and (21.9) in the Bose system theory.

We separate the contribution in action (32.4) from the vortex tubes and then make the change of variables (32.5). We obtain the expression

$$
-\int \left(\frac{1}{2} (\nabla \varphi)^2 + \frac{1}{2} \sqrt{\frac{2}{\rho_0}} \pi (\nabla \varphi)^2 + \right.
$$

$$
\left. + \frac{1}{2} \left(\frac{(\nabla \pi)^2}{1 + \sqrt{2/\rho_0}\, \pi} + 2\lambda \pi^2 \right) \right) dx - \qquad (32.19)
$$

$$
- \sum_i m_B(r_0) \int ds_i + \frac{\lambda^2}{g} \int dx.
$$

Functional $\exp S$ has to be integrated with respect to fields φ, π and also over the trajectories of the vortex centres. Function $\varphi(x)$ in action (32.19) is not unique and has the increment

$$
\pm 2\pi \sqrt{2\rho_0} \equiv \pm g. \qquad (32.20)
$$

The quantity q has the meaning of an electrical charge. In order to prove it we have to use a functional integral with respect to a new variable which has the meaning of a vector potential of the electromagnetic field. The passage to the new variable is analogous to that made in Section 21 for the Bose gas and will not be repeated here. Note that charge q is reciprocal to the coupling constant of the $\varphi-\pi$ and the $\pi-\pi$ interactions.

The quantum vortices exist as independent particles. The obvious conservation law of the difference of the number of vortices rotating in positive and negative directions can be interpreted as the conservation law of the difference of the number of particles and antiparticles.

The mass of an individual vortex is, to be exact, infinite due to the energy of the φ field surrounding the volume. For instance, the mass (energy) $m_B(r)$ inside the circle of radius with the centre coinciding with the vortex centre is given by formula (32.18) with the substitution $r_0 \rightarrow r$.

The Goldstone model with the quantum vortices can be called the simplest model of strong + electromagnetic interactions in a $(2+1)$-dimensional space-time. Here the quantum vortices have the meaning of protons, the π particles the meaning of π mesons, the φ particles the meaning of photons. The analogy is based on the properties of the particles and their corresponding fields. In fact, the interactions among the quantum vortices are mediated on large distances by field φ and on small ones also by field π analogously to the interactions among protons which are mediated by photons on large distances but on small ones also by pions. Similarly, the decay of the massive particle on two massless ones can be considered as an analogue of the π meson decay on 2γ quanta. Finally, the difference of the number of 'particles' and 'antiparticles' is conserved for the quantum vortices. This difference has the meaning of the electric charge coinciding with the baryonic one in this model.

33. ON VORTEX-LIKE SOLUTIONS IN QUANTUM FIELD THEORY

We shall consider the possibility of the existence of *vortex-like excitations* in some field theoretical models in four-dimensional space-time. First we shall take a generalization of the Goldstone model studied in the previous section that is the model with three real scalar fields φ_a and an action functional of the form

$$S = -\frac{1}{2}\int d^4x \left(\sum_a (\nabla \varphi_a)^2 - \lambda \sum_a \varphi_a^2 + \frac{g}{2}\left(\sum_a \varphi_a^2 \right)^2 \right), \qquad (33.1)$$

which is expressed here in the Euclidean variables. The condition $\delta S = 0$ is given by the equation

$$-\Delta_4 \varphi_a - \lambda \varphi_a + g\left(\sum_a \varphi_a^2 \right)\varphi_a = 0. \qquad (33.2)$$

This equation has the constant solution, $\varphi_a = $ constant, with the constraint

$$\sum_a \varphi_a^2 = \frac{\lambda}{g}, \qquad (33.3)$$

as well as the solution describing vortex-like excitations which does not depend on the 'temporal' coordinate x_4 and is of the form

$$\varphi_a = \frac{x_a}{r}f(r). \qquad (33.4)$$

Here $r = (x_1^2 + x_2^2 + x_3^2)^{1/2}$ is the distance from the origin of coordinates in the three-dimensional space. Equation (33.3) reduces to the second-order equation for the function f

$$f'' + \frac{2}{r}f' - \frac{2f}{r^2} + \lambda f - gf^3 = 0. \qquad (33.5)$$

Let us consider a solution of the equation which behaves as $\sim r$ for $r \to 0$ and goes to constant $(\lambda/g)^{1/2}$ for $r \to \infty$. It is possible to show that this solution of the equation exists. The functional

$$\frac{1}{2}\int_{r<r_0}\left[\sum_a(\nabla\varphi_a)^2 - \lambda\sum_a\varphi_a^2 + \frac{g}{2}\left(\sum_a\varphi_a^2\right)^2\right]d^3x, \qquad (33.6)$$

giving the excitation energy in volume $r < r_0$ is, however, proportional to r_0 in the limit $r_0 \to \infty$. Thus the vortex-like excitation has infinite energy and cannot be interpreted as a new particle.

There are more complicated vortex-like solutions in models with Yang–Mills fields. For instance, for the system with the action

$$-\frac{1}{2}\int\left[\sum_{a,\mu}(\partial_\mu\varphi_a + \varepsilon\varepsilon_{abc}b_\mu^b\varphi_c)^2 - \lambda\sum_a\varphi_a^2 + \frac{g}{2}\left(\sum_a\varphi_a^2\right)^2\right]dx -$$
$$-\frac{1}{2}\int\sum_{a,\mu}(\partial_\mu b_\nu^a - \partial_\nu b_\mu^a + \varepsilon\varepsilon_{abc}b_\mu^b b_\nu^c)^2\,d^4x \qquad (33.7)$$

we can look for a solution of the form

$$\varphi_a(x) = x_a u(r)r^{-1}; \qquad b_\mu^a(x) = \varepsilon_{\mu ab}x_b(a(r) - (\varepsilon r^2)^{-1}). \qquad (33.8)$$

Such a solution was suggested and studied by 't Hooft [129] and independently by A. M. Polyakov [128]. It was shown that it has a finite energy functional and consequently can be treated as a new particle. Attempts to find out other, more realistic field theoretical models with vortex-like solutions belong to very actual problems of the present time. It is not excluded that the key to the construction of a successive strong interaction theory lies just on that path [151–158].

REFERENCES

1. Wiener, N.: 'Differential Space', *J. Math. Phys.* **2** (1923), 131–174; 'The Average Value of a Functional', *Proc. Lond. Math. Soc.* **22** (1924), 454–467.
2. Feynman, R. P.: 'Space-time Approach to Nonrelativistic Quantum Mechanics', *Rev. Mod. Phys.* **20** (1948), 367–387.
3. Feynman, R. P.: 'Mathematical Formulation of the Quantum Theory of Electromagnetic Interaction', *Phys. Rev.* **80** (1950), 440–457.
4. Feynman, R. P.: 'An Operator Calculus Having Applications in Quantum Electrodynamics', *Phys. Rev.* **84** (1951), 108–128.
5. Heisenberg, W. and Pauli, W.: 'Quantum Dynamics of Wave Fields', *Z. Phys.* **56** (1929), 1–61.
6. Schwinger, J.: 'Quantum Electrodynamics, I: A Covariant Formulation', *Phys. Rev.* **74** (1948), 1439–1461; II: 'Vacuum Polarization and Self-energy', *Phys. Rev.* **75** (1949), 651–679.
7. Bogoliubov, N. N.: 'On the Representation of Green–Schwinger Functions by Means of Functional Integrals', *Dokl. Akad. Nauk S.S.S.R.* **99** (1954) 225–226 (in Russian).
8. Gel'fand, I. M. and Minlos, R. A.: 'Solution of the Quantum Field Equations', *Dokl. Akad. Nauk S.S.S.R.* **97** (1954), 209–212 (in Russian).
9. Matthews, P. T. and Salam, A.: 'The Green's Functions of Quantized Fields', *Nuovo Cim.* **12** (1954), 563–565.
10. Khalatnikov, I. M.: 'On a Method of Calculation of the Statistic Sum', *Dokl. Akad. Nauk S.S.S.R.* **87** (1952), 539–542; 'Representation of Green Functions in Quantum Electrodynamics in the Form of Functional Integrals', *Zh. Eksp. Teor. Fiz.* **28** (1955), 633–636 (in Russian).
11. Fradkin, E. S.: 'The Green Function Method in Quantum Field Theory and in Quantum Statistics', *Trudy. Fiz. Inst. Akad. Nauk S.S.S.R.* **29** (1965), 7–138 (in Russian).
12. Feynman, R. P.: 'Quantum Theory of Gravitation', *Acta Phys. Polon.* **24** (1963), 697–722.
13. De Witt, B. S.: 'Quantum Theory of Gravity', *Phys. Rev.* **160** (1967), 1113; **162** (1967), 1195, 1239.
14. Faddeev, L. D. and Popov, V. N.: 'Feynman Diagrams for Yang–Mills Field', *Phys. Lett.* **B25** (1967), 30–31.
15. Popov, V. N. and Faddeev, L. D.: 'Perturbation Theory of Gauge Invariant Fields', *Inst. Teor. Fiz. Akad Nauk Ukr. S.S.R.* preprint, Kiev, 1967 (in Russian).
16. Mandelstam, S.: 'Feynman Rules for Electromagnetic and Yang–Mills Fields from the Gauge-inependent Field-theoretic Formalism', *Phys. Rev.* **175** (1968), 1580; 'Feynman Rules for the Gravitational Fields from Coordinate-independent Field theoretic Formalism', *Phys. Rev.* **175** (1968), 1604.
17. Fradkin, E. S. and Tyutin, I. V.: 'S-matrix for Yang–Mills and Gravitational Fields', *Phys. Lett.* **B30** (1969), 562–563; *Phys. Rev.* **D2** (1970), 2841–2856.

290

18. 't Hooft, G.: 'Renormalization of Massless Yang–Mills Fields', *Nucl. Phys.* **B33** (1971), 173–199.

19. De Witt, B.: 'Gravity: A Universal Regulator', *Phys. Rev. Lett.* **13** (1964), 114–118.

20. Khriplovich, I. B.: 'Gravity and Finite Renormalizations in QED', *Yad. Fiz.* **3** (1966), 575–581 (in Russian).

21. Weinberg, S.: 'A Model of Leptons', *Phys. Rev. Lett.* **19** (1967), 1264–1266.

22. Lee, B. W.: 'Renormalizable Massive Vector-meson Theory – Perturbation Theory of the Higgs Phenomenon', *Phys. Rev.* **5** (1972), 823–835.

23. Faddeev, L. D.: 'A Gauge Invariant Model of Electromagnetic and Weak Interactions of Leptons', Leningrad. Otd. Mat. Inst. Akad Nauk S.S.S.R. preprint, Leningrad, 1972 (in Russian).

24. Feynman, R. P.: 'Atomic Theory of the Transition in Helium', *Phys. Rev.* **91** (1953), 1921; 'Atomic Theory of Liquid Helium Near Absolute Zero', *Phys. Rev.* **91** (1953), 1301.

25. Feynman, R. P.: 'Slow Electrons in Polar Crystals', *Phys. Rev.* **97** (1954), 660–665.

26. Gelfand, I. M. and Yaglom, A. M.: 'Integration in Functional Spaces and its Application in Quantum Physics', *Usp. Mat. Nauk* **11** (1956), 77–114 (in Russian).

27. Kovalchik, I. M.: 'Wiener Integral', *Usp. Mat. Nauk* **18** (1963), 97–134 (in Russian).

28. Kac, M.: 'Probability and Related Topics in Physical Sciences', *Proc. Summer Seminar, Boulder, Col.*, Interscience, London, New York, 1957.

29. Berezin, F. A.: *Method of Second Quantization*, Academic Press, New York, London, 1966.

30. Berezin, F. A.: 'Non-Wiener Functional Integrals', *Teor. Mat. Fiz.* **6** (1971), 194–212 (in Russian).

31. Daletzki, Yu. L.: 'Functional Integrals Connected with the Operator Evolution Equations', *Usp. Mat. Nauk* **17** (1962), 3–115 (in Russian).

32. Evgraphov, M. A.: 'On a Formula for Representation of Differential Equation Fundamental Solution by the Functional Integral', *Dokl. Akad. Nauk S.S.S.R.* **191** (1970), 979 (in Russian).

33. Alimov, A. L. and Buslaiev, V. S.: 'On the Functional Integral of the Second-order Parabolic Equation', *Vestnik Leningr. Univ. Ser. Mat.* No. 1 (1972), 5–14 (in Russian); Alimov, A. L.: 'On the Connection Between Functional Integrals and Differential Equations', *Teor. Mat. Fiz.* **11** (1972), 182–189 (in Russian).

34. Faddeev, L. D.: 'Feynman Integral for Singular Lagrangians', *Teor. Mat. Fiz.* **1** (1969), 3–17 (in Russian).

35. Bogoliubov, N. N. and Shirkov, D. V.: *Introduction to the Theory of Quantized Fields*, Interscience, New York, 1959.

36. Schweber, S. S.: *Relativistic Quantum Field Theory*, Row, Peterson, Evanston, 1961.

37. Akhiezer, A. I. and Berestecki, V. B.: *Quantum Electrodynamics*, Nauka, Moscow, 1969 (in Russian).

38. Abrikosov, A. A., Gorkov, L. P., and Dzyaloshinski, I. E.: *Methods of Quantum Field Theory in Statistical Physics*, Fizmatgiz, Moscow, 1962 (in Russian).

39. Bialynicki-Birula, I. T.: 'On the Gauge Covariance of Quantum Electrodynamics', *J. Math Phys.* **3** (1962), 1094–1098.

40. Ward, T. C.: 'The Scattering of Light by Light', *Phys. Rev. Lett.* **77** (1949), 293; 'An Identity in Quantum Electrodynamics', *Phys. Rev. Lett.* **78** (1950), 182.

41. Yang, C. N. and Mills, R. L.: 'Conservation of Isotopic Spin and Isotopic Gauge Invariance', *Phys. Rev.* **96** (1954), 191–198.

42. Schwinger, J.: 'Non-abelian Gauge Fields', *Phys. Rev.* **125** (1962), 1043; **127** (1962), 324.

43. Gross, D. T. and Wilczek, F.: 'Ultraviolet Behavior of Non-abelian Gauge Theories', *Phys. Rev. Lett.* **30** (1973), 1343–1346.

44. Politzer, H. D.: 'Reliable Perturbation Results for Strong Interactions', *Phys. Rev. Lett.* **30** (1973), 1346–1349.

45. Weyl, H.: 'A Remark on the Coupling of Gravitation and Electron', *Phys. Rev.* **77** (1950), 699–705.

46. Umezawa, H. and Takahashi, Y.: 'The General Theory of the Interaction Representation', *Prog. Theor. Phys.* **9** (1953), 14–32, 501–523.

47. Lee, T. D. and Yang, C. N.: 'Theory of Charged Vector Mesons Interacting with Electromagnetic Field', *Phys. Rev.* **128** (1962), 885.

48. Dirac, P. A. M.: 'Generalized Hamiltonian Dynamics', *Proc. Roy. Soc.* **A246** (1958), 326–332; 'The Theory of Gravitation in Hamilton Form', *Proc. Roy. Soc.* **A246** (1958), 333–343.

49. Arnowitt, R., Deser, S., and Misner, C. W.: 'Canonical Variables for General Relativity', *Phys. Rev.* **117** (1960), 1595–1602.

50. Schwinger, J.: 'Quantized Gravitational Field', *Phys. Rev.* **130** (1963), 1253; **132** (1963), 1317.

51. Bergman, P.: 'Observables in General Relativity', *Rev. Mod. Phys.* **33** (1961), 510–514.

52. Andersen, T. L.: 'Coordinate Conditions and Canonical Formulism in Gravitational Theory', *Rev. Mod. Phys.* **36** (1964), 929–938.

53. Faddeev, L. D.: 'Hamilton Form of the Theory of Gravitation', *Proc. 5th Int. Conf. Gravity and Relativity, Tbilisi*, 1968 (in Russian).

54. Favard, J.: *Cours de la geometrie différentielle locale*, Gauthier-Villars, 1957.

55. Fock, V. A.: *Theory of Space, Time and Gravity*, Fizmatgiz, Moscow, 1961.

56. Higgs, P. W.: Spontaneous Symmetry Breakdown Without Massless Bosons', *Phys. Rev.* **145** (1956), 1156–1163.

57. 't Hooft, G.: 'Renormalizable Lagrangians for Massive Yang–Mills Fields', *Nucl. Phys.* **B35** (1971), 167–188.

58. Schwinger, J.: 'On the Euclidean Structure of Relativistic Field Theory', *Proc. Nat. Acad. Sci. U.S.A.* **44** (1958), 956–965.

59. Abrikosov, A. A., Landau, L. D., and Khalatnikov, I. M.: 'On the Elimination of Infinities in Quantum Electrodynamics', *Dokl. Akad. Nauk S.S.S.R.* **95** (1954), 497; 'Asymptotic Expression for the Electron Green's Functions in Quantum Electrodynamics', *Dokl. Akad. Nauk S.S.S.R.* **95** (1954), 773; 'Asymptotic Expression for the Photon Green's Function in Quantum Electrodynamics', *Dokl. Akad. Nauk S.S.S.R.* **95 (1954)**, 1177 (all in Russian).

60. Milekhin, G. A. and Fradkin, E. S.: 'Double Logarithmic Approximation in Quantum Electrodynamics', *Zh. Eksp. Teor. Fiz.* **45** (1963), 1926–1939 (in Russian).

61. Yennie, D. R., Frautschi, S. C., and Suura, H.: 'The Infrared Divergence Phenomena and High Energy Processes', *Ann. Phys.* **13** (1961), 379.

62. Chung, V.: Infrared Divergence in Quantum Electrodynamics', *Phys. Rev.* **B140** (1965), 1110–1122.

63. Kulish, P. P. and Faddeev, L. D.: 'Asymptotic Conditions and Infrared Divergences in Quantum Electrodynamics', *Teor. Mat. Fiz.* **4** (1970). 153–170 (in Russian).

64. Gerdzhikov, V. S. and Kulish, P. P.: 'On Low-energy Theorems for Photons and Infrared Divergences', *Teor. Mat. Fiz.* **18** (1974), 51–55 (in Russian).

65. Soloviev, L. D.: 'Low-energy Theorems and Dispersion Relations in Quantum Electrodynamics', *Teor. Mat. Fiz.* **15** (1973), 59–69 (in Russian).

66. Nordstrom, D. L.: 'Some Applications of Coherent States in Quantum Electrodynamics', thesis, Iowa State Univ., Ames, U.S.A., 1970.

67. Barbashov, B. M.: 'Functional Integrals in Quantum Electrodynamics and Infrared Asymptotics of the Green's Functions', *Zh. Eksp. Teor. Fiz.* **48** (1965), 607 (in Russian).

68. Pervushin, V. N.: 'Functional Integration Method and Eikonal Approximation to the Potential Scattering Amplitudes', *Teor. Mat. Fiz.* **4** (1970), 22–31 (in Russian).

69. Abarbanel, H. D. I. and Itzykson, C.: 'Relativistic Eikonal Expansion', *Phys. Rev. Lett.* **23** (1969), 53–56.

70. Barbashov, B. M., Kuleshov, S. P., Matveev, V. A., Pervushin, V. N., Sissakian, A. N., and Tavhelidze, A. N.: 'Straight-line Paths Approximation for Studying High Energy Elastic and Inelastic Hadron Collisions in Quantum Field Theory', *Phys. Lett.* **B33** (1970), 484–488.

71. Barbashov, B. M. and Nesterenko, V. V.: 'On an Approximation of Virtual Particle Propagators and High-energy Approximation Behaviour of the Feynman Diagrams', *Teor. Mat. Fiz.* **4** (1970), 93; 'Functional Integration and Redge-Eikonal Representation of the Scattering Amplitude', *Teor. Mat. Fiz.* **10** (1972), 196 (in Russian).

72. Landau, L. D.: 'Theory of Superfluid Helium', *Zh. Eksp. Teor. Fiz.* **1** (1941), 592 (in Russian).

73. Bogoliubov, N. N.: 'On the Theory of Superfluidity', *Izv. Akad. Nauk S.S.S.R.* **11** (1947), 77 (in Russian).

74. Bogoliubov, N. N. and Zubarev, D. N.: 'The Ground State Wave Function of a System of Interacting Bose Particles', *Zh. Eksp. Teor. Fiz.* **8** (1955), 129–139 (in Russian).

75. Tserkovnikov, Yu. A.: 'On the Theory of Ideal Bose Gas at Temperatures Different From Zero', *Dokl. Akad. Nauk S.S.S.R.* **143** (1962), 832; 'Random Phase Approximation of Non-ideal Bose Gas', *Dokl. Akad. Nauk S.S.S.R.* **159** (1964), 1023; 'The Second Sound in a Weakly Non-ideal Bose Gas', *Dokl. Akad. Nauk S.S.S.R.* **159** (1964), 1264 (all in Russian).

76. Tolmachev, V. V.: 'Connection Between the Statistic Variation Principle and the Method of Partial Summation of Thermodynamical Perturbation Theory Diagrams in Modified Formulation of the Problem of Non-ideal Bose–Einstein System', *Dokl. Akad. Nauk S.S.S.R.* **134** (1960), 1324; 'Construction of Weak-coupling Asymptotic Expansions from the Formal Thermodynamical Perturbation Theory in Modified Formulation of the Problem of Non-ideal Bose–Einstein System', *Dokl. Akad. Nauk S.S.S.R.* **135** (1960), 41; 'Elementary Temperature Excitation on Non-ideal Bose–Einstein System', *Dokl. Akad. Nauk S.S.S.R.* **135** (1960), 825–828 (in Russian).

77. Beliaev, S. T.: 'Application of Quantum Field Theory Methods to the System of Bose Particles', *Zh. Eksp. Teor. Fiz.* **34** (1958), 417; 'Energy Spectrum of Non-ideal Bose Gas', *Zh. Eksp. Teor. Fiz.* **34** (1958), 433–446 (in Russian).

78. Hugenholtz, N. M. and Pines, D. : 'Ground State Energy and Excitation Spectrum of a System of Interacting Bosons', *Phys. Rev.* **116** (1959), 489–506.

79. Lee, T. D. and Yang, C. N.: 'Many-body Problem in Quantum Mechanics and Quantum Statistical Mechanics', *Phys. Rev.* **105** (1957), 1119.

80. Lee, T. D. and Yang, C. N.: 'Low Temperature Behaviour of a Dilute Bose System of Hard Spheres', *Phys. Rev.* **112** (1958), 1419; **113** (1959), 1406. 'Many-body Problem in Quantum Statistical Mechanics, I–V', *Phys. Rev.* **113** (1959), 1165; **116** (1959), 25; **117** (1960), 12, 897.

81. Bogoliubov, N. N.: 'Quasi-expectation Values in Problems of Statistical Mechanics, JINR preprint D-781, 1961 (in Russian).

82. Bogoliubov, N. N.: 'On the Problem of Hydrodynamics of Superfluid Liquid', JINR preprint R-1495, 1963 (in Russian).

83. Galasevich, Z.: 'Asymptotic Calculation of the Green's Function in Viscose-liquid Approximation for Superfluid Bose Systems', JINR preprint R-1517, 1964 (in Russian).

84. Girardeau, M., Arnowitt, R.: 'Theory of Many-boson Systems. Pair Theory', *Phys. Rev.* **113** (1959), 755.

85. Luban, M.: 'Statistical Mechanics of a Nonideal Boson Gas. Pair Hamiltonian Model', *Phys. Rev.* **128** (1962), 965–987.

86. Huang, K. and Yang, C. N.: 'Quantum Mechanical Many-body Problem with Hardsphere Interactions', *Phys. Rev.* **105** (1957), 767.

87. Weller, W.: 'Zur Superfluidität eines Bose-Systems', *Z. Naturforsch.* **18a** (1963), 279; 'Zur Begründung des Zweiflüssigkeitenmodelles zur Helium II aus der Mikroskopischen Theorie', *Z. Naturforsch.* **19a** (1964), 410.

88. Gavoret, T. and Nozieres, P.: 'Structure of the Perturbation Expansion for the Bose Liquid of Zero Temperature', *Ann. Phys.* **28** (**1964**), 349.

89. Sonin, E. B.: 'Fluctuations, Long-range Correlations and Superfluidity', *Zh. Eksp. Teor. Fiz.* **59** (1970), 1416–1428 (in Russian).

90. Reatto, L. and Chester, Q. V.: 'Phonons and the Properties of a Bose System', *Phys. Rev.* **155** (1967), 88–100.

91. Lasher, G.: 'Coherent Phonon States and Long-range Order in Two Dimensional Bose Systems', *Phys. Rev.* **172** (1968), 224–229.

92. Berezinski, V. L.: 'Long-range Correlation Breaking in One-dimensional and Two-dimensional Systems with Continuous Symmetry Group', *Zh. Eksp. Teor. Fiz.* **59** (1970), 907; **61** (1971), 1144 (in Russian).

93. Popov, V. N.: 'On the Theory of Superfluidity of two-dimensional and One-dimensional Bose Systems', *Teor. Mat. Fiz.* **11** (1972), 354–365 (in Russian).

94. Popov, V. N. and Faddeev, L. D.: 'On an Approach to Bose gas Theory at Low Temperatures', *Zh. Eksp. Teor. Fiz.* **47** (1964), 1315–1321 (in Russian).

95. Popov, V. N.: 'Green's Functions and Thermodynamical Functions of a Non-ideal Bose Gas', *Zh. Eksp. Teor. Fiz.* **47** (1964), 1759–1764 (in Russian).

96. Popov, V. N.: 'Green's Functions and Thermodynamical Functions of a Nonideal Bose Gas (Second Approximation)', *Vestn. LGU Ser fiz. i. chim* **22** (1965), 58–64 (in Russian).

97. Popov, V. N.: 'Application of Functional Integration to the Derivation of Low-energy Asymptotics of Green's Functions and Kinetic Equations for Non-ideal Bose Gas', *Teor. Mat. Fiz.* **6** (1971), 90–108 (in Russian).

98. Popov, V. N.: 'Hydrodynamical Hamiltonian of a Non-ideal Bose Gas', *Teor. Mat. Fiz.* **11** (1972), 236–247 (in Russian).

99. Chapman, S. and Cowling, T. G.: *The Mathematical Theory of Non-uniform Gases*, Cambridge Univ. Press, Cambridge, 1952.

100. Khalatnikov, I. M.: *Theory of Superfluidity*, Nauka, Moscow, 1971 (in Russian).

101. Lieb, E. H. and Liniger, W.: 'Exact Analysis of an Interacting Bose Gas, I: The General Solution and the Ground State', *Phys. Rev.* **130** (1963), 1605–1616.

102. Lieb, E. H.: 'Exact Analysis of an Interacting Gas, II: The Excitation Spectrum', *Phys. Rev.* **130** (1963), 1616–1624.

103. Yang, C. N. and Yang, C. P.: 'Thermodynamics of a One-dimensional System of Bosons

with Repulsive Delta Function Interaction', *J. Math Phys.* **10** (1969), 1115–1122.

104. Abrikosov, A. A.: 'Influence of Dimensions on the Critical Field of the Second Class Superconductors', *Dokl. Akad. Nauk S.S.S.R.* **86** (1952), 489; 'On the Magnetic Properties of the Second Class Superconductors', *Zh. Eksp. Teor. Fiz.* **32** (1957), 1442 (in Russian).

105. Popov, V. N.: 'Quantum Vortices and Phase Transitions in Bose Systems', *Zh. Eksp. Teor. Fiz.* **64** (1973), 647–680 (in Russian).

106. Pitayevski, L. P.: 'Vortex Strings in a Non-ideal Bose Gas', *Zh. Eksp. Teor. Fiz.* **40** (1961), 646–651 (in Russian).

107. Bycling, E.: 'Vortex Lines and the λ Transition', *Ann. Phys.* **32** (1965), 367–376.

108. Wiegel, F. W.: 'Vortex-ring Model of Bose Condensation', *Physica* **65** (1973), 321–336.

109. Andronikashvili, E. L. and Mamaladze, Yn. G.: 'Quantization of Macroscopic Motions and Hydrodynamics of Rotating Helium II', *Rev. Mod. Phys.* **38** (1966), 567–625.

110. Landau, L. D.: 'On the Electron Plasma Oscillations', *Zh. Eksp. Teor. Fiz.* **16** (1946), 574 (in Russian).

111. Kapitonov, V. S. and Popov, V. N.: 'Longwave Plasma Oscillation Damping', *Zh. Eksp. Teor. Fiz.* **63** (1972), 143–149 (in Russian).

112. Perel, V. I. and Eliashberg, G. M.: 'Absorption of Electromegnetic Waves in Plasma', *Zh. Eksp. Teor. Fiz.* **41** (1961), 886 (in Russian).

113. Onsager, L.: 'Crystal Statistics 1: A Two-dimensional Model with an Order-disorder Transition', *Phys. Rev.* **65** (1944), 117–149.

114. Yang, C. N.: 'The Spontaneous Magnetization of a Two-dimensional Ising Model', *Phys. Rev.* **85** (1952), 808.

115. Kac, M. and Ward, T. C.: 'A Combinatorial Solution of the Two-dimensional Ising Model', *Phys. Rev.* **88** (1951), 1332–1337.

116. Potts, R. B. and Ward, T. C.: 'The Combinatorial Method and the Two-dimensional Ising Model', *Prog. Theor. Phys.* **13** (1955), 38–46.

117. Kaufman, B.: 'Crystal Statistics, II: Partition Function Evaluated by Spinor Analysis', *Phys. Rev.* **76** (1949), 1232–1252.

118. Rumer, Yu. B.: 'Thermodynamical Expectation Values for Infinite Flat Ising Lattice', *Zh. Eksp. Teor. Fiz.* **47** (1964), 278–293 (in Russian).

119. Vdovichenko, N. V.: 'Calculation of a Statistic Sum of a Flat Dipole Lattice', *Zh. Eksp. Teor. Fiz.* **47** (1964), 715; 'Spantaneous Magnetization of a Flat Dipole Lattice', *Zh. Eksp. Teor. Fiz.* **48** (1965), 526 (in Russian).

120. Ryazanov, G. V.: 'Correlation Functions of the Finite Ising Model', *Zh. Eksp. Teor. Fiz.* **54** (1968), 1010–1015 (in Russian),

121. Molokanov, A. A.: Diploma Thesis, LGU, 1971 (in Russian).

122. Berezin, F. A.: 'Flat Ising Model', *Usp. Mat. Nauk* **24** (1969), 3–22 (in Russian).

123. Fradkin, E. S. and Kalashnikov, O. K.: 'On the Ising Model Theory', FI Akad. Nauk S.S.S.R. preprint No. 93, 1968 (in Russian).

124. Wilson, K. and Kogut, G.: 'The Renormalization Group and the ε-expansion', *Phys. Rep.* **C12** (1974), 75–200.

125. Larkin, A. I. and Khmelnitski, I. E.: 'Phase-transitions in Axialsymmetric Segnetoelectrics', *Zh. Eksp. Teor. Fiz.* **56** (1969), 2087–2098 (in Russian).

126. Wilson, K. and Fisher, M. E.: 'Critical Exponents in 3.99 Dimensions', *Phys. Rev. Lett.* **28** (1972), 240–243.

127. Goldstone, J.: 'Field Theories with Superconductor Solutions', *Nuovo Cim.* **19** (1961), 154–164.

128. Polyakov, A. M.: 'Particle Spectrum in Quantum Field Theory', *Pisma v Zh. Eksp. Teor. Fiz.* **20** (1974), 430–433 (in Russian).

129. 't Hooft, G.: 'Magnetic Monopoles in Unified Gauge Theories', *Nucl. Phys.* **B79** (1974), 276–284.

130. Furry, W.: 'A Symmetry Theorem in the Positron Theory', *Phys. Rev.* **51** (1937), 125–129.

131. Faddeev, L. D. and Popov, V. N.: 'Covariant Quantization of Gravitation Field', *Usp. Fiz. Nauk* **111** (1973), 428–450 (in Russian).

132. Konopleva, N. P. and Popov, V. N.: *'Gauge Fields'*, Atomizdat, Moscow, 1972 (in Russian).

133. Hasert, F. J. *et al.*: 'Search for Elastic Muon–Neutrino Electron Scattering', *Phys. Lett.* **B46** (1973), 121–124.

134. Hasert, F. J. *et al.*: 'Observation of Neutrino-like Interactions without Muon or Electron in the Gargamelle Neutrino Experiment', *Phys. Lett.* **B46** (1973), 138–140.

135. Kibble, T. W. B.: 'Coherent Soft Photon States and Infrared Divergences', *J. Math. Phys.* **9** (1968), 315–324.

136. Kibble, T. W. B.: 'Coherent Soft Photon States and Infrared Divergences, II: Mass-shell Singularities of Green's Functions', *Phys. Rev.* **173** (1968), 1527; 'III: Asymptotic States and Reduction Formulas', *Phys. Rev.* **174** (1968), 1882; 'IV: The Scattering Operator', *Phys. Rev.* **175** (1968), 1624–1640.

137. Abrikosov, A. A.: 'On the Infrared Catastrophe in Quantum Electrodynamics', *Zh. Eksp. Teor. Fiz.* **30** (1956), 96; 'Compton Effect at High Energies', *Zh. Eksp. Teor. Fiz.* **30** (1956), 386; 'On the High-energy Electron–Electron and Electron–Positron Scattering', *Zh. Eksp. Teor. Fiz.* **30** (1956), 544 (all in Russian).

138. Sudakov, V. V.: 'Vertex Functions for Ultra-high Energies in Quantum Electrodynamics', *Zh. Eksp. Teor. Fiz.* **30** (1956), 87–95 (in Russian).

139. Onsager, L.: 'Statistical Hydrodynamics', *Nuovo Cim. Sup.* **6** (1949), 279–287.

140. Feynman, R. P.: *Application of Quantum Mechanics to Liquid Helium*, Prog. Low. Temp. Phys. Vol. 1, North Holland, Amsterdam, 1955, Ch. 2.

141. Bardeen, T., Cooper, L. N., and Schrieffer, T. R.: 'Theory of Superconductivity', *Phys. Rev.* **108** (1957), 1175–1204.

142. Bogoliubov, N. N.: 'New Method in Superconductivity Theory, I', *Zh. Eksp. Teor. Fiz.* **34** (1958), 58; Tolmachev, V. V. and Tyablikov, C. V.: 'New Method in Superconductivity Theory, II', *Zh. Eksp. Teor. Fiz.* **34** (1958), 66; Bogoliubov, N. N.: 'New Method in Superconductivity Theory, III', *Zh. Eksp. Teor. Fiz.* **34** (1958), 73 (all in Russian).

143. Bogoliubov, N. N., Tolmachev, V. V., and Shirkov, D. V.: *New Method in Superconductivity Theory*, Izd. Akad. Nauk, S.S.S.R., Moscow, 1958 (in Russian).

144. Ginzburg, V. L. and Landau, L. D.: 'On the Superconductivity Theory', *Zh. Eksp. Teor. Fiz.* **20** (1950), 1064–1082 (in Russian).

145. Gell-Mann, M. and Brueckner, K. A.: 'Correlation Energy of an Electron Gas at High Density', *Phys. Rev.* **106** (1957), 354–372.

146. Maleyev, S. V.: 'Analytic Continuation of Temperature Diagrams and Unitary Conditions at Finite Temperatures', *Teor. Mat. Fiz.* **4** (1970), 86–100 (in Russian).

147. Stanley, E. H.: *Introduction to the Transition and Critical Phenomena*, Clarendon Press, Oxford, 1971.

148. Berezin, F. A.: 'Quantization', *Izv. Akad. Nauk S.S.S.R. Set Mat.* **38** (1974), 1116–1175 (in Russian).

149. Svidzinski, A. V.: 'Functional Integration Method in Superconductivity Theory', *Teor. Mat. Fiz.* **9** (1971), 273–290 (in Russian).

150. Andrianov, V. A. and Popov, V. N.: 'Low-frequency Asymptotics of Green's Functions and Kinetic Equations for a Fermi – Bose Gas', Vestn. Leningr. Inst. No. 16 (1974), pp. 7–15 (in Russian).

151. Kapitonov, V. S. and Popov, V. N.: 'Hydrodynamical Action for Plasma' *Teor. Mat. Fiz.* **26** (1976), 246–255 (in Russian).

152. Faddeev, L. D.: 'Hadrons from Leptons?', *Pisma v Zh. Eksp. Teor. Fiz.* **21** (1975), 25 (in Russian).

153. Arefyeva, I. Ya and Korepin, V. E.: 'Scattering in Two-dimensional Model with the Lagrangian $\gamma^{-1}[2^{-1}(\partial_\mu u)^2 + \cos u - 1]$', *Pisma v Zh. Eksp. Teor. Fiz.* **20** (1974), 312 (in Russian).

154. Korepin, V. E., Kumin, P. P., and Faddeev, L. D.: 'Quantization of Solitons', *Pisma v Zh. Eksp. Teor. Fiz.* **21** (1975), 138 (in Russian).

155. Korepin, V. E. and Faddeev, L. D.: 'Quantization of Solitons', *Teor. Mat. Fiz.* **25** (1975), 147 (in Russian).

156. Dashen, R. E., Haslacher, B., and Neveau, A.: Nonperturbation Methods and Extended-Hadron Models in Field Theory', *Phys. Rev.* **D10** (1974), 4125.

157. Christ, N. and Lee, T. D.: 'Quantum Expansion of Soliton Solutions', *Phys. Rev.* **D12** (1975), 1607.

158. Tomboulis, E.: 'Canonical Quantization of Nonlinear Waves', *Phys. Rev.* **D12** (1975), 1678.

159. Jackiw, R. and Woo, N.: 'Semiclassical Scattering of Quantized Nonlinear Waves', *Phys. Rev.* **D12** (1975), 1643.

160. Faddeev, L. D.: Vortex-like Solutions of a Unified Model of Electromagnetic and Weak Interactions of Leptons', München preprint MPI-PAE/Pth 16, 1974.

161. Fradkin, E. S. and Vilkovisky, G. A.: *Phys. Rev.* **D8** (1974), 4242.

162. Kaku, M.: *Nucl. Phys.* **B91** (1975), 99.

163. Aragone, C. and Chela-Flores, J.: *Nuovo Cim.* **B25** (1975), 225.

164. Leggett, A. J.: *Rev. Mod. Phys.* **47** (1975), 331.

165. Wheatley, J. C.: *Rev. Mod. Phys.* **47** (1975), 415.

166. Alonso, V. and Popov, V. N.: *Zh. Eksp. Teor. Fiz.* **72** (1977), 1445 (in Russian).

167. Brusov, P. N. and Popov, V. N.: *Zh. Eksp. Teor. Fiz.* **78** (1980), 234 (in Russian).

168. Brusov, P. N. and Popov, V. N.: *Zh. Eksp. Teor. Fiz.* **78** (1980), 2419. (in Russian).

169. Brusov, P. N. and Popov, V. N.: *Zh. Eksp. Teor. Fiz.* **79** (1980), 1871 (in Russian).

170. Brusov, P. N. and Popov, V. N.: *Zh. Eksp. Teor. Fiz.* **80** (1981), 1564 (in Russian).

171. Brusov, P. N.: *Zap. Nauch. Semin. LOMI* **101** (1981), 28 (in Russian).

172. Kapitonov, V. S.: *Zap. Nauch. Semin. LOMI* **77** (1978) 84 (in Russian).

SUBJECT INDEX

298